Microprobe Techniques in the Earth Sciences

Mineralogical Society Series

Series editor
Dr A.P. Jones

The aim of the series is to provide up-to-date reviews through the selected but specialized contri██████████████████████ A particularly attractive feature of the series i███████████████████ it suitable for third year undergraduates and ██████████████████ solely within the Earth Sciences. Each volume is purpose-designed, highly illustrated and serves as an excellent reference tool.

TITLES AVAILABLE

1 Deformation Processes in Minerals, Ceramics and Rocks
Edited by D.J. Barber and P.G. Meredith

2 High-temperature Metamorphism and Crustal Anatexis
Edited by J.R. Ashworth and M. Brown

3 The Stability of Minerals
Edited by Geoffrey D. Price and Nancy L. Ross

4 Geochemistry of Clay-Pore Fluid Interactions
Edited by D.A.C. Manning, P.L. Hall and C.R. Hughes

5 Mineral Surfaces
Edited by D.J. Vaughan and R.A.D. Pattrick

6 Microprobe Techniques in the Earth Sciences
Edited by Philip J. Potts, John F.W. Bowles, Stephen J.B. Reed and Mark R. Cave

Microprobe Techniques in the Earth Sciences

Edited by Philip J. Potts
Department of Earth Sciences, The Open University, Milton Keynes MK7 6AA, UK

John F.W. Bowles
Mineral Science Ltd, Chesham HP5 2PZ, UK

Stephen J.B. Reed
Department of Earth Sciences, University of Cambridge CB2 3EQ, UK

and

Mark R. Cave
Analytical Geochemistry Group, British Geological Survey, Keyworth, Nottingham NG12 5GG, UK

Published in association with
The Mineralogical Society of Great Britain and Ireland

CHAPMAN & HALL
London · Glasgow · Weinheim · New York · Tokyo · Melbourne · Madras

Published by Chapman & Hall, 2–6 Boundary Row, London SE1 8HN, UK

Chapman & Hall, 2–6 Boundary Row, London SE1 8HN, UK

Blackie Academic & Professional, Wester Cleddens Road, Bishopbriggs, Glasgow G64, 2NZ, UK

Chapman & Hall GmbH, Pappelallee 3, 69469 Weinheim, Germany

Chapman & Hall USA, 115 Fifth Avenue, New York, NY 10003, USA

Chapman & Hall Japan, ITP-Japan, Kyowa Building, 3F, 2-2-1 Hirakawacho, Chiyoda-ku, Tokyo 102, Japan

Chapman & Hall Australia, 102 Dodds Street, South Melbourne, Victoria 3205, Australia

Chapman & Hall India, R. Seshadri, 32 Second Marin Road, CIT East, Madras 600 035, India

First edition 1995

© 1995 The Mineralogical Society

Typeset in 10/12 Times by Best-set Typesetter Ltd., Hong Kong
Printed in Great Britain at the University Press, Cambridge

ISBN 0 412 55100 4

Apart from any fair dealing for the purposes of research or private study, or criticism or review, as permitted under the UK Copyright Designs and Patents Act, 1988, this publication may not be reproduced, stored, or transmitted, in any form or by any means, without the prior permission in writing of the publishers, or in the case of reprographic reproduction only in accordance with the terms of the licences issued by the Copyright Licensing Agency in the UK, or in accordance with the terms of licences issued by the appropriate Reproduction Rights Organization outside the UK. Enquiries concerning reproduction outside the terms stated here should be sent to the publishers at the London address printed on this page.

The publisher makes no representation, express or implied, with regard to the accuracy of the information contained in this book and cannot accept any legal responsibility or liability for any errors or omissions that may be made.

A catalogue record for this book is available from the British Library

Library of Congress Catalog Card Number: 94-74691

∞ Printed on permanent acid-free text paper, manufactured in accordance with ANSI/NISO Z39.48-1992 and ANSI/NISO Z39.48-1984 (Permanence of Paper).

Contents

Colour plates appear between pages 20 and 21; other plates appear between pages 68 and 69, and 292 and 293.

List of contributors		ix
Preface		xi
1	**Microanalysis from 1950 to the 1990s**	1
	James V.P. Long	
1.1	Introduction	1
1.2	Microscope methods with image-plane selection	1
1.3	Microprobe methods	3
1.4	The possibilities: primary and secondary beams	3
1.5	Some history	5
1.6	Quantitative analysis	11
1.7	Precision, accuracy and resolution	14
1.8	Other techniques using electrons and/or X-rays	25
1.9	Selected-area mass spectrometry with direct excitation	33
1.10	Selected-area mass spectrometry with indirect excitation	39
1.11	Laser heating	39
1.12	Detection of molecular species	41
1.13	Which technique? Analytical strategy and tactics	42
	References	46
2	**Electron microprobe microanalysis**	49
	Stephen J.B. Reed	
2.1	Introduction	49
2.2	X-ray spectroscopy	52
2.3	Wavelength-dispersive spectrometers	57
2.4	Energy-dispersive spectrometers	60
2.5	The electron column	65
2.6	Scanning and mapping	70
2.7	Qualitative analysis	73
2.8	Quantitative analysis – experimental	75
2.9	Quantitative analysis – data reduction	81
	References	87
	Further reading	88

3	**Analytical electron microscopy**	91
	Pamela E. Champness	
3.1	Introduction	91
3.2	The X-ray analysis of thin specimens	96
3.3	Electron energy-loss spectroscopy	124
3.4	EDS versus EELS	135
3.5	Future directions of AEM instrumentation	135
	Acknowledgements	136
	References	136
4	**The nuclear microprobe – PIXE, PIGE, RBS, NRA and ERDA**	141
	Donald G. Fraser	
4.1	Introduction	141
4.2	The proton microprobe	142
4.3	The PIXE technique	144
4.4	Geological applications of PIXE	151
4.5	Nuclear reaction analysis (NRA) and PIGE	156
4.6	Rutherford backscattering (RBS)	157
4.7	Elastic recoil detection analysis (ERDA)	159
4.8	Conclusions	159
	Acknowledgements	160
	References	160
5	**Synchrotron X-ray microanalysis**	163
	Joseph V. Smith and Mark L. Rivers	
5.1	Introduction	163
5.2	Synchrotron storage rings	173
5.3	Beamline and experimental stations	186
5.4	Specimen preparation	201
5.5	Analysis	204
5.6	Illustrative applications and future advances	213
	Acknowledgements	221
	Invitation	222
	Appendix 5A Notation and abbreviations	222
	References	223
	Further reading	227
6	**Ion microprobe analysis in geology**	235
	Richard W. Hinton	
6.1	Introduction	235
6.2	Instrumentation	235
6.3	Secondary ion production	246
6.4	Analytical procedures	255

6.5	Applications: elemental analysis	262
6.6	Applications: isotopic analysis	273
6.7	Future developments	281
	Acknowledgements	282
	References	282

7 Mineral microanalysis by laserprobe inductively coupled plasma mass spectrometry — 291
William T. Perkins and Nicholas J.G. Pearce

7.1	Introduction	291
7.2	Inductively coupled plasma mass spectrometry	291
7.3	Laser ablation and ICP-MS	300
7.4	Analytical rationale	302
7.5	History of laser ablation analysis	303
7.6	Bulk sampling with LA-ICP-MS: an alternative to solution preparation	304
7.7	Analytical methodology	304
7.8	Detection limits	313
7.9	Applications	316
7.10	Future trends	321
7.11	Conclusions	323
	References	323

8 Ar–Ar dating by laser microprobe — 327
Simon P. Kelley

8.1	Introduction	327
8.2	The ^{40}Ar–^{39}Ar dating technique	329
8.3	Estimation of errors	334
8.4	Instrumentation	336
8.5	Methodology	347
8.6	Applications of the laser microprobe technique	349
	References	356

9 Stable isotope ratio measurement using a laser microprobe — 359
Ian P. Wright

9.1	Introduction	359
9.2	Stable isotope geochemistry	360
9.3	Conventional preparation of gases for stable isotope studies	361
9.4	Stable isotope ratio mass spectrometry	362
9.5	Historical development of the laser microprobe	364
9.6	Application of the laser microprobe to stable isotope studies	365
9.7	Instrumentation	367
9.8	Carbon and oxygen isotopic analyses of carbonates	369
9.9	Sulphur isotopic measurements of sulphides	373

9.10	Oxygen isotopic analyses of silicates and oxides	375
9.11	Nitrogen isotope analyses	380
9.12	Conclusions	382
	Acknowledgements	383
	References	383
10	**Micro-Raman spectroscopy in the Earth Sciences**	387
	Stephen Roberts and Ian Beattie	
10.1	Introduction	387
10.2	Principles	387
10.3	Instrumentation	392
10.4	Sample handling and routine operation	394
10.5	Advantages and disadvantages of the Raman technique	394
10.6	Some applications of micro-Raman spectroscopy in the Earth Sciences	395
10.7	Other applications of micro-Raman spectroscopy in the Earth Sciences	405
10.8	Other Raman techniques and closing remarks	406
	Acknowledgements	406
	References	406
Index		409

Contributors

Ian Beattie	Department of Geology, University of Southampton, Highfield, Southampton SO9 5NH, UK.
Pamela E. Champness	Department of Geology, University of Manchester, Oxford Road, Manchester M13 9PL, UK.
Donald G. Fraser	Department of Earth Sciences, University of Oxford, Parks Road, Oxford OX1 3PR, UK.
Richard W. Hinton	Department of Geology and Geophysics, University of Edinburgh, West Mains Road, Edinburgh EH9 3JW, UK.
Simon P. Kelley	Department of Earth Sciences, The Open University, Walton Hall, Milton Keynes MK7 6AA, UK.
James V.P. Long	Department of Earth Sciences, Bullard Laboratories, University of Cambridge, Madingley Rise, Madingley Road, Cambridge CB3 0EZ, UK.
Nicholas G.B. Pearce	Institute of Earth Studies, University of Wales – Aberystwyth, Aberystwyth SY23 3DB, UK.
William T. Perkins	Institute of Earth Studies, University of Wales – Aberystwyth, Aberystwyth, SY23 3DB, UK.
Stephen J.B. Reed	Department of Earth Sciences, University of Cambridge, Downing Street, Cambridge CB2 3EQ, UK.
Mark L. Rivers	Department of Geophysical Sciences and Center for Advanced Radiation Studies, University of Chicago, 5734 South Ellis Avenue, Chicago IL 60637, USA
Stephen Roberts	Department of Geology, University of Southampton, Highfield, Southampton SO9 5NH, UK.
Joseph V. Smith	Department of Geophysical Sciences and Center for Advanced Radiation Studies, University of Chicago, 5734 South Ellis Avenue, Chicago IL 60637, USA.

LIST OF CONTRIBUTORS

Ian P. Wright — Planetary Sciences Unit, Department of Earth Sciences, The Open University, Walton Hall, Milton Keynes MK7 6AA, UK.

Preface

The opportunity to compile contributions to this book arose from a two-day meeting held at The Open University in October 1992, sponsored by The Mineralogical Society and the Royal Society of Chemistry and convened by three of the present editors. The second day of this meeting was entitled 'Microanalysis Techniques in the Earth Sciences' and included a range of invited speakers who were asked to present a paper reviewing the current state of development in their microanalytical technique of interest, as well as indicating the direction in which the science might be heading in forthcoming years. These invited speakers (with one or two additional authors) readily agreed to develop their paper into a chapter to be contributed to the present volume.

The scope of this book includes all the most common microprobe techniques used in geological research covering instrumentation, details of analytical practice and representative applications. The treatment is designed to be of interest to the graduate student, as well as the research scientist, whose work involves the use of microprobe techniques or the interpretation of analytical results. In the course of writing, the title of the book was changed from that used for the research meeting to *Microprobe Techniques in the Earch Sciences*, to emphasize that it is about techniques wherein selected areas of samples are excited by a focused beam of particles (electrons, protons, ions) or electromagnetic radiation.

In dedicating this book, we should like to recognize the pioneers of microprobe instrumentation, one of whom has contributed the first chapter, and those who have carried forward the science, including the contributors to this volume.

PJP, JFWB, SJBR, MRC
May 1994

CHAPTER ONE
Microanalysis from 1950 to the 1990s

James V.P. Long

This chapter contains a review of fundamentals and the development of selected-area techniques.

1.1 Introduction

Before 1950, the term 'microanalysis' was used to denote the analysis of a small sample, sometimes removed mechanically from a larger specimen. While this definition is still valid today, this monograph concentrates upon techniques which are capable of spatial discrimination as well as analysis. Because the region selected for analysis usually lies at the surface of a polished mount or polished thin section of the specimen, the term 'selected-area' technique is often used although, in fact, a small **volume** of the specimen is examined. In essence, we are concerned with techniques that answer the questions 'What is it? Where is it? How much of it is present?'

There are basically two ways in which such analyses may be performed: either by using an instrument which is essentially a microscope or with a device which has come to be known as a 'microprobe'. Let us first distinguish between these methods.

1.2 Microscope methods with image-plane selection

In this approach, the specimen is illuminated by a broad beam of incident radiation, which may be visible light, electrons, ions, etc. Some form of lens is then used to focus either the reflected or transmitted primary radiation, or alternatively secondary radiation emitted by the specimen, in order to produce an enlarged image in which the contrast derives from the variation of chemical composition within the specimen itself (Figure 1.1(a)). An aperture placed in the image plane will then select radiation associated with a small volume in the specimen. With the aid of a suitable spectrometer located behind the aperture, this radiation may be analysed and the spectrum used to give information on the composition of perhaps a few cubic

Microprobe Techniques in the Earth Sciences. Edited by P.J. Potts, J.F.W. Bowles, S.J.B. Reed and M.R. Cave. Published in 1995 by Chapman & Hall, London. ISBN 0 412 55100 4

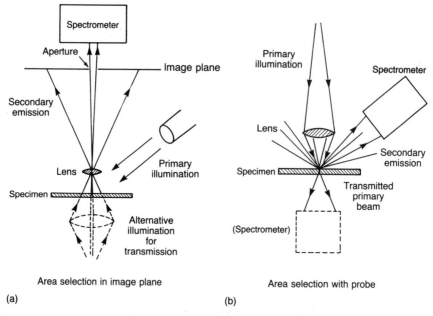

Figure 1.1 Area selection by means of (a) a microscope with an aperture in the image plane, and (b) focused illumination or 'probe'.

micrometres (10^{-10}–10^{-12} g) of material. The most familiar apparatus using this arrangement is the polarized-light microscope applied to petrological thin sections: crystal structure, refractive index and birefringence are all dependent on chemical composition and provide the means of identifying mineral grains and, by reference to tabulated data, of assigning to them a general chemical formula. Sometimes, it is also possible to determine their composition within a solid-solution series. The final image plane of the apparatus is in the eye of the observer and the processes of aperturing and spectrometry are performed by the retina and brain.

It is perhaps worth noting that the petrologist is more fortunate than the metallurgist in that rocks, unlike metals, may be crushed and separated into their component minerals which, in turn, may be subjected to bulk chemical analysis. As a result, it has been possible to assemble extensive tables correlating optical properties and chemical composition, so that petrologists have for decades been able to use the optical microscope as a qualitative or semiquantitative microanalytical tool. The limitations of the optical technique are, of course, that except in special cases, variations in composition within a solid-solution series, or the presence of minor and trace elements, produce either no measurable variation in the optical properties, or alternatively a variation which cannot be interpreted uniquely in terms of composition.

1.3 Microprobe methods

In the second method, only the area of interest, often less than 10 μm across, is illuminated by a focused or collimated beam of electromagnetic radiation or charged particles, as shown in Figure 1.1(b). The term 'probe' (or 'microprobe') is often used to describe such a beam. Compositional information is then obtained by observing secondary radiation or particle emissions emerging from the small volume excited by the primary beam, by measuring the attenuation of the primary beam if it passes right through the specimen, or by observing the scattering of the primary radiation.

Figure 1.1(b) illustrates schematically the essential features of any probe system used for analysis. The beam emerging from the source (in the case of ions or electrons usually referred to as the 'gun') is divergent and, although collimating pinholes may be employed to produce a fine beam, much higher intensities can be obtained in the final probe by using a suitable lens or lenses to collect a larger fraction of the total emission from the source. The lens system is normally arranged to produce a demagnified image of the source on the surface of the specimen. More than one stage of demagnification is frequently used, the intermediate image formed by the first lens (the condenser) providing the object for the second lens. The final lens is often referred to as the objective.

The specimen is mounted on an $x-y(-z)$ stage capable of fine incremental movements. In order that the area to be analysed may be precisely located with respect to the probe, most instruments incorporate an optical microscope, adjusted so that the crosswires at the centre of the field of view coincide with the point of impact of the probe on the specimen. Figure 1.2 shows a number of alternative ways of arranging the optical microscope, the most commonly employed being the reflecting objective (Figure 1.2(a)) with axial holes drilled through the mirrors to allow passage of the primary beam. In a probe system using charged particles, electrostatic deflection plates or electromagnetic deflection coils allow the beam to be moved rapidly and precisely over small distances on the surface of the specimen. As an alternative to the use of an optical microscope for locating the probe on the specimen, the deflection plates or coils may be used to scan the beam over the surface and the resulting signal used to form an image; this technique has many other applications and is discussed in section 1.5.

1.4 The possibilities: primary and secondary beams

In general, the interaction of the primary beam with the specimen gives rise to a number of observable effects: for example, Figure 1.3 shows diagram-

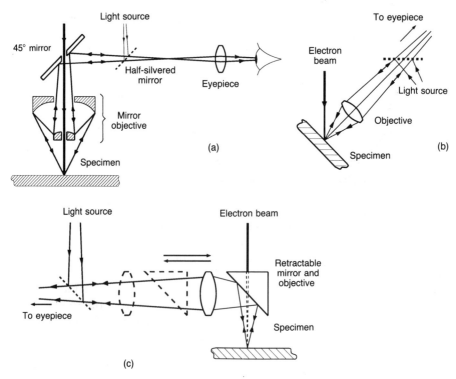

Figure 1.2 Alternative arrangements for an optical microscope in a probe instrument: (a) reflecting objective with axial holes in mirrors to allow passage of the incident bombarding beam; (b) refracting objective with axis inclined to the bombarding beam and normal to the specimen; (c) retractable objective and plane mirror.

matically the results of electron bombardment. Note that in many cases, much of the energy contained in the primary beam is converted into heat and that at high beam intensities the resultant temperature rise may be sufficient to melt or volatilize a small part of the specimen. Thus, in addition to producing observable secondary emission, a high-intensity probe may be used as a means of releasing material from the small volume of interest into the vacuum system for subsequent analysis.

There are many possible combinations of incident probe and resultant secondary signal, some of which are illustrated in Figure 1.4. When the bombardment results only in the production of electromagnetic radiation or of electrons, no material is removed from the specimen and the technique is essentially non-destructive. On the other hand, bombardment by energetic ions or by intense laser beams physically removes material from the point of impact, leaving a small crater. One obvious difference between these two types of interaction is that in the former, the process and any observations

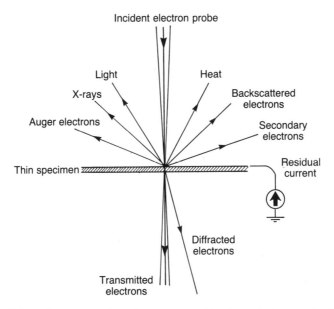

Figure 1.3 Schematic representation of processes resulting from electron bombardment.

associated with it may be continued indefinitely, whereas in the latter the duration of the measurement is limited by the survival of the material.

Subsequent chapters in this book deal in some detail with specific techniques. It is the purpose of this introduction to examine fundamental principles and practices common to all such analytical methods, to review the development of selected-area methods and to summarize the characteristics and applicability of individual techniques. To set the present into perspective, it is useful first to look briefly at the developments which have taken place over the past 40–50 years. Central to this theme is the electron microprobe, the first such technique to enter the field.

1.5 Some history

The physical principles of the electron probe, in which characteristic X-ray line spectra are produced by a fine, focused, electron beam impinging on the specimen, were established well before the Second World War. In 1913, Moseley discovered the relationship between the wavelength of the characteristic emissions of individual elements and their atomic number (Moseley, 1913, 1914) using, for this purpose, the apparatus shown in Figure 1.5. This arrangement, which used a single crystal plate and photographic emulsion as an X-ray spectrograph, was really an electron macroprobe and during the 1920s and 1930s, developments of this type of demountable X-ray tube were

MICROANALYSIS FROM 1950 TO THE 1990S

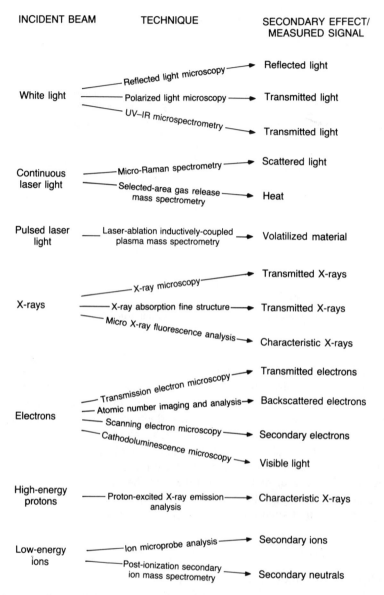

Figure 1.4 Some combinations of incident probe and measured signal and the resultant analytical techniques.

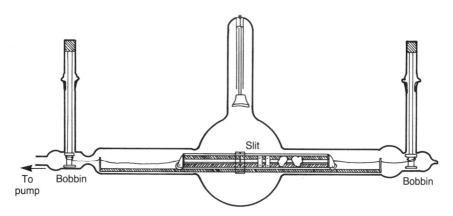

Figure 1.5 Moseley's apparatus for examination of X-ray emission spectra. In this apparatus, specimens were carried on a trolley drawn along rails by silk fishing line wound on brass bobbins that could be rotated by means of two greased cone joints. X-rays excited by electron bombardment from the gas discharge passed through the slit to a Bragg crystal spectrometer with photoplate recording. (Moseley, 1914).

used for analysis. Because of the regular spacing of the X-ray emission lines of neighbouring elements, it was easy to detect a gap in the sequence and, for example, the then unknown element hafnium was found by examining the X-ray emission spectrum of zircon (Coster and von Hevesy, 1923). Much of this early work has been summarized by von Hevesy (1932).

The remaining requirement for the electron microprobe, a means of focusing the electron beam, was also to hand as a result of the work of Knoll and Ruska (1932) in Germany on the design of magnetic lenses for the newly invented electron microscope. Thus in 1951, Raymond Castaing, a research student at the University of Paris, supervised by A. Guinier, presented his PhD dissertation (Castaing, 1951) in which he not only described the theory, design and construction of the instrument itself, but also discussed in detail the theory of the process of X-ray production and practical methods of converting measured intensities of characteristic lines into weight concentrations of the elements in the specimen. Castaing's first instrument was converted from an electron microscope with electrostatic lenses; all subsequent instruments have used magnetic lenses. Castaing was not able to provide detailed answers to all the problems of quantitative analysis, but he set up a comprehensive framework to which subsequent workers have added (and are still adding) refinements and which remains valid to the present day.

Castaing's 'sonde electronique' was able to perform quantitative analysis with a spatial resolution of $1-3 \mu m$, a typical measurement occupying a few minutes. Thus, for example, complete concentration profiles could be obtained across diffusion couples without the need for tedious stepwise removal of layers for analysis. Indeed, this advance in analytical capability was so great that it took several years for the full implications to be realized and for commercial instruments to become available. The first of these was that produced by the French Cameca Company in 1958.

A significant contribution to the development of the electron probe resulted from the independent work of two groups in Cambridge in the early 1950s on electron-optical arrangements almost identical to those used in Castaing's instrument.

In the Engineering Laboratory at Cambridge, Oatley and McMullen (McMullen, 1953) developed a successful scanning electron microscope in which a fine electron probe was scanned in a square raster pattern on the specimen while the spot on a cathode-ray tube was scanned in synchronism over an identical but much larger square. By collecting secondary (low-energy) electrons ejected from the specimen by the scanning probe and using this signal to control the brightness of the spot on the cathode-ray tube, they formed micrographs of the specimen in which the contrast was due to the variations in the secondary electron emission from point to point on the specimen, arising principally from the topography of the surface.

While the concept was not new, this microscope was the first to achieve high resolution and formed the prototype for many commercial instruments over the next 40 years. Its particular relevance to the present story is that information on the technique of scanning travelled rapidly the few hundred metres to the Cavendish Laboratory, where W.C. Nixon, a research student under V.E. Cosslett in the Electron Microscope Section, had already developed the point-projection X-ray microscope, described in section 1.8.5. With the in-house expertise in the formation of fine electron beams, it was a natural step to apply the scanning technique to microanalysis. This was done by another of Cosslett's research students, P. Duncumb, who constructed a scanning electron-probe analyser by converting an elderly RCA transmission electron microscope and equipping it with an X-ray spectrometer. Micrographs could then be formed, showing not only topography but also the distribution of particular elements at the surface of the specimen (Duncumb and Cosslett, 1957). The schematic arrangement of such a scanning instrument is shown in Figure 1.6.

In the same laboratory, the present author constructed a second instrument equipped with scanning and a polarized light microscope specifically for use with geological material. Figure 1.7(a), (b) show the first (and by present-day standards very primitive) X-ray scanning images to be obtained on this type of specimen (Long, 1958; Agrell and Long, 1960).

In the 1960s a number of commercial instruments appeared: the Cambridge Instrument Company's Microscan I, derived from the design by Duncumb and Melford (1960) found many applications in the metallurgical field. Also in the UK, Associated Electrical Industries produced an instrument based

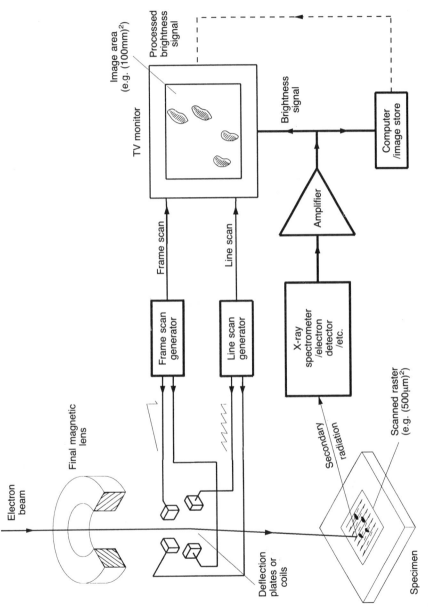

Figure 1.6 Schematic arrangement of an electron probe equipped with beam scanning.

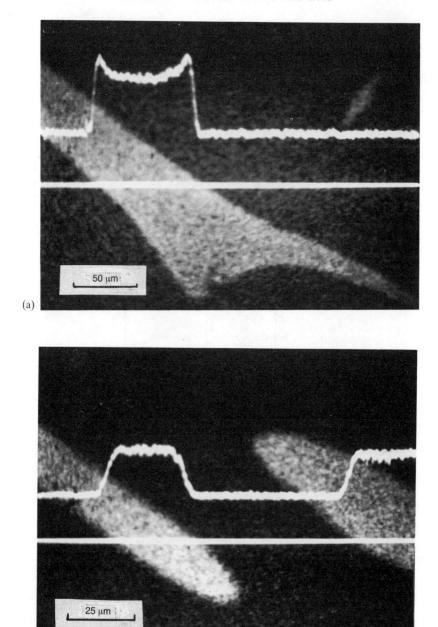

Figure 1.7 X-ray scanning micrographs showing (a) the distribution of nickel in kamacite and taenite of the octahedrite, Canyon Diablo; and (b) distribution of calcium between augite and hypersthene in inverted pigeonite from the Bushveld intrusion. Superimposed traces show variation of Ni Kα and Ca Kα respectively along tracks of probe marked. (Long, 1958; Agrell and Long, 1960).

on a prototype designed by Mulvey (1960), while in the USA, Applied Research Laboratories marketed the EMX with three X-ray spectrometers, in collaboration with D. Wittry. The first commercial instrument designed specifically for geological applications and equipped with high-quality polarized light optics was the Cambridge Instruments 'Geoscan', designed in collaboration with the present writer.

All of these instruments incorporated beam scanning and optical microscopes, the latter for many years being the hallmark of a true electron probe as distinct from a scanning electron microscope with an added X-ray spectrometer.

1.6 Quantitative analysis

In general, qualitative analyses are derived from the identification of the wavelength, energy, mass, etc., of the components of the secondary emission. Quantitative analysis requires that the intensities of these components be measured. In general, both the 'peak' and 'background' need to be recorded (Figure 1.8) to give the net intensity, usually measured and expressed in photons, ions, etc. per second.

In order to convert this measured intensity into a concentration in weight percent, it is necessary to have a knowledge of:

1. The primary beam intensity;
2. the fractional loss of primary beam energy from the specimen by scattering;
3. the efficiency of conversion of primary beam energy into secondary radiation;
4. the effect of the matrix of the specimen on secondary intensity, e.g. attenuation of the secondary radiation in escaping from the surface;
5. the solid angle of collection of the spectrometer and its efficiency;
6. the efficiency of the final detector.

Several of these quantities are very difficult to estimate and in some analytical techniques we simply do not have a sufficient understanding or control of the detailed physical processes to make calculations possible. In practice, therefore, the worst problems are circumvented by making comparative measurements of the specimen and a standard under constant conditions of excitation and with the spectrometer geometry fixed. Since the secondary radiation is the same for both, the detector efficiency also remains constant. We are thus left with only the factors arising from differences between the composition of the standard and the unknown, for example, between a pyroxene being analysed for iron and a known olivine used as standard.

The effect of the matrix varies with the technique used but in general, the following relationship applies:

$$C_{spec} = C_{std} \frac{I_{spec}}{I_{std}} \cdot \frac{f_{spec}}{f_{std}} \tag{1.1}$$

where C denotes concentration in wt%, I is the measured intensity and f a 'matrix factor' embodying all the effects of the matrix on the intensity, i.e. 2, 3 and 4 above.

1.6.1 'Spiking'

There are a number of possible ways of determining f_{spec}/f_{std}. One of the simplest, used in trace-element of bulk analysis, e.g. of a rock powder, is that of 'spiking', in which several samples of the specimen are prepared, each with a different known standard addition of the element or elements to be determined. All samples, including the unspiked specimen, are measured, the secondary radiations from the added spikes being subject to the same matrix effects as those from the elements occurring naturally in the sample. A graph of intensity *versus* concentration of added spike for any element then gives a straight line with an intercept equal to the concentration in the original sample. The technique is mainly applicable to methods of bulk analysis and is clearly very difficult to apply when the sample is only a few cubic micrometres in volume, although some experiments have been made on the use of ion implantation (Leta and Morrison, 1980).

1.6.2 Use of matching standards

As the composition of the standard approaches that of the specimen, so the ratio f_{spec}/f_{std} approaches 1. The use of matching standards is a very simple method of eliminating the matrix correction and may be applied to phases with a limited variation in composition. For example, in a binary solid-solution series such as forsterite–fayalite, a few homogeneous olivines, the compositions of which have been determined by bulk analysis, may be used to construct an empirical calibration curve. When the intensity of the signals arising from individual constituent elements is plotted against their respective concentrations, any departure from linearity is due to the matrix effect.

The use of empirical standards clearly becomes less practicable with complex multi-element phases, e.g. amphiboles, owing to the difficulty of assembling well-characterized and homogeneous standards covering the required range of composition.

1.6.3 Calculation of matrix factors

If the physics of the interaction between the primary beam and the specimen is sufficiently well understood, the matrix factors may be calculated, either

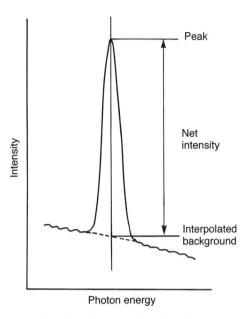

Figure 1.8 Peak, background and net intensity. Note that in many spectra it is necessary to interpolate background from measurements on either side of the peak.

from first principles or partly by semi-empirical modelling. This approach was used by Castaing in the case of the electron probe, where the individual processes in the loss of energy of the electron beam, the production of X-rays and their escape from the specimen can be identified. Details of the nature of the calculations for the electron-probe case are discussed in Chapter 2, where it is shown that in typical analyses, the factor lies in the range 0.80–1.20, depending on the combination of specimen and standard. In most cases, the accuracy of the calculation is such that in the final analysis, the error in concentration is less than 2% of the measured quantity.

Even though the physics of X-ray production is well established, the refinement of methods for calculating parameters associated with compound matrices has occupied the attention of many workers for the past 40 years and is still an active field of investigation. In other techniques, for example the ion microprobe (Chapter 6), the development of matrix-factor corrections is much less advanced and resort to more empirical methods is necessary.

To calculate matrix factors, it is necessary to know the composition of both standard and 'unknown'. This apparent impasse is overcome in practice by a process of iteration. The first estimate of the composition of the specimen is obtained by assuming initially that the matrix factors are all equal to 1. (This method was used by Castaing and is sometimes referred to as 'Castaing's first approximation'.) Using this first estimate of the composition, the matrix factors may then be calculated and applied to give a 'second approximation'. The process is then continued until successive approximations converge to a final result which changes insignificantly with further iteration. There are a number of computer programs available for the electron probe which perform such calculations for a multi-element specimen in a few seconds: prior to their development, the calculation often took much longer than the actual measurements.

1.6.4 Calibrated 'micro-standards'

If the concentration of a major element is changed, for example in the solid-solution series melilite–gehlenite, the complementary changes in the concentrations of the remaining elements result in a significant change in matrix composition and hence in the matrix factor. However, the composition of the matrix is essentially unaffected by the presence of trace elements or variation in their concentrations. The relationship between intensity and concentration is therefore almost always linear below about 1 wt%. It is thus possible to use the electron probe, with its well-established methods of calculating matrix corrections, to measure the concentrations of trace elements at, say, 500–10 000 ppm within a mineral grain and then to use the same grain as an intermediate standard for trace-element analysis of minerals of the same major element composition by another technique, e.g. the ion probe, for which calculation of matrix factors is not possible.

1.7 Precision, accuracy and resolution

Just as microanalysis may be summarized as 'what?, where? and how much?', so may the performance of a given technique be defined in terms of its ability to detect given elements or species, the spatial resolution, and the precision, accuracy and sensitivity for quantitative measurements. It is convenient to consider these attributes in the reverse order.

1.7.1 Precision

The precision of an analytical technique is a measure of the closeness of agreement within a set of results, irrespective of whether the result is correct in absolute terms. The accuracy is a measure of how closely the result represents the true value. Thus, in measurements of the Fe and Mg concentrations in olivine grains within a chondritic meteorite for the purpose of determining the degree of homogeneity, we require a high precision but accuracy is of secondary importance. On the other hand, if we wish to compare the measured olivine compositions with those in published literature, a knowledge of the accuracy is essential, both for the new measurements and also for the published data. In assessing analytical techniques, it is common practice to circulate well-characterized specimens of undisclosed composition to a number of laboratories and then to compare the resultant data. A very frequent finding in such exercises is that duplicate results submitted by any one laboratory show very good agreement, i.e. good precision, but that a much larger spread of results is apparent when data from different laboratories are compared, i.e. the accuracy is poor.

Precision is affected by a number of factors: the stability of the primary beam and of the spectrometer and detector are clearly important. These are determined not only by the design of the instrument but also by the expertise of the operator in choosing and setting the optimum operating conditions. The specimen itself may be a source of fluctuations in the measured intensity, particularly if it contains inhomogeneities on a scale comparable with the probe diameter. Small movements, either of the probe or the specimen can then give rise to variations in the recorded signal, depending on the size and the extent of the inhomogeneity.

Defects in specimen preparation may also be important: when electrically insulating specimens are examined by bombardment with a beam of charged particles, it is necessary to make a conducting path for the current by coating the surface with a very thin layer of evaporated metal or carbon. Defects in this film, or failure to ensure that it is properly grounded, can lead to charging and either deflection of the incident beam or a change in its effective energy. Occasionally, oil used as a lubricant during polishing becomes trapped in cracks in the specimen and under vacuum is released to form a film on the surface which can cause the conducting coating to lift; this too can cause instability or a reduction of intensity.

Errors arising from deficiencies in experimental technique such as those listed above are avoidable and it is useful to have available the means of predicting the errors which are fundamental and which represent the limit of precision when all parts of the equipment are working perfectly.

(a) Counting statistics and error prediction

The variation among a number of repeat measurements of any quantity may be expressed in the form of the standard deviation (σ_{exp}) or 'sample variance' $(\sigma_{\text{exp}})^2$

$$\sigma_{\text{exp}}^2 = \sum_{i=1}^{n} \frac{(x_i - \bar{x})^2}{n - 1} \qquad (1.2)$$

where x_i is an individual observation, $\bar{x}$ is the arithmetic mean and n is the number of observations. This expression is programmed on most scientific pocket calculators.

Even with perfect stability of specimen and analytical instrumentation, it is necessary to take into account the quantum nature of the secondary signal, which will consist of X-ray photons, light photons, ions, electrons, etc. In every case, repeated measurements of intensity will show variations about some mean value due to the random nature of the emission process, no matter how constant the behaviour of the apparatus.

The magnitude of this variation is governed by Poisson statistics, which apply in those cases where the observed events are the result of a very large number of possible events, each with only a very small probability of occurrence. (A classic example is the number of deaths in ten Prussian Army Corps each year between 1875 and 1894 as a result of mule kicks.)

For the Poisson distribution, σ is given very simply by:

$$\sigma = (\bar{x})^{1/2} \tag{1.3}$$

where $\bar{x}$ is average number of events recorded in each measurement.

When x exceeds about 20, the Poisson distribution approximates very closely to the familiar Gaussian distribution or normal error curve, for which the probability of observing any particular result x, is given by

$$P(x) = \frac{\bar{x}^2 e^{-\bar{x}}}{x!} \tag{1.4}$$

As for the Poisson distribution, $\sigma = (\bar{x})^{1/2}$.

These are extremely important results in that they enable us to predict the behaviour of the ideal photon- or particle-measuring system. Thus, Table 1.1 is obtained from equation 1.3, and shows the standard deviation and relative standard deviation ($100.\sigma/\bar{x}$) associated with measurements in which given average numbers of counts are recorded.

Another way of expressing these results is in terms of confidence levels. Figure 1.9 shows the areas under the Gaussian curve between limits expressed in units of σ. Thus, a single measurement with a total count of 10 000 has approximately a two-thirds chance of being within 1% of the correct result and a 95% chance of being within 2%.

The predictable nature of the variations attributable to counting statistics

Table 1.1 Standard deviation and relative standard deviation associated with given levels of counts

Counts	σ	Relative deviation (%)
100	±10	±10
400	20	5
1 000	32	3.2
10 000	100	1
10^6	1000	0.1

PRECISION, ACCURACY AND RESOLUTION

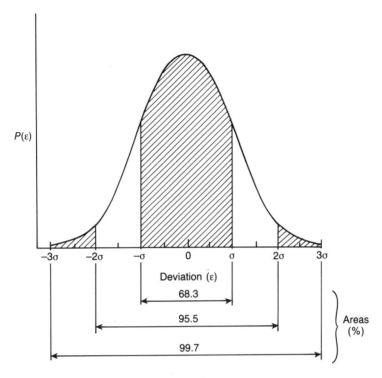

Figure 1.9 The normal distribution curve showing areas corresponding to probability of a measurement lying within $\pm 1\sigma$, $\pm 2\sigma$ and $\pm 3\sigma$, respectively, of the mean.

may be used to check the behaviour of an analytical instrument by repeating the same measurement perhaps 20 or 30 times. If no other sources of instability are present, the sample variance given by equation 1.2 should equal σ^2 calculated from equation 1.3. (Note that we should not expect the two quantities to be identical since there is always an uncertainty in the value of the measured standard deviation, $\sigma_\sigma = \sigma/(2(n-1))^{1/2}$, where n is the number of measurements). If the experiment is repeated a number of times with increasing total counts, the value of $\sigma_{exp}/\bar{x}$ will start to exceed $\sigma/\bar{x}$ when other sources of error and instability, e.g. drift in primary beam current, become significant. It should be emphasized that calculation of σ must be performed on total counts and not on the count rate.

(b) Combination of errors

A number of useful relationships, which apply irrespective of whether σ derives from counting statistics alone or in combination with other sources of error, are noted below. For a full treatment of the errors of measurement the reader is referred to texts by Topping (1972), Taylor (1982) and Lyons (1986) or, particularly in relation to counting errors, to Knoll (1979).

If measurements are repeated, then the standard error of the mean is given by:

$$\sigma_{\text{mean}} = \frac{\sigma}{(n)^{1/2}} \qquad (1.5)$$

where n is the number of observations.

In many techniques of analysis, two essential steps in the reduction of experimental data are the subtraction of background to obtain the net count on a peak, and the ratioing of two net peak counts for specimen and standard respectively (equation 1.1). Using A and B to denote two separate measurements in each case we have for the former:

$$\sigma_{A \pm B} = (\sigma_A^2 + \sigma_B^2)^{1/2} \qquad (1.6)$$

and for the latter:

$$\frac{\sigma_{A/B}}{A/B} = \left[\left(\frac{\sigma_A}{A}\right)^2 + \left(\frac{\sigma_B}{B}\right)^2 \right]^{1/2} \qquad (1.7)$$

(c) Sensitivity and limit of detection

The limit of detection of any analytical technique is in general determined by the confidence with which the measured signal due to a given element or analysed species can be distinguished from the background signal. The confidence level is commonly set at 99.7%, corresponding to 3σ. Thus, if the observation at the peak position exceeds the background by more than 3σ of the background then the statement that the species sought is present has a probability of 0.997 of being correct.

Given the basic characteristics of any analytical method it is thus possible to make an estimate of the ultimate limit of detection. Thus, if in an electron probe operated under conditions appropriate to trace analysis, the count rate for Ni Kα obtained with a terrestrial olivine containing 5000 ppm Ni is 5420 per second, and the background is 1000 counts per second, we can estimate the minimum concentration of nickel which could be determined with two measurements each of 400 seconds duration on peak and background respectively. (This estimate would be useful, for example, in deciding the applicability of the method to the analysis of meteoritic olivine where, owing to the strong partitioning of nickel into the metal phase, concentrations are generally a few tens of ppm).

In 400 seconds, the total background count = 4×10^5 counts.

Therefore, $3 \times \sigma_{\text{background}} = 3.(4 \times 10^5)^{1/2} = 1896$ counts (equivalent to $1896/400 = 4.7$ cps).

Using the calibration obtained with the known terrestrial olivine and ignoring matrix factors, this count rate is equivalent to a detection limit of:

$$\frac{\sqrt{2} \cdot 4.7 \times 5000}{5420} \sim 6\,\text{ppm}$$

The factor of $\sqrt{2}$ is introduced to account for the additional uncertainty in subtracting a background signal from a peak-plus-background signal in accordance with equation 1.6. The magnitudes of peak and background (and therefore the corresponding values of σ_A and σ_B in equation 1.6) can be assumed to be approximately equal at the detection limit if equal counting times are used in these two measurements.

Note that 4.7 counts per second is only about 0.5% of the background rate so that in order to achieve the limit calculated above it would be necessary for the stability of the equipment to be substantially better than 0.5%. A practical way of minimizing the effects of instrumental drift is to alternate short measurements of peak and background, a method which may be readily applied by suitable programming of the operating routine.

(d) Summary of requirements for high precision and sensitivity

It is clear that high counting rates and a high peak-to-background ratio are required to achieve maximum sensitivity. Frequently it is necessary to compromise, since in almost all types of spectrometer, attainment of high peak-to-background generally involves reduction of slit widths with a consequent reduction of count rate. It may be shown (e.g. Knoll, 1979) that the optimum conditions are obtained when P^2/B is a maximum: this expresses the fact that there is no point in striving for infinite peak-to-background if, by so doing, the peak is made infinitely small. At low concentrations, when the peak and background are nearly equal, the total measuring time should be divided equally between the two.

1.7.2 Accuracy and standards

Accuracy is not amenable to a simple mathematical treatment: nearly all analytical techniques depend on comparison with a calibration standard and as a result, any error in the assumed composition of the latter is reflected directly as an unpredictable error in that of the unknown. For this reason the choice and preparation of the calibration standard are crucial and often present the most difficult stages of an analysis. In those techniques where the physics underlying matrix effects is well understood, it is possible to use either pure elements or simple stoichiometric compounds, e.g. oxides, as standards of high integrity. Considerable problems arise when characterization of the standard relies on a bulk analysis of the material: not only are errors in the bulk analytical technique carried over, but also it is necessary to ensure that the bulk composition is representative of the composition on a microscale, i.e. that the material is homogeneous. Another hazard arising if the standard is analysed for trace elements by a bulk technique is the effect of any minor contaminant phase containing a high concentration of the trace element. For example, the bulk analysis of quartz often shows titanium at trace concentrations whereas a selected-area technique reveals that the majority of the Ti is present not in solid solution but as minute inclusions of rutile.

1.7.3 Spatial resolution

The resolution of an analytical technique may be defined in terms of the smallest volume from which analytical information may be obtained, unaffected by the surrounding material. The parameters and effects which determine the size and shape of this volume are:

1. The diameter of the focused probe at the surface of the specimen;
2. the penetration depth of the primary beam in the specimen and the extent to which scattering causes a lateral spread;
3. the depth from which the secondary radiation can escape from the specimen;
4. effects produced outside the volume excited by the primary beam, e.g. the production by primary electrons of X-rays which in turn excite surrounding material by virtue of their greater penetrating power.

These effects are illustrated in Figure 1.10 which shows that the shape of the analysed volume is very dependent on the nature and energy of the primary radiation. With electrons, the beam energy is dictated by the need to excite the characteristic radiation of the target atoms and is generally in the range 10–30 keV, resulting in a penetration of the order of $0.8-7\,\mu m$ in a silicate specimen. With low-energy ions, on the other hand, the penetration of the primary beam is less than 10 nm, and secondary ions are generated very close to the surface. Thus, even with a wide primary ion beam, the 'depth resolution' is very high. This characteristic is exploited in the analysis of surface layers and, most importantly, in the measurement of very short-range concentration profiles perpendicular to the surface. The broad ion beam slowly erodes the specimen and, provided precautions are taken to avoid recording the signal from the edges of the resultant crater, resolution in depth can be of the order of 10 nm. At the other extreme, a fine beam of high-energy X-rays would irradiate a thin cylindrical volume, the resolution in depth being determined largely by the escape depth of the characteristic fluorescence X-rays. Since the energy and the absorption of characteristic X-rays vary considerably with atomic number and with the composition of the matrix, it is possible for significant 'spatial discrimination' to be introduced. For example, if the cylindrical volume depicted in Figure 1.10(d) is inhomogeneous and contains inclusions rich in, say, iron, the apparent concentration of iron will depend on the position of the inclusions: Fe $K\alpha$ radiation from an inclusion at the surface would be recorded with relatively

Plate 1.1 **(a)** Andesite (West Indies) showing plagioclase with bytownite core and labradorite rim (x Polars; width of field: 1mm); **(b)** Cathodoluminescence micrograph of the thin section in (a) showing change in luminescence colour at outer rim, almost certainly associated with change in trace-element concentration.

Plate 1.2 Cathodoluminescence micrograph of a thin section of a calcite-filled vug from the Dalradian (Islay, Scotland); zoned calcite (orange-black) and dolomite (red) (width of field: 1mm).

Micrographs courtesy of Dr J. A. D. Dickson.

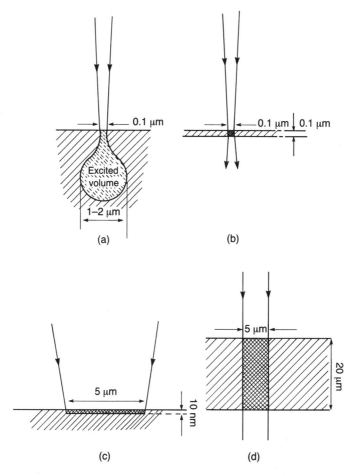

Figure 1.10 Penetration and scattering of the primary beam in the specimen and the associated analysed volume: (a) electron probe and thick specimen; (b) electron probe and thin specimen; (c) ion probe; and (d) X-ray probe. Dimensions shown are order of magnitude only and depend on beam energy and the nature of the specimen.

high efficiency, whereas that from one at the base of the cylinder would be subject to significant attenuation.

It is important to recognize that the resolution, whether measured laterally or in depth, does not define a region with sharp boundaries. The primary beam itself often has a non-uniform distribution of intensity across its diameter and the penetration and scattering in the specimen result in gradients in the level of excitation, extending from the surface and from the axis of the beam respectively. It is often convenient to make a very rough approximation to the analysed volume by assuming Gaussian distributions for individual effects and to add these in quadrature. Thus, if the diameter of the beam striking the specimen is D and the diameter of the volume excited by an infinitesimally narrow beam (i.e. allowing for scatter within the sample) is D_p, then the effective resolution is given by $(D^2 + D_p^2)^{1/2}$.

1.7.4 The diameter of the incident probe

The lateral resolution can never be less than the diameter of the incident probe. In those systems in which the probe is formed by demagnifying a source, the size of the image at the surface of the specimen is determined by D_0, the product of the size of the source and the demagnification factor, to which must be added any enlargement caused by aberrations in the optical system.

A serious defect of lenses used for focusing electrons or ions is spherical aberration, which results in a shorter focal length for marginal rays than for paraxial rays, as shown in Figure 1.11. The only practical method of restricting the diameter of the disc of least confusion, D_s, is by limiting the aperture of the lens. Assuming a Gaussian distribution in both the source and in D_s, we may write he effective diameter D as

$$D = (D_0^2 + D_s^2)^{1/2} \qquad (1.8)$$

If we wish to restrict the enlargement due to spherical aberration to 10%, i.e. $D = 1.1D_0$, we obtain from equation 1.8 that

$$D_s = \frac{D_0}{2} \qquad (1.9)$$

D_s is proportional to the cube of the semi-angular aperture α:

$$D_s = \frac{C_s \alpha^3}{2} \qquad (1.10)$$

where the constant C_s is the spherical aberration coefficient of the lens. In an electron microprobe C_s is typically a few centimetres.

As illustrated by Figure 1.12, the object distance for a lens used to produce a demagnified image of a source is proportional to the demagnification if the image distance remains constant. The current passing through the aperture of the lens is therefore given by

$$i = kD_0^2 \alpha^2 \qquad (1.11)$$

where k is a constant. Combining equations 9, 10 and 11 then gives the important relationship:

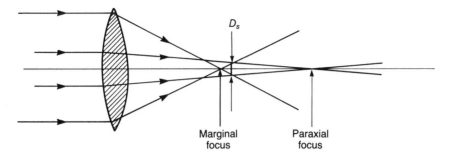

Figure 1.11 Ray paths through a lens showing the effect of spherical aberration. D_s is the minimum diameter of the image, i.e. the 'disc of least confusion'.

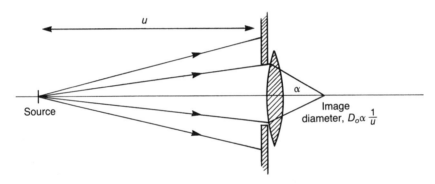

Figure 1.12 Demagnified image of source produced by a single lens with an aperture. The current in the probe is proportional to the area of the aperture and inversely proportional to D_0^2 and hence to α^2.

$$i \propto D_0^{8/3} \tag{1.12}$$

This very strong dependence of probe current upon probe diameter (a reduction of a factor of 10 in D_0 is obtained at the expense of a drop in probe current by a factor of 460) often imposes a limitation on the useful analytical resolution even when penetration, scattering, etc. are not important. Sensitivity and precision have already been shown to depend on the square root of the total count and hence, for a given counting time, on the square root of the probe current. There is thus an inevitable trade-off between precision and resolution which needs to be taken into account in the planning of experimental measurements.

1.7.5 Resolution in scanning images

Another aspect of the relationship between counting rate and resolution appears in the recording of scanning images. The 'brightness' or grey level of each individual picture element (or 'pixel') is determined by the total number of counts recorded while the scanning probe is on the corresponding point on the specimen. In a region of constant composition, fluctuations in the brightness of neighbouring pixels will be given by equation 1.3, and will determine not only the minimum compositional change that can be detected at a phase boundary but also the precision with which the position of the boundary can be delineated.

In Figure 1.13, the three X-ray scanning images from an electron microprobe show the distribution of silicon between mineral grains and a glassy host phase. They illustrate clearly the reduction in statistical noise as the recording time is increased.

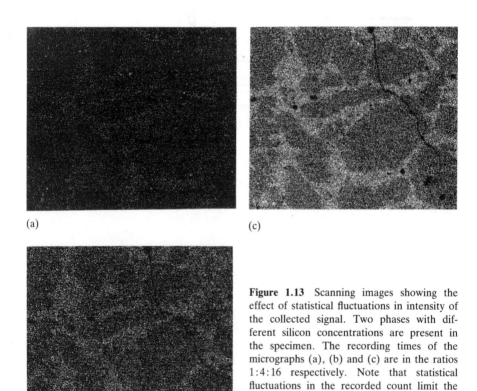

Figure 1.13 Scanning images showing the effect of statistical fluctuations in intensity of the collected signal. Two phases with different silicon concentrations are present in the specimen. The recording times of the micrographs (a), (b) and (c) are in the ratios 1:4:16 respectively. Note that statistical fluctuations in the recorded count limit the ability to distinguish both concentration differences and the position of phase boundaries. (Courtesy of Dr S.J.B. Reed).

1.7.6 Qualitative applicability

Thhe microanalytical methods of interest to the Earth scientist are generally those which detect and measure the concentration of some constituent species – element, isotope or molecule. The wide range in the efficiencies of excitation, spectrometry and ultimate detection of the measured species results in a corresponding range in precision, sensitivity and resolution, not only from one technique to another but also for different species measured with the same technique. For example, the ion microprobe has exceptional sensitivity (in the ppb range) for the alkali elements on account of their low ionization potentials, but is almost useless in its normal configuration for the next-neighbour rare gases.

It is useful to set the performance of the electron probe against the criteria developed in this section in order to see where the scope for alternative techniques lies. The performance of the electron probe is summarized in Table 1.2.

The particular advantages of the electron probe lie in the ease of specimen preparation, the simple and reliable excitation by electron bombardment and the well-understood mechanism of the excitation process, which permits the use of calculated matrix factors.

The main deficiencies of the technique lie in its inability to measure the light elements with high precision or sensitivity, a limited sensitivity for geochemically important elements such as the REE (>100 ppm) and the limitation on spatial resolution imposed by penetration and lateral scattering of the incident electrons. The technique does not, of course, provide any information on isotopic composition, nor the presence of molecular species, although in some cases coordination or valence state may be deduced from observations of the shape, or small changes in the energy, of emission lines.

1.8 Other techniques using electrons and/or X-rays

There is no doubt that Castaing's realization of a working electron probe provided the stimulus for devising new selected-area micro-analytical methods with alternative bombarding and emitted species. Other important influences were, of course, advances in the technology of spectrometers, lasers and detectors and, in the 1970s and 1980s, the impact of the microcomputer in instrument control and data recording.

Whilst attempting to present some aspects of the historical setting, the following sections are organized primarily according to the nature of the physical interaction between bombarding beam and specimen and the interrelationships of the techniques discussed. Many of these techniques are treated in much more detail in subsequent chapters. Where this is not the case, leading references are given. Thus initially, we follow the theme which includes electron bombardment and/or detection of X-rays.

Table 1.2 Summary of performance of the electron probe

Parameter	Performance
Element range	$Z = 4$ (Be) upwards. Quantitative analysis becomes increasingly difficult for atomic numbers below $Z = 11$ (Na) owing to the decreasing energies of the characteristic photon emissions, increasing matrix factors and problems of spectrometry and detection of low-energy X-rays
Precision	Dependent on concentration and measuring time but typically ±1% (relative) and with care ±0.1–0.4%
Accuracy	Dependent upon element and matrix and on quality of calibration standards. Above $Z = 11$, generally within ±3% (relative); major elements in silicates ±1% (relative)
Detection limit	Again dependent on element, matrix and measuring time; also on type of spectrometer used. (Bragg crystal spectrometer gives peak-to-background for a pure element of 1000, cf. 100–150 for energy-dispersive spectrometer, with corresponding detection limits of 10–200 ppm and 500–5000 ppm respectively)
Analysis time	Typically seconds to tens of seconds per element but dependent on concentration and required precision
Resolution	Limited by electron penetration to approximately 1–3 μm and in some cases by excitation of secondary fluorescence

1.8.1 Analytical electron microscopy

One of the simplest modifications of the standard electron-probe technique, which overcame the restriction on spatial resolution imposed by electron scattering, was made by Duncumb (1959) using very thin specimens, prepared for the transmission electron microscope (TEM).

In a solid specimen, electrons from the probe are scattered into a volume which, depending upon the mean atomic number of the target area, is more or less pear-shaped, as shown in Figure 1.10(a). The extreme range of the incident electrons is given approximately by the Thomson–Whiddington law:

$$p = \frac{0.025V^2}{\rho} \qquad (1.13)$$

where p = extreme range (or penetration) in μm; V = beam accelerating voltage in kV; ρ = density of target in g cm^{-3}.

Thus, for a typical silicate with $\rho = 3$ g cm^{-3} and a beam voltage of 20 kV, the electron range is approximately 3 μm.

If the thickness of the specimen is restricted to about 100 nm or less, very little lateral scattering of the electron beam occurs and the analysed volume approximates to a cylinder, the diameter of which approaches that of the incident probe. The intensity of emitted X-rays, and hence the concentration sensitivity, is, of course, much reduced because the incident beam expends only a small portion of its energy in the specimen. However, the transmitted electrons may be used for electron microscopy and microdiffraction and Duncumb (1963) designed and constructed a new instrument referred to as EMMA (Electron Microscope Micro-Analyser) for this purpose. A subsequent version (EMMA3) formed the basis of the first commercial design by AEI Scientific Apparatus Ltd of Manchester, which appeared in 1969. Today, facilities for analysis of sub-micrometer areas are a standard option on most TEMs and the technique is reviewed in detail in Chapter 3.

1.8.2 Selected-area X-ray fluorescence analysis

Conventional X-ray fluorescence analysis is performed on bulk specimens such as powdered rocks or glass discs prepared by fusing rock powder with a suitable flux, and typically examines between 0.5 and 10 g of material. Characteristic X-ray spectra, essentially identical with those generated by electron bombardment, are excited by illuminating the specimen with X-rays whose photon energy is higher than the binding energy of the K (or L) electrons in the elements to be determined in the sample. Under these circumstances, a K or L electron may be ejected by photoelectric absorption of the incoming photon, leaving the atom ionized just as it would be following electron excitation. An important difference, however, is the absence of the 'white' radiation or 'bremsstrahlung' produced by the slowing down of electrons within the sample, noted in section 1.8.4 and discussed in more detail in Chapter 2. As a result, the peak-to-background ratio is much higher than in the electron probe (10^5-10^6, cf. 10^3) and the detection limit is typically of the order of 0.1–10 ppm.

The first attempt to use X-ray fluorescence analysis to reveal the distribution of elements within a specimen was made by von Hamos (1934), who used a curved X-ray monochromator to image the surface of a specimen on a photographic emulsion by means of the secondary X-rays excited by a primary X-ray beam (i.e. with an imaging rather than a probe technique). However, resolution and sensitivity were limited and the technique was not quantitative. In the 1950s Zeitz and Baez (1957) and Long and Cosslett (1957) used microfocus X-ray tubes together with small apertures to illuminate a small area of a thin specimen placed immediately above the aperture. Although it was possible to detect the fluorescence radiation from a mass of the order of 10^{-9} g of some elements (Long and Röckert, 1963), the technique suffered from lack of intensity despite the high specific loading of the microfocus tubes, and was not applied to mineralogical problems.

The situation has been transformed in recent years by the development of very intense synchrotron X-ray sources which have been used both with collimating apertures and with X-ray focusing optics to illuminate areas of the order of 10–20 μm diameter on polished specimens. With the aid of energy-dispersive X-ray detectors, which permit relatively high collection efficiencies, detection limits of the order of 1 ppm have been obtained, corresponding to a limiting mass of the order of 10^{-15} g. A full review of this method is given in Chapter 5.

1.8.3 Cathodoluminescence

As indicated in Figure 1.3, electron bombardment often results in the emission of light. 'Cathodoluminescence' of minerals was studied more than a hundred years ago by Crookes (1879), who observed that 'substances known to be phosphorescent shine with great splendour when subjected to the negative discharge in high vacuum'. Despite many applications of this phenomenon and much research which showed that luminescence is frequently the result of small concentrations of certain trace element activators, the possibilities afforded by the microscopical examination of minerals under electron bombardment do not seem to have been explored until the 1960s when the luminescence was observed during the use of electron microprobes. Smith and Stenstrom (1965) and Long and Agrell (1965) independently obtained luminescence micrographs by illuminating the surface of thin sections with a broad (approximately 1 mm diameter) beam of electrons. The results showed clearly the correlation between colour and brightness of the luminescence and variations in concentration of certain trace elements, particularly in zoned calcites and also in quartz.

The present author demonstrated a simple cold cathode source of electrons (in fact a miniature version of the tubes used by Crookes), which could be mounted on a small specimen chamber carried on the stage of a normal petrological microscope. Refinements of this apparatus are now in common use in sedimentary petrology for distinguishing growth zoning in diagenetic minerals. Typical luminescence micrographs are shown in Plates 1.1 and 1.2.

As an analytical technique, cathodoluminescence imaging has obvious limitations: in almost all cases the luminescence is produced by trace concentrations of elements such as manganese and the rare earths. It is a sensitive method for revealing growth zones and secondary overgrowths which are not distinguishable by normal optical examination. However, it is not easy to make quantitative estimates of concentration of the activating element or, by visual observation alone, to determine its identity unambiguously. On the other hand, the technique is very simple to use and in addition to the optical microscope, the average equipment cost is less than 1% of that of a modern electron microprobe.

For further details of the technique and its applications the reader is referred to a recent review by Marshall (1988).

1.8.4 Particle-induced X-ray emission: the proton microprobe

The efficiency of X-ray production with proton bombardment is much lower than that with electrons and in order to obtain useful characteristic intensities it is necessary to accelerate the proton beam to energies of the order of 2–3 MeV by means of a small van de Graaff generator. However, as with X-ray fluorescence spectra, the 'white' continuum radiation is of very low intensity and hence the peak-to-background ratio is high. The first experiments in the development of an analytical technique were made by Johansson, Akselsson and Johansson (1970) at the University of Lund. They were able to demonstrate a well-defined Ti Kα peak obtained in a 1-hour run on a sample containing 4×10^{-11} g Ti. The technique was also applied to the detection of atmospheric pollutants collected on thin carbon foils exposed out-of-doors (Figure 1.14).

More recent instruments use magnetic quadrupole lenses to focus the beam into probes of the order of a few micrometres in diameter. They still employ, as did Johansson and his co-workers, energy-dispersive solid-state X-ray detectors to maximize the efficiency of collection and spectrometry of the emitted X-rays.

The technique suffers from the same limitations in the case of the light elements as do the electron probe and X-ray fluorescence methods. In common with all X-ray analytical techniques, however, the excitation and emission processes are well understood so that calibration is much more straightforward than is the case with techniques involving emission of secondary ions, where the physical processes are very complex. Despite its advantages for trace-element analysis, the practice of the technique is restricted to relatively few laboratories where the instruments in use have generally been constructed in-house, albeit with a commercially-designed accelerator.

1.8.5 X-ray microscopy and absorption spectrometry

X-ray microscopy or 'microradiography' (Goby, 1913) has advantages for examining internal features in optically opaque material and for revealing three-dimensional relationships without the need for sectioning. The simplest method of obtaining microradiographs is by placing the specimen on the surface of a high-resolution photographic emulsion and exposing it to radiation from a conventional X-ray tube, as in Figure 1.15(a). After development, the microradiograph is enlarged photographically. Using this method, Hooper (1960) was able to examine the internal structure of large

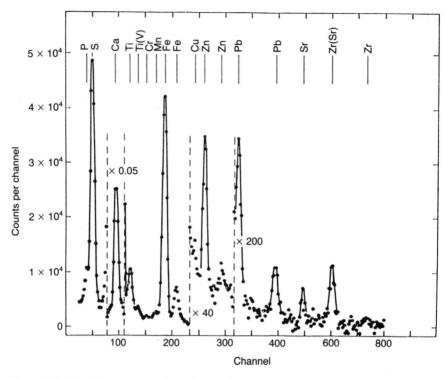

Figure 1.14 The X-ray spectrum from dust particles collected on carbon film excited by proton bombardment. (Johansson, Akselsson and Johansson, 1970).

numbers of foraminifera with a view to classification on the basis of internal dimensions. The optimum contrast in such micrographs is obtained when the wavelength of the X-ray radiation is chosen so that it is just shorter than the characteristic absorption edge of a major element contained within a textural feature of the specimen.

The point-projection X-ray microscope, developed by Cosslett and Nixon (1951) in the Cavendish Laboratory, Cambridge, provides another method for producing microradiographs. This instrument uses an electron-optical system identical with that of the electron probe; the essential difference is that the electron beam is focused on a thin X-ray-transparent foil instead of the normal thick target. A thin section or other specimen to be examined is placed close to the opposite side of the foil and a magnified radiograph is recorded on a high-contrast photographic emulsion a few centimetres away, as in Figure 1.15(c). An ultra-high resolution is no longer required in the emulsion because the primary magnification of the instrument can be of the order of 20 times. A typical micrograph is shown in Figure 1.16. Many electron probe instruments could be used for microradiography by a suitable

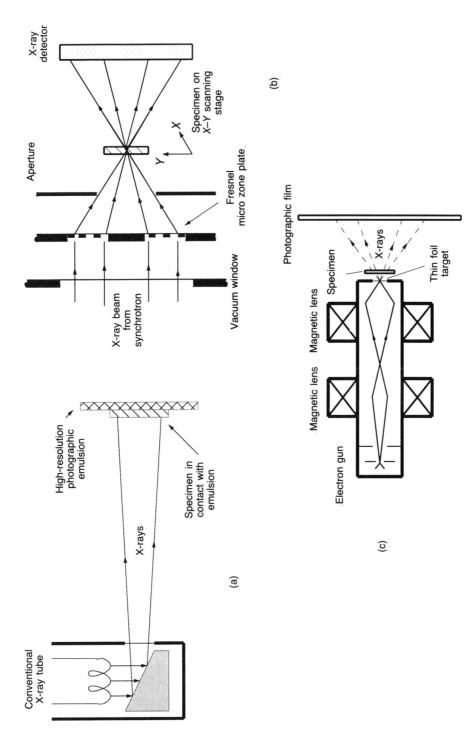

Figure 1.15 Techniques for X-ray microscopy: (a) contact microradiography; (b) zone-plate scanning microscope; and (c) point-projection microscope.

Figure 1.16 Point-projection microradiograph of 30 μm section of garnetiferous schist, taken with a copper target. Note the strong absorption of garnets and biotite where the iron and manganese absorption edges are of lower energy than Cu Kα. (Long, 1958.)

adaption of the specimen stage area although, due to the longer working distance and larger spherical aberration coefficient of the final lens, X-ray intensities will generally be lower than in purpose-built instruments. These techniques are fully reviewed by Cosslett and Nixon (1960).

It is also possible to use the point-projection microscope for absorption microanalysis by placing a fluorescent screen in the plane of the photographic emulsion in order to view the image and then substituting a small aperture to select X-rays that have passed through a small region of the specimen (similar to Figure 1.1(b)). Spectrometry of the X-ray continuum transmitted by the aperture enables the transmission of the specimen to be measured for wavelengths just above and below the absorption edge of an element, the mass thickness of which in the specimen is to be determined. For a general review see Cosslett and Nixon (1960). The technique as originally developed for the point-projection microscope is tedious and slow, but the availability of synchrotron X-ray sources has opened up new possibilities in this field, as illustrated by the high-resolution microradiographs shown in Figure 1.17 (Morrison *et al.*, 1990). In addition, it is now possible to make measurements of the fine structure of the transmitted spectrum close to X-ray absorption edges and so to obtain information on chemical bonding of

SELECTED-AREA MASS SPECTROMETRY WITH DIRECT EXCITATION

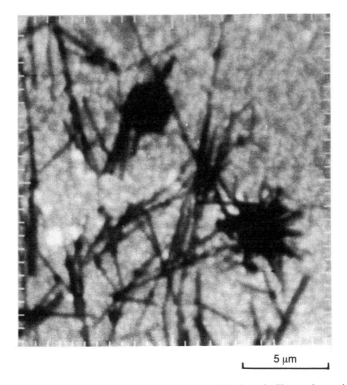

Figure 1.17 High-resolution X-ray microradiograph of hydrated silicates in setting cement (48 h) taken with synchrotron radiation with energy (351 eV) close to the Ca L absorption edge. (Courtesy of Dr G.R. Morrison; Morrison *et al.*, 1990.)

elements within small volumes of a thin specimen. For further information on this topic, the reader is referred to Chapter 5.

1.9 Selected-area mass spectrometry with direct excitation

1.9.1 *Secondary ion mass spectrometry: the ion microprobe*

The detection and measurement of ions generated from the specimen offers a new set of possibilities. In the conventional mass spectrometric technique for isotopic analysis of metallic elements, e.g. Sr, Nd and Pb, the sample is chemically processed and a small drop of the solution containing the separated element is evaporated onto a ribbon filament. Positive ions are generated when the filament is heated *in vacuo* in the source of the mass spectrometer. This thermal ionization technique clearly possesses no intrinsic spatial resolution other than that which may be associated with the original sampling, e.g. in the selection of individual zircon grains.

Similarly, in the case of non-metallic elements, e.g. C, N, S, etc., mass spectrometry is conventionally performed on ions produced when a gaseous sample is ionized by an electron beam. Techniques of stepwise heating and stepwise combustion of the sample have, particularly for extraterrestrial material, provided much information on the isotopic characteristics of specific phases (Wright and Pillinger, 1988).

These methods, however, require very careful attention to the problems of contamination of the sample during processing and, although very powerful in the appropriate application, do not fall directly within the scope of this book.

In secondary ion mass spectrometry, the sample is bombarded with a beam of primary ions which erode the surface and eject both atomic and molecular species, some of which emerge as ions (Figure 1.18). This technique was explored as a means of analysis by a number of workers prior to 1960 but the first development of a selected-area technique was due to Slodzian (1964), working in Paris as a research student under R. Castaing. Their first apparatus was essentially a secondary ion microscope capable of mass resolution and so in the general category illustrated in Figure 1.1(a). This remarkable instrument produced images of the specimen surface with a resolution of the order of $1\,\mu$m (Figure 1.19). An aperture in an image plane enabled a quantitative measurement of the intensity of a particular ion species to be obtained from a selected area of the surface.

Just as Castaing's electron probe was developed commercially by the Cameca Company, so did Slodzian's work lead to the production of a series of instruments by the same company. The IMS 300 was followed by the IMS 3F in 1980 and later by 4F and 5F versions. Recently, the IMS 1270, equipped with a large mass spectrometer, has come into production.

The possibility of constructing an ion analogue of the electron probe (i.e.

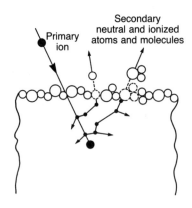

Figure 1.18 Diagrammatic representation of the process of ion sputtering, illustrating the production of secondary ions and neutral species.

with the configuration of Figure 1.1(b)) was explored independently by two groups. In the UK the present author investigated theoretically the design of a probe instrument (Long, 1965) leading subsequently to the development of a prototype instrument by Drummond (1968), followed eventually by the commercial AEI IM20 in the mid-1970s. Meanwhile, in the USA the work of Liebl (1967) resulted in the production of the Applied Research Laboratories (ARL) IMMA instrument.

For the formation of ion images, the 'microscope' approach has considerable advantages provided that the resolution required is not less than about $3\,\mu$m. This is because information is being recorded simultaneously from all points in the field of view whereas in the case of an image produced with a scanning probe each pixel is addressed sequentially. Thus, unless there is a corresponding increase in the current density of the primary beam and/or the collection efficiency of the secondary ions in the scanning mode, the information rate will be lower. In fact, as shown by Slodzian (1988), these compensating factors exceed unity below about $3\,\mu$m so that scanning becomes more efficient at high spatial resolution.

In many applications in the Earth Sciences, however, we require quantitative information from small grains rather than images; in these circumstances the probe configuration is always more efficient, not only on grounds of secondary-ion intensity, but also because the probe erodes only the area of interest, thus conserving material and largely eliminating contamination of the measured signal by ions from other parts of the field of view. Most measurements are, therefore, made in the probe mode and all Cameca instruments from the IMS 3F onwards are able to operate in either configuration.

The important characteristics of the ion probe are:

1. It is a destructive technique: i.e. sample is removed by the beam;
2. the spectrometer records individual isotopes;
3. the peak-to-background ratio is very high;
4. hydrogen and the light elements (but not the rare gases) can be measured;
5. the physical processes of secondary ion formation are very complex.

We may show by examining the five characteristics listed above in a little more detail that the role of the ion probe is very different from that of the electron probe and that the two techniques are essentially complementary.

As noted in section 1.4, there is a limit to the duration of a measurement imposed by the rate of removal of material by the primary beam. Thus, except for certain electron-sensitive materials, electron probe measurements may be continued indefinitely, whereas in the ion probe the number of atoms of the element of interest actually present in the volume of interest sets a fundamental limit on precision. However, the destructive nature of the ion bombardment, coupled with the fact that secondary ions are generated very close to the surface, may be turned to advantage in studying

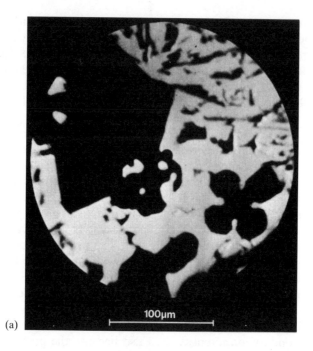

(a)

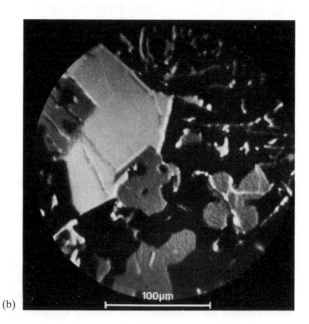

(b)

Figure 1.19 Mass-selected images of an alloy showing the distribution of (a) ^{27}Al and (b) ^{28}Si. (Slodzian, 1964. Reproduced with permission, Annales de Physique.)

the variation of composition with depth. With suitable control, a depth resolution of approximately 10 nm can be obtained and this type of measurement has been of great importance in the study of short-range diffusion profiles in minerals and in semiconductors.

The background between peaks in secondary ion spectra is generally very low so that peak/background for many elements can exceed 10^7 (cf. $\sim 10^3$ for the electron probe). For those elements for which the secondary-ion yield is also high, the detection limit is often well below 1 ppm and it is worth noting that lithium, which is beyond the range of the electron probe, can be measured at the 1 ppb level. However, to attain such detection limits with elements of higher atomic number, it is usually necessary to take into account the effect of spectral interferences caused by molecular peaks which are isobaric (i.e. have the same integer mass numbers as the atomic peak of interest). Typical examples are the interference of $^{28}Si^{16}O_2$ with ^{60}Ni and $^{174}Hf^{16}O_2$ with ^{206}Pb, the latter encountered in the U–Pb dating of zircon. Such molecules constitute a background which can severely limit sensitivity for the atomic species but fortunately there are techniques, detailed in Chapter 6, by which their effect may be much reduced or even eliminated. We may refer here briefly to one important method of eliminating interferences because of its influence on the development of ion probe instrumentation for the geological sciences.

Actual atomic masses differ from the sum of the masses of the constituent protons and neutrons by small fractions of a mass unit. For a given isotopic species this difference, which arises from the nuclear binding energy, is known as the 'mass defect' of the species. On the accepted scale, ^{12}C has a mass of 12.0000 and a mass defect of zero. For the first of the examples given above the actual masses of the two species are ^{60}Ni: 59.931 and $^{28}Si^{16}O_2$: 59.967 mass units (daltons) respectively. There is thus a mass difference of 0.036 mass units and, if the spectrometer has an adequate resolution, the peak at mass 60 will appear as a doublet with the SiO_2 separated from the atomic Ni.

The mass resolution of a spectrometer is given by $M/\Delta M$ where ΔM is the mass difference of two peaks at mass M which are separated by a valley, usually defined at 10% of the peak height. Thus, a resolution of 60/0.036 = 1667 would be required to separate Ni from SiO_2. In terms of instrumental parameters, the resolution for a given design of magnetic spectrometer is given by:

$$M/\Delta M = k.R/s \qquad (1.14)$$

where R is the radius of curvature of the ion flight path in the magnet, s is the width of the entrance slit and k is a constant. Hence, for a given instrument it is necessary to reduce s to obtain high resolution, generally at the expense of signal intensity.

Some measurements, notably U–Pb dating of zircons, need both high mass resolution (separation of the HfO_2–Pb doublet above requires an $M/\Delta M$ of ~5000) and high transmission to give good statistical counting precision. To this end, W. Compston at the Australian National University in Canberra (Compston, Williams and Clement, 1982) constructed a large instrument ('SHRIMP') with $R = 100$ cm, thus permitting a large value for s and consequently high ion transmission. This instrument became operational in the early 1980s and for 10 years has dominated work in the field. Early work on the high-resolution dating of zircon with the IM20 instrument operated in the present author's laboratory (Hinton and Long, 1979) achieved specific count rates of the order of 0.01 counts/s/ppm Pb/nA incident probe current. The ANU instrument improved upon this figure by a factor of about 1000!

The spectra shown in Figure 1.20 were obtained with a new instrument (Long and Gravestock, 1988; Coath, 1990), which has been designed to combine high transmission and resolution using a relatively small mass spectrometer ($R' = 30$ cm). Figure 1.20 illustrates a number of points from the above discussion: the very low background and high sensitivity of the ion probe, the resolution of molecular species from isobaric atomic species and the localized measurement of isotopic abundance.

The high sensitivity, low detection limits and isotopic capability of the ion probe thus make it a valuable complement to, but by no means a substitute for, the electron probe. The final characteristic ((5) above) imposes a severe restriction on its use as a general analytical tool: the complex relationship

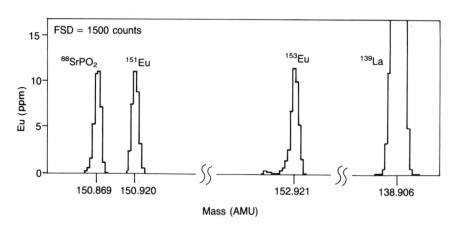

Figure 1.20 High-resolution mass spectra of rare-earth elements in Durango apatite taken with the P7-Concept ion microprobe: primary beam O^-, 5 μm diameter, current 2 nA mass resolution, $M/\Delta M$, ~8000; specific intensity for Eu ~2 counts per second/ppm/nA. Note the very low inter-peak background and resolution of molecular species. The isotopic abundance ratio, $^{153}Eu/^{151}Eu$ of 0.905 ± 0.02, is determined on ~10^{-16} g Eu.

between the secondary ion intensity and concentration is discussed fully in Chapter 6.

1.10 Selected-area mass spectrometry with indirect excitation

There are considerable advantages in decoupling the process of sample removal from the subsequent ionization prior to mass spectrometry. An example of such decoupling is the ionization of sputtered material by a suitably directed electron beam. One obvious attraction is that the conditions for the two physical processes may be optimized independently. For example, selective excitation may be employed to enhance the yield and hence the sensitivity for one particular element.

Even at thermal energies, sputtered material spreads rapidly away from the point of impact of the primary probe, and thus the volume within which indirect ionization takes place will inevitably be larger than that within which sputtered ions are formed in the normal ion probe. The resulting plasma is therefore more dilute and interelement effects correspondingly reduced, thus, in principle, simplifying the task of establishing calibration curves appropriate to different sample matrices.

The technique is, however, by no means a panacea for the problems of the ion probe: no improvement in sensitivity would be obtained for those elements, such as the alkalis, which in any case have a very high ion yield in sputtering. Furthermore, in calculating the overall gain, it is necessary to take into account factors such as the fraction of the total ablated plume irradiated by the ionizing beam, the ionization cross section and the proportion of ions generated that can actually be collected at the entrance slit of the spectrometer. In comparison with the very high efficiency (often 100%) with which directly produced secondary ions may be collected in the conventional ion probe, the efficiency of the indirect process, given by the product of the factors noted above, may be disappointingly small. Nevertheless, techniques have emerged over the past 10–20 years in which indirect excitation has been used with a net advantage. For further information on indirect excitation and selective ionization by focused laser light (resonance ion mass spectrometry or RIMS) the reader is referred to papers in recent proceedings of the series of international conferences on secondary ion mass spectrometry, e.g. De Bisschop and Vandervorst (1988).

1.11 Laser heating

As an alternative to ion bombardment or mechanical excavation, a very effective method of removing a small volume of a specimen for subsequent mass spectrometry is by local heating. The most convenient method of producing the necessary concentration of energy for this purpose is by use of a focused laser beam. Focused electron beams could also be used but inevitably involve an extension to the vacuum system and possibly also problems of specimen charging. A focused laser beam, on the other hand, may be directed onto the sample via a glass or quartz window, so giving a very clean system with only a small increase in volume.

A full discussion, both of the technique and of the history of the use of lasers for this purpose, is given in Chapters 7 and 8. We may note here that although the beam may be focused to a few micrometres in diameter the volume of specimen actually heated is generally significantly larger and typically of the order of 20–50 μm. Lasers may be operated either to give a steady output (continuous wave or CW operation) or in a pulsed mode in which a much higher peak power is produced for a very short period, often of only a few nanoseconds. Table 1.3 shows qualitatively the effect of increasing power density.

Under continuous low-power irradiation, the specimen reaches thermal equilibrium with a hot spot at the focus of the beam and generally a logarithmic radial temperature gradient away from the central region. With a pulsed laser, thermal equilibrium is not established, partly because the pulse duration may be only of the order of nanoseconds and also because at high power, material is almost instantaneously ablated.

1.11.1 Localized release of gases

At low power densities, maintained continuously but gradually increased in steps with time, heating of the specimen may be controlled so as to produce a corresponding stepwise increase in temperature. This technique is particularly applicable to the controlled release of gases from the irradiated volume for subsequent isotopic analysis by a sensitive version of a conventional gas mass spectrometer and stems from the first use by Megrue (1967) for noble-gas analysis. In addition to release of gas by the effect of heat alone, localized combustion and fluorination can be carried out by introducing oxygen and fluorine respectively into the sample chamber.

1.11.2 Laser ablation inductively coupled plasma-mass spectrometry

The highly ionized gas produced in an electrical discharge such as an arc struck between two electrodes is referred to as a plasma. Optical emission

Table 1.3 Heating effect of focused laser probes

Power density ($W\,m^{-2}$)	Power input for $100\,(\mu m)^2$ ($11.3\,\mu m$ diameter) (W)	Effect
10^{10}	0.1	Heating
10^{13}	100 (peak)	Melting
$>10^{13}$	>100	Ablation

spectroscopy, one of the earliest physical techniques used in the Earth Sciences, employs a DC arc to volatilize 20–100 mg of rock or mineral powder and to excite optical emission spectra from atoms excited above the ground state or in some cases from ions. The technique was widely used for trace element analysis (e.g. Nockolds and Mitchell, 1948).

A 'plasma torch' may be constructed by placing a copper coil carrying a radio-frequency current around a silica tube through which argon gas is flowing. Once ionization is initiated (by means of a spark discharge), the conducting gas forms the secondary of the transformer of which the copper coil is the primary. A continuous power dissipation of the order of 1 kW is readily achieved, resulting in a core temperature in the gas of up to 10 000 K. A plasma at this temperature is a very efficient means of vaporizing and ionizing fine particulate material introduced into the gas stream and this arrangement forms the basis of a number of analytical techniques.

In inductively coupled plasma-atomic emission spectrometry (ICP-AES), the sample is usually taken into solution which, in turn, is introduced into the plasma as a fine spray. This arrangement has the advantage that the concentration of sample atoms in the plasma is much lower than in the DC arc method so that troublesome inter-element matrix effects associated with the latter are largely eliminated. The technique is used extensively for the routine analysis of major and trace elements, including the rare-earth elements as described by Thompson and Walsh (1983).

Inductively coupled plasma-mass spectrometry (ICP-MS), first developed by Gray (1974), uses the same technique except that, instead of recording the optical spectra of atoms in the plasma, ions are extracted from the plasma stream by way of a small aperture and a differentially pumped inlet system and then accelerated into a mass spectrometer. For further information on the development and application of this technique, the reader is referred the text by Date and Gray (1988).

The above techniques, which all examine bulk samples, strictly have no place in this volume except in tracing the evolution of the latest variant, in which a pulsed laser is used to volatilize the material in a small volume of the specimen directly into the gas entering the plasma torch. This technique, known as laser ablation ICP-MS, again has the advantage of decoupling sampling and ionization and is treated fully in Chapter 7.

1.12 Detection of molecular species

The absorption of electromagnetic radiation in the ultraviolet, visible and infrared regions of the spectrum takes place as a result of interaction with electrons occupying electronic, vibrational and rotational energy levels within atoms and molecules of the absorbing specimen. Spectrometry of the transmitted beam can thus give information on electronic configurations and molecular associations. Infrared spectrometry, for example, has been used for many years to distinguish, by their characteristic absorption frequencies, hydrogen present in hydroxyl groups from that in water molecules within minerals. For a review of the theory and technique as used on a 'macro' scale (typically on samples of the order of a few milligrams), the reader is referred to the monograph edited by Farmer (1974).

1.12.1 Optical absorption microspectrometry

In many cases, however, useful measurements can be made only by isolating much smaller samples such as single mineral grains or separate phases, e.g. fluid inclusions, contained within individual grains. For anisotropic minerals, the absorption characteristics may vary with orientation, which thus needs to be set during the measurement. The first measurements of this type were made in California by Burns, who coupled a petrological microscope equipped with a universal stage to a commercial spectrophotometer (Burns, 1966(a)). This equipment was used in various applications, including a study of the origin of optical pleochroism in orthopyroxenes (Burns, 1966(b)). The growing interest in fluid inclusions over the past two decades has led to the use of infrared spectrometry for the identification of the molecular constituents of individual inclusions down to $\sim 10\,\mu$m in diameter (Vry, Brown and Beauchaine, 1987).

1.12.2 Micro laser-Raman spectrometry

A technique complementary to micro infrared spectrometry is that of micro-Raman spectrometry. This makes use of the effect first observed by Raman and Krishnan (1928) in which part of the light scattered by a solid, liquid or gas suffers a shift in frequency by an amount which is characteristic of the vibrational energy levels of the scattering molecules. The proportion of the light scattered in this way is very small and the development of a microtechnique has only come about since the advent of lasers that provide very intense sources of sharply defined wavelengths. It is described in detail in Chapter 10.

1.13 Which technique? Analytical strategy and tactics

Cost limits the number of geochemical techniques that can be supported in any one laboratory and it is always tempting to try to tailor an analytical problem to those techniques available locally. A better approach is to define the objective of the analysis, select the best technique from those available nationally or even internationally, and then attempt to obtain access to the appropriate facility. Many university departments provide a 'service' on a more-or-less informal basis which makes this possible.

1.13.1 Sampling

First, it is essential to describe the information required in as much detail as possible. Almost inseparable from this exercise will be a decision on the size of sample necessary to provide a meaningful result. For example, if the objective is to determine how a particular trace element is distributed between different rocks in a differentiation sequence, then it follows that the samples must be sufficiently large to be representative of each rock. The amount of material required will depend on factors such as grain size and whether or not the trace element is strongly concentrated in a particular, perhaps rare, mineral phase. In such a case, the sample size will be of the order of grams, or perhaps kilograms, and the techniques discussed in this volume will not be appropriate.

Broadly, therefore, the choice of technique will be determined by 'what and where'. When the nature of the problem indicates a selected-area technique, a simple sequence of initial measurements will provide answers to a number of important questions.

1. Repeated measurements on homogeneous standard material:	What is the reproducibility of the technique?
2. Repeated measurements at a single point on the unknown:	Is the reproducibility on the unknown worse than that of the standard?
3. Measurements at different points within single mineral grain:	Is the grain zoned?
4. Measurements on different grains:	Are all grains from the same population?

The observed variability at stage 2 requires careful evaluation. When due account has been taken of the effect of counting statistics, any increase over that observed with a homogeneous standard can arise for a variety of reasons. In the case of a 'destructive' technique, e.g. the ion probe, the observed variation may simply represent a compositional change with depth and, if associated with a concentration gradient normal to the surface, should be repeatable at a neighbouring point. For a nominally non-destructive technique, e.g. microbeam X-ray fluorescence, excess variability may be observed if the specimen is inhomogeneous on a scale comparable with the probe diameter, so that any relative movement of specimen and probe results in sampling of compositionally different material; such a situation could arise, for example, from the presence of small inclusions or at phase boundaries.

The problem of analysing material where the scale of the inhomogeneity is comparable with the resolution of the technique does not have a simple solution. It is sometimes possible to effect an improvement of the resolution by deconvolution of the smeared concentration/distance profile if the instrumental excitation function is known. More often, however, there are only two possibilities. The technique may be modified so as to improve the spatial resolution, e.g. by changing from the normal thick specimen electron probe technique to the thin specimen TEM method discussed earlier. Alternatively, the resolution may be degraded to the point where the exciting beam is large enough to excite a representative sample of the inhomogeneous material. Scanning the probe over a sufficiently large area and integrating the resultant signal is often a convenient way of achieving such sampling.

1.13.2 Measurement efficiency

It is a fundamental characteristic of selected-area techniques that the potential volume of data is very large. For example, while the bulk XRF analysis of a rock powder can only yield one true answer, the number of individual, potentially different electron-probe analyses of a $1\,\text{mm}^2$ grain of a mineral in a rock section exceeds 10^5. An important part of measurement strategy is therefore the efficient use of instrument time and selection of the best method for presenting the data. Thus, a preliminary investigation as discussed in section 1.13.1 may reveal inhomogeneity so presenting a number of options:

1. To investigate the inhomogeneity on a point-by-point basis at high precision and/or accuracy;
2. to show the distribution of the inhomogeneity, using a scanning technique, almost certainly at lower precision, the result being presented as a distribution map, contour plot or as a three-dimensional x–y–concentration plot;
3. to determine the variation of composition along one or more line traverses at intermediate precision;
4. to degrade the spatial resolution (by enlarging the probe or by scanning) and so to obtain an average for a known area;
5. on the basis of the probe data, to excise part of the sample and subject it to 'bulk' analysis by a technique offering greater precision or accuracy, e.g. by thermal ionization mass spectrometery after preliminary measurement with an ion probe.

Often a combination of these procedures will be appropriate and when working with new, unknown material it is essential for the operator to review progress and to be prepared to make tactical changes during the course of an investigation (while always bearing in mind the implications for instrument running time).

A very important point emerges from this discussion: in the case of techniques using a bulk sample and consequently yielding only one result, the work of the scientist is essentially that of choosing the bulk sample, specifying the technique of analysis and interpreting the result. Given that the analytical method is fully developed, the actual analysis is a technological exercise, possibly complex, but nevertheless capable of giving only one result if competently executed. Any feedback will involve either the selection and preparation of a new sample or the decision to subject the same sample to some alternative analytical procedure.

The situation in the case of selected-area techniques is generally very different for two reasons: first, as noted above, the number of possible individual measurements is very large, and secondly, the speed of analysis is often very high, so that preliminary results, at least, may appear in minutes if not seconds. As a result, except where exploratory work has already defined an exact course of action, it is essential, if maximum efficiency is to be achieved, that the owner of the specimen should sit at, and preferably drive, the instrument and so be able to dictate the tactics of the investigation. No arrangement where a thin section is 'posted' to the resident instrument operator can achieve the same ends unless the operator is as fully acquainted with the material and its background as the owner himself.

The increasing sophistication of the control of many analytical instruments also has a number of important consequences. As already indicated, there are often several different ways of configuring the analytical procedure and of presenting the results. Further, individual operations may be initiated by means of a few keystrokes, so that, in great contrast to the situation in the pre-computer age, the user is frequently isolated from the actual functions of the instrument itself. For such reasons, all but very experienced users require guidance from a specialist who has a full understanding, not only of the physics of the instrument and the interaction of the exciting beam with the specimen, but also of the limitations imposed by the methods of data reduction and of their current state of development. Only by constant awareness of the whole situation can the quality of the resulting data be assured.

The specialist and user can interact in two ways: the specialist may act as 'chauffeur' and drive the instrument as directed by the user; alternatively, he may sit in the passenger seat and act as a driving instructor to the user. If the user only makes very infrequent use of the technique, the first arrangement is clearly more efficient. On the other hand, there are many advantages of 'hands-on' control. To pursue the analogy, only then does the user become fully aware of the 'feel of the road' – in practice, to sense when a measurement is not totally reliable or that there is an incipient malfunction of the instrument. With this first-hand knowledge, he or she is much more likely to countenance appropriately cautious corrective action, for example the remeasurement of a standard, than when the roles are reversed. Then, the impatience of a dominant (and possibly ill-informed) user, anxious only for results, can inhibit the performance of essential confirmatory measurements, to the detriment of the end result. It is worth remembering that although the capability of analytical equipment has never been greater, never has it been so easy to produce bad data at such high speed and with so little effort.

References

Agrell, S.O. and Long, J.V.P. (1960) The application of the scanning X-ray microanalyser to mineralogy, in *X-ray Microscopy and Microanalysis* (eds A. Engström, V.E. Cosslett and H.H. Pattee), Elsevier, Amsterdam, pp. 391–400.

Burns, R.G. (1966a) Apparatus for measuring polarized absorption spectra of small crystals. *J. Sci. Instrum.*, **43**, 58–60.
Burns, R.G. (1966b) Origin of optical pleochroism in orthopyroxenes. *Mineral. Mag.*, **35**, 715–19.

Castaing, R. (1951) Application des sondes électroniques à une methode d'analyse ponctuelle chimique et crystallographique. PhD Thesis, University of Paris.
Coath, C.D. (1990) A study of ion optics for micro-beam SIMS. PhD Thesis, University of Cambridge.
Compston, W., Williams, I.S. and Clement, S.W.J. (1982) U–Pb ages within single zircons using a sensitive high-resolution ion microprobe. 30th Ann. Conf. Am. Soc. Mass Spectrometry, pp. 593–5.
Cosslett, V.E. and Nixon, W.C. (1951) X-ray shadow microscope. *Nature*, **168**, 24–5.
Cosslett, V.E. and Nixon, W.C. (1960) *X-ray Microscopy*, Cambridge University Press.
Coster, D. and von Hevesy, G. (1923) On the missing element of atomic number 72. *Nature*, **111**, 79.
Crookes, W. (1879) Contributions to molecular physics in high vacuo. *Phil. Trans.*, **170**, 641–62.

Date, A.R. and Gray, A.L. (eds) (1988) *Applications of Inductively Coupled Plasma Mass Spectrometry*, Blackie, Glasgow and London.
De Bisschop, P. and Vandervorst, W. (1988) Investigation of some characteristics of a laser-based post-ionization scheme, in *Secondary Ion Mass Spectrometry, SIMS VI* (eds A. Benninghoven, A.M. Hüber and H.W. Werner), John Wiley, Chichester, pp. 809–12.
Drummond, I.W. (1968) An investigation into the design of an ion-probe analyser. PhD Dissertation, University of Cambridge.
Duncumb, P. (1959) The X-ray scanning microanalyser. *Brit. J. Appl. Phys.*, **10**, 420–7.
Duncumb, P. (1963) X-ray micro-analysis of elements in the range Z = 4–92 combined with electron microscopy and electron diffraction, in *X-ray Optics and X-ray Microanalysis* (eds H.H. Pattee, V.E. Cosslett and A. Engström), Academic Press, NY, pp. 431–9.
Duncumb, P. and Cosslett, V.E. (1957) A scanning microscope for X-ray emission pictures, in *X-ray Microscopy and Microradiography* (eds V.E. Cosslett, A. Engström and H.H. Pattee), Academic Press, NY, pp. 374–80.
Duncumb, P. and Melford, D.A. (1960) Design considerations of an X-ray scanning microanalyser used mainly for metallurgical applications, in *X-ray Microscopy and Microanalysis* (eds A. Engström, V.E. Cosslett and H.H. Pattee), Elsevier, Amsterdam, pp. 358–64.

Farmer, V.C. (ed.) (1974) The infra-red spectra of minerals. *Mineral. Soc. Monograph*, **4**, Mineralogical Society, London.

Goby, P. (1913) Une application nouvelle des rayons X: la microradiographie. *Compt. Rend.*, **156**, 686–8.

REFERENCES

Gray, A.L. (1974) A plasma source for mass analysis. *Proc. Soc. Anal. Chem.*, **11**, 182–3.

Hinton, R.W. and Long, J.V.P. (1979) High-resolution ion-microprobe measurement of lead isotopes: variations within single zircons from Lac-Seul, northwestern Ontario. *Earth Planet. Sci. Lett.*, **45**, 309–25.

Hooper, K. (1960) Microradiography in quantitative micropaleontology: techniques, in *X-ray Microscopy and Microanalysis* (eds A. Engström, V.E. Cosslett and H.H. Pattee), Elsevier, Amsterdam, pp. 216–23.

Johansson, T.B., Akselsson, R. and Johansson, S.A.E. (1970) Elemental trace analysis at the 10^{-12} g. level. *Nucl. Instr. and Methods*, **84**, 141–3.

Knoll, G.F. (1979) *Radiation Detection and Measurement*, Wiley, NY.
Knoll, M. and Ruska, E. (1932) Das Elektronenmikroscop. *Z. Physik*, **78**, 318–39.

Leta, D.P. and Morrison, G.H. (1980) Ion implantation for in-situ quantitative ion-microprobe analysis. *Anal. Chem.*, **52**, 277–80.
Liebl, H.J. (1967) Ion microprobe mass analyzer. *J. Appl. Phys.*, **38**, 5277–83.
Long, J.V.P. (1958) Microanalysis with X-rays, Thesis, University of Cambridge.
Long, J.V.P. (1965) A theoretical assessment of the possibility of selected-area mass-spectrometric analysis using a focused ion beam. *Brit. J. Appl. Phys.*, **16**, 1277–84.
Long, J.V.P. and Agrell, S.O. (1965) The cathodoluminescence of minerals in thin section. *Miner. Mag.*, **34**, 318–26.
Long, J.V.P. and Cosslett, V.E. (1957) Some methods of X-ray microchemical analysis, in *X-ray Microscopy and Microradiography* (eds V.E. Cosslett, A. Engström and H.H. Pattee), Academic Press, NY, pp. 435–42.
Long, J.V.P. and Gravestock, D.C. (1988) An ion microprobe system for geological and other applications, in *Secondary Ion Mass Spectrometry SIMS VI* (eds A. Benninghoven, A.M. Hüber and H.W. Werner), John Wiley, Chichester, pp. 161–4.
Long, J.V.P. and Röckert, H.O.E. (1963) X-ray fluorescence microanalysis and the determination of potassium in nerve cells, in *X-ray Optics and X-ray Microanalysis* (eds H.H. Pattee, V.E. Cosslett and A. Engström), Academic Press, NY.
Lyons, L. (1986) *Statistics for Nuclear and Particle Physicists*, Cambridge University Press.

Marshall, D.J. (1988) *Cathodoluminescence of Geological Materials*, Unwin Hyman, London.
McMullen, D. (1953) An improved scanning electron microscope for opaque specimens. *Proc. Instn. Elec. Engrs.*, **100**, 254–9.
Megrue, G.M. (1967) Isotopic analysis of rare gases with a laser microprobe. *Science*, **157**, 1555–6.
Morrison, G.R., Beswetherick, J.T., Browne M.T. *et al.* (1990) Development and applications of the Kings's College/Daresbury X-ray microscope, in *X-ray Microscopy in Biology and Medicine* (eds Shinohara *et al.*), Japan Sci. Soc. Press, Tokyo/Springer-Verlag, Berlin, pp. 99–108.
Moseley, H.G.J. (1913) The high-frequency spectra of the elements. *Phil. Mag.*, **26**, 1024–34.
Moseley, H.G.J. (1914) The high-frequency spectra of the elements; part II. *Phil. Mag.*, **27**, 703–13.
Mulvey, T. (1960) A new microanalyser, in *X-ray microscopy and microanalysis* (eds A. Engström, V.E. Cosslett and H.H. Pattee), Elsevier, Amsterdam, pp. 372–8.

Nockolds, S.R. and Mitchell, R.L. (1948) The geochemistry of some Caledonian plutonic rocks: a study in the relationships between the major and trace elements of igneous rocks and their minerals. *Trans. Roy. Soc. Edinburgh*, **61**, 535–75.

Raman, C.V. and Krishnan, R.S. (1928) A new class of spectra due to secondary radiation. *Indian J. Phys.*, **2**, 399–419.

Raman, C.V. (1928) A new radiation. *Indian J. Phys.*, **2**, 387–98.

Slodzian, G. (1964) Étude d'une methode d'analyse locale chimique et isotopique utilisant l'émission ionique secondaire. Thesis, University of Paris. *Ann. Phys.*, **9**, 591–648.

Slodzian, G. (1988) Introduction to fundamentals in direct and scanning secondary-ion microscopy, in *Secondary Ion Mass Spectrometry SIMS VI* (eds A. Benninghoven, A.M. Hüber and H.W. Werner), John Wiley, Chichester, pp. 3–12.

Smith, J.V. and Stenstrom, R.C. (1965) Electron-excited luminescence as a petrologic tool. *J. Geol.*, **73**, 627–35.

Taylor, J.R. (1982) *An Introduction to Error Analysis*, University Science Books, Mill Valley, California.

Thompson, M. and Walsh, J.N. (1983) *A Handbook of ICP Spectrometry*, Blackie, Glasgow and London.

Topping, J. (1972) *Errors of Observation and their Treatment*, Chapman & Hall, London.

von Hamos, L. (1934) Micro-chemical analysis of plane polished surfaces by means of monochromatic X-ray images. *Nature*, **134**, 181–2.

von Hevesy, G. (1932) *Chemical Analysis by X-rays and its Applications*, McGraw Hill, NY.

Vry, J., Brown, P.E. and Beauchaine, J. (1987) Application of micro FTIR spectroscopy to the study of fluid inclusions. *Fluid Inclusion Res.*, **20**, 396.

Wright, I.P. and Pillinger, C.T. (1988) New frontiers in stable isotope research: laser probes, ion probes and small-sample analysis. *US Geol. Survey Bull.*, **1890**.

Zeitz, L. and Baez, A.V. (1957) Microchemical analysis by emission spectrographic and absorption methods, in *X-ray Microscopy and Microradiography* (eds V.E. Cosslett, A. Engström and H.H. Pattee), Academic Press, NY, pp. 417–34.

CHAPTER TWO
Electron probe microanalysis
Stephen J.B. Reed

2.1 Introduction

2.1.1 Principles of the technique

Electron probe microanalysis makes use of the X-ray spectrum emitted by a solid sample bombarded with a focused beam of electrons to obtain a localized chemical analysis. All elements from atomic number 4 (Be) to 92 (U) can be detected in principle, though not all instruments are equipped for 'light' elements ($Z < 10$). Qualitative analysis involves the identification of the lines in the spectrum and is fairly straightforward owing to the simplicity of X-ray spectra. Quantitative analysis (determination of the concentrations of the elements present) entails measuring line intensities for each element in the sample and for the same elements in calibration standards of known composition.

By scanning the beam in a television-like raster and displaying the intensity of a selected X-ray line, element distribution images or 'maps' can be produced. Also, images produced by electrons collected from the sample reveal surface topography or mean atomic number differences according to the mode selected. The scanning electron microscope (SEM), which is closely related to the electron probe, is designed primarily for producing electron images, but can also be used for element mapping, and even point analysis, if an X-ray spectrometer is added. There is thus a considerable overlap in the functions of these instruments.

2.1.2 Accuracy and sensitivity

X-ray intensities are measured by counting photons and the precision obtainable is limited by statistical error. For major elements it is usually not difficult to obtain a precision (defined as 2σ) of better than $\pm 1\%$ (relative), but the overall analytical accuracy is commonly nearer $\pm 2\%$, owing to other factors such as uncertainties in the compositions of the standards and errors in the various corrections which need to be applied to the raw data.

As well as producing characteristic X-ray lines, the bombarding electrons also give rise to a continuous X-ray spectrum (section 2.2.4), which limits

Microprobe Techniques in the Earth Sciences. Edited by P.J. Potts, J.F.W. Bowles, S.J.B. Reed and M.R. Cave. Published in 1995 by Chapman & Hall, London. ISBN 0 412 55100 4

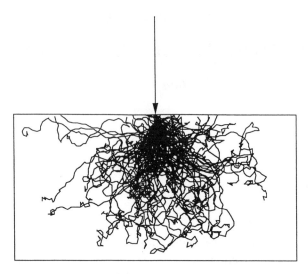

Figure 2.1 Simulated trajectories of electrons (initial energy 20 keV) in silicon (rectangle = 1 × 2 μm).

the detectability of small peaks, owing to the presence of 'background'. Using routine procedures, detection limits are typically about 100 ppm (by weight) but can be reduced by using long counting times. Lower detection limits are obtainable with proton bombardment (Chapter 4) because the continuous X-ray spectrum is less intense. The background in X-ray fluorescence analysis (Chapter 5) is also lower.

2.1.3 Spatial resolution

Spatial resolution is governed by the penetration and spreading of the electron beam in the specimen (Figure 2.1). Since the electrons penetrate an approximately constant mass, spatial resolution is a function of density. In the case of silicates (density about $3 \, \text{g cm}^{-3}$), the nominal resolution is about of 2 μm under typical conditions, but for quantitative analysis a minimum grain size of several micrometres is desirable.

Better spatial resolution is obtainable with ultra-thin (~100 nm) specimens, in which the beam does not have the opportunity to spread out so much. Such specimens can be analysed in a transmission electron microscope (TEM) with an X-ray spectrometer attached, also known as an analytical electron microscope, or AEM (Chapter 3).

2.1.4 Sample preparation

Since the electron probe analyses only to a shallow depth, specimens should be well polished so that surface roughness does not affect the results.

INTRODUCTION

Sample preparation is essentially as for reflected light microscopy, with the proviso that only vacuum compatible materials must be used. Opaque samples may be embedded in epoxy resin blocks. For transmitted light viewing, polished thin sections on glass slides are prepared.

In principle, specimens of any size and shape (within reasonable limits) can be analysed. Holders are commonly provided for 25 mm (1") diameter round specimens and for rectangular glass slides. Standards are either mounted individually in small mounts or in batches in normal-sized mounts.

Many geological samples are electrically nonconducting and a conducting surface coat must be applied to provide a path for the incident electrons to flow to ground. The usual coating material is vacuum-evaporated carbon (~20 nm thick), which has a minimal influence on X-ray intensities on account of its low atomic number, and (unlike gold, which is commonly used for SEM specimens) does not add unwanted peaks to the X-ray spectrum. However, steps should be taken to maintain as constant a thickness as possible.

2.1.5 Applications

Electron probe analysis is a standard technique in mineralogy and petrology, thanks to its capability for rapid non-destructive *in situ* analysis with high spatial resolution, good accuracy and reasonably low detection limits, and the fairly simple specimen preparation requirements. Other microanalytical techniques described elsewhere in this volume can offer better performance in specific aspects (e.g. spatial resolution or detection limits), but none compares for all round usefulness and convenience.

Applications may be summarized under the following headings:

1. Mineral identification. The electron probe is invaluable for this purpose (together with optical microscopy and diffraction methods). It is especially useful for opaque minerals and for small or rare grains.
2. Descriptive petrology. Electron probe analysis is widely used to provide analytical data for the description and classification of rocks.
3. Geothermometry and geobarometry. Compositions of coexisting phases determined with the electron probe can be used to derive temperatures and pressures of rock formation.
4. Experimental petrology. High spatial resolution is particularly useful in experimental studies aimed at determining phase relationships and element partitioning (though the latter application is constrained by the attainable detection limits; section 2.9.10).
5. Cosmochemistry. The electron probe has played a vital role in the study of meteorites and lunar samples, where its nondestructive nature is especially important.
6. Zoning. The spatial resolution of the technique makes it a powerful tool for studying zoned minerals.

7. Diffusion studies. The high spatial resolution is also useful for measuring experimental diffusion profiles in systems of petrological interest.
8. Particle analysis. The ability to analyse small individual particles is valuable in various fields, including sedimentology and environmental studies.
9. Rare-phase location. Automated searching for areas of a specific composition enables grains of rare phases to be located.

2.2 X-ray spectroscopy

2.2.1 Atomic structure

According to the Rutherford–Bohr model of the atom, electrons orbit around the positive nucleus. In the normal state the number of orbital electrons equals the number of protons in the nucleus (given by the atomic number, Z). Only certain orbital states with specific energies exist and these are defined by quantum numbers (see standard texts). With increasing Z, orbits are occupied on the basis of minimum energy, those nearest the nucleus, and therefore the most tightly bound, being filled first.

Orbital energy is determined mainly by the principal quantum number (n). The shell closest to the nucleus ($n = 1$) is known as the K shell; the next is the L shell ($n = 2$), then the M shell ($n = 3$), etc. The L shell is split into three subshells designated L1, L2 and L3, which have different quantum configurations and slightly different energies (whereas the K shell is unitary). Similarly, the M shell has five subshells. This model of the inner structure of the atom is illustrated in Figure 2.2.

The populations of the inner shells are governed by the Pauli exclusion principle, which states that only one electron may possess a given set of quantum numbers. The maximum population of a shell is thus equal to the number of possible states possessing the relevant principal quantum number. In the case of the K shell this is 2, for the L shell 8, and for the M shell 18. Thus for $Z \geq 2$ the K shell is full, and for $Z \geq 10$ the L shell is full. Electrons occupying outer orbits are usually not directly involved in the production of X-ray spectra, which are therefore largely unaffected by chemical bonding etc.

2.2.2 Origin of characteristic X-rays

'Characteristic' X-rays result from electron transitions between inner orbits, which are normally full. An electron must first be removed in order to create a vacancy into which another can 'fall' from an orbit further out. In electron probe analysis vacancies are produced by electron bombardment, which also applies to X-ray analysis in the TEM (Chapter 3). Other mechanisms can be used, including proton bombardment, as in 'PIXE'

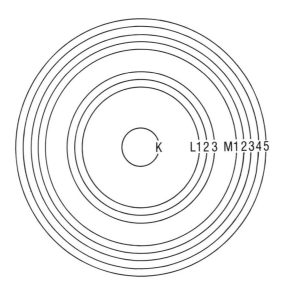

Figure 2.2 Schematic diagram of inner atomic electron shells.

(Chapter 4), and excitation with X-rays, as in X-ray fluorescence analysis (Chapter 5). Characteristic X-ray spectra produced by all these processes are essentially the same.

X-ray lines are identified by a capital Roman letter indicating the shell containing the inner vacancy (K, L or M), a Greek letter specifying the group to which the line belongs in order of decreasing importance (α, β, etc.), and a number denoting the intensity of the line within the group in descending order (1, 2, etc.). Thus the most intense K line is $K\alpha_1$. (The less intense $K\alpha_2$ line is usually not resolved, and the combined line is designated $K\alpha_{1,2}$ or just $K\alpha$). The most intense L line is $L\alpha_1$. Because of the splitting of the L shell into three subshells, the L spectrum is more complicated than the K spectrum and contains at least 12 lines, though many of these are weak.

Characteristic spectra may be understood by reference to the energy level diagram (Figure 2.3), in which horizontal lines represent the energy of the atom with an electron removed from the shell (or subshell) concerned. An electron transition associated with X-ray emission can be considered as the transfer of a vacancy from one shell to another, the energy of the X-ray photon being equal to the energy difference between the levels concerned. For example, the $K\alpha$ line results from a K–L3 transition (Figure 2.3). Energies are measured in electron volts (eV), 1 eV being the energy corresponding to a change of 1 V in the potential of an electron (=1.602×10^{-19} J). This unit is applicable to both X-rays and electrons. X-ray energies of interest in electron probe analysis are mostly in the range 1–10 keV.

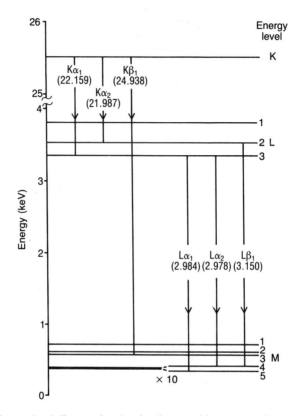

Figure 2.3 Energy level diagram for Ag showing transitions responsible for main K and L emission lines (arrows show direction of vacancy movement). Energy of emission line indicated in brackets.

A nomenclature system for characteristic lines based on the energy levels involved has been recommended by the International Union of Pure and Applied Chemistry (Jenkins *et al.*, 1991), but has not yet found widespread use. In this system, the Kα_1 line is known as K-L$_3$, for example.

The 'critical excitation energy' (E_c) is the minimum energy which bombarding electrons (or other particles etc.) must possess in order to create an initial vacancy. Figure 2.4 shows the dependence of E_c on Z for the principal shells. In electron probe analysis the incident electron energy (E_0) must exceed E_c and should preferably be at least twice E_c to give reasonably high excitation efficiency. For atomic numbers above about 35 it is usual to change from K to L lines to avoid the need for an excessively high electron beam energy (which has undesirable implications with respect to the penetration of the electrons in the sample, and in any case may exceed the maximum available accelerating voltage).

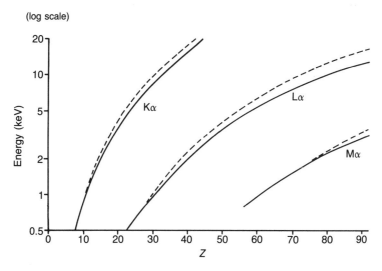

Figure 2.4 Energies of principal characteristic lines (—) and their excitation energies (---).

2.2.3 Wavelengths, energies and intensities of X-ray lines

The preceding discussion treated X-rays as photons possessing a specific energy (E). Sometimes it is more appropriate to describe X-rays by their wavelength (λ), which is related to energy by the expression:

$$E\lambda = 12396 \tag{2.1}$$

where E is in electron volts and λ is in Å, where $1\,\text{Å} = 10^{-10}\,\text{m}$ (this is the traditional unit for X-ray wavelength and is still commonly used in preference to the 'correct' SI unit, the nanometre, which is equal to $10\,\text{Å}$).

Since X-ray lines originate in transitions between inner shells, the energy of a particular line shows a smooth dependence on atomic number, varying approximately as Z^2 (Moseley's law). The energies of the $K\alpha_1$, $L\alpha_1$ and $M\alpha_1$ lines are plotted against Z in Figure 2.4.

The total X-ray intensity for a particular shell is divided between several lines. In the case of the K shell, more than 80% of the total intensity is contained in the combined $K\alpha_{1,2}$ line (Figure 2.5). The relative intensity of the $K\beta$ line decreases with decreasing atomic number, in accordance with the electron occupancies of the relevant energy levels. The intensity is more widely spread in the L spectrum – the combined $L\alpha_{1,2}$ line contains about 55% of the total intensity (Figure 2.6). The $L\beta$ lines have greater relative intensity and are split into two or more components. For elements of relatively low atomic number (e.g. Sr, $Z = 38$) the L spectrum is simplified,

ELECTRON PROBE MICROANALYSIS

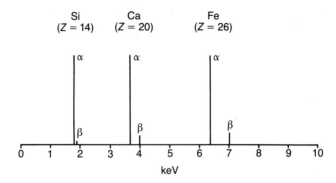

Figure 2.5 Typical K spectra.

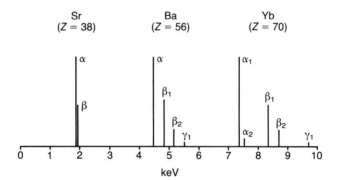

Figure 2.6 Typical L spectra.

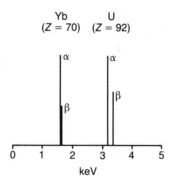

Figure 2.7 Typical M spectra.

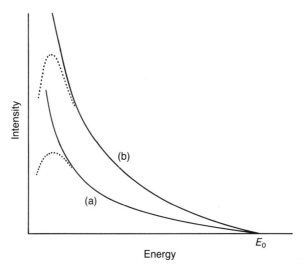

Figure 2.8 Continuum intensity as a function of X-ray energy (E_0 = incident electron energy): (a) low atomic number, (b) high atomic number; dotted lines show intensity after allowing for absorption in detector window, etc.

owing to certain energy levels being unoccupied. M spectra (Figure 2.7) are confined to relatively low energies and are quite simple, because of the limited occupancy of the outermost shells.

2.2.4 The continuous spectrum

Electron bombardment not only produces characteristic X-ray lines resulting from electron transitions between inner atomic shells but also a continuous X-ray spectrum or 'continuum', covering all energies from zero to E_0 (the incident electron energy). This continuum arises from interactions between incident electrons and atomic nuclei. The intensity of the continuum decreases monotonically with increasing X-ray energy (Figure 2.8), and is approximately proportional to Z. The main significance of the continuum in the present context is that it contributes the 'background' upon which characteristic elemental lines are superimposed.

2.3 Wavelength-dispersive spectrometers

2.3.1 Bragg reflection

X-ray spectrometers of the wavelength-dispersive (WD) type make use of crystal diffraction for wavelength selection. X-rays of wavelength λ are

Table 2.1 Crystals used in WD spectrometers

Name	Abbreviation	d (Å)	Wavelength range (Å)
Lithium fluoride	LIF	2.013	0.8–3.2
Pentaerythritol	PET	4.371	1.8–7.0
Thallium acid phthallate	TAP	12.95	5.1–21.0

strongly 'reflected' by a crystal of interplanar spacing d when the following condition (Bragg's law) is satisfied:

$$n\lambda = 2d \sin \theta \qquad (2.2)$$

where θ is the glancing angle of incidence and reflection (the 'Bragg angle') and n is the order of reflection (usually 1). Strong reflection occurs because rays diffracted by successive layers are in phase, the difference in path length being λ.

In the WD spectrometer (WDS) Bragg reflection is utilized to select a single wavelength at a time, different wavelengths being obtained by varying the angles of incidence and reflection simultaneously. The reflected ray is defined by a slit in front of the detector.

2.3.2 Choice of crystal

In a typical spectrometer the available range of angles corresponds to $\sin \theta$ = 0.2 to 0.8 approximately. For a given d the range of wavelengths is thus $0.4d$ to $1.6d$. Crystals of different d value are therefore needed to cover the required wavelength range (Table 2.1). These crystals cover the Kα lines of elements of atomic number 9–39 (F–Y) and Lα lines for $Z > 24$ (Cr). (The range varies somewhat for different instruments).

Crystals with large enough d values to cover the Kα lines of 'light' elements, i.e. $Z < 10$ (Ne), are difficult to obtain and for long wavelengths their place is taken by synthetic diffracting structures such as lead stearate (d = 50 Å) and other 'pseudo-crystals' made of soap film layers. Relatively recently, multilayers consisting of alternating evaporated layers of heavy and light elements (e.g. W and C) have been introduced. These give intense, though broad, reflections.

2.3.3 Focusing geometry

A flat crystal would be unsuitable for use with the point X-ray source of the electron microprobe since the angle of incidence would vary across the face of the crystal. WDSs in this application, therefore, use curved crystals,

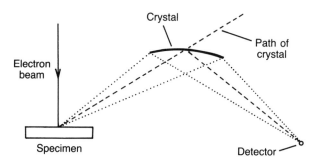

Figure 2.9 Geometry of WD spectrometer.

which give an approximately constant Bragg angle over the whole surface. The reflected X-rays form a focus at a distance from the crystal equal to the distance of the crystal from the source. The wavelength is varied by moving the crystal along a line passing through the source while the spectrometer mechanism maintains the correct crystal angle and detector position (Figure 2.9).

WD spectrometers are sensitive to displacement of the source, as this changes the angle of incidence at the crystal. This factor is important for quantitative analysis and in elemental mapping, where the beam is scanned.

2.3.4 Multiple spectrometers

As noted previously, several crystals are needed to cover the full range of relevant wavelengths. Since changing crystals is inconvenient and time-consuming, it is very desirable to have several WDSs fitted to a microprobe instrument, three or four being typical. SEMs often have only an EDS (section 2.4), though sometimes a WDS (normally a single one with several interchangeable crystals) is fitted.

Usually, WDSs are oriented vertically and disposed radially around the electron column. With this arrangement the apparent intensity of a peak is affected strongly by vertical displacement of the source, though this is not a problem if the specimen is flat and well polished and its height relative to the focal plane of the spectrometer is set with the aid of a high-power optical microscope. Horizontally oriented spectrometers are insensitive to vertical displacement of the source, but fewer of these can be fitted, due to space restrictions.

2.3.5 Proportional counters

In a WDS, X-rays are detected by means of a 'proportional counter', consisting of a gas-filled tube with a coaxial anode wire and a 'window' through

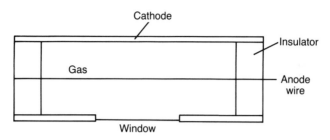

Figure 2.10 Proportional counter.

which X-rays can enter (Figure 2.10). Incoming X-rays ionize gas atoms, producing free electrons which move to the anode while the positive ions move out to the cathode (the counter body). Each absorbed X-ray photon thus produces an electronic pulse, its size reflecting the number of ions produced, which is in turn proportional to the photon energy. Pulses are counted to measure X-ray intensities, which are expressed in counts per second.

A commonly used counter gas is argon with 10% methane added to improve its properties. However, this is not dense enough to absorb short-wavelength X-rays effectively and xenon is sometimes used instead for this purpose. Alternatively, the required absorption can be obtained by using a counter filled with argon at higher pressure.

For long wavelengths (>4 Å, approximately) a very thin window (e.g. 1 μm polypropylene) is used to minimize absorption. Since such windows are slightly permeable, gas that escapes must be replaced from a cylinder supply. The gas flows continuously through the counter, which is thus known as a 'flow counter'. The 'sealed counter' used for shorter wavelengths has a thicker window and is sealed for life.

2.3.6 Pulse height analysis

As noted previously, the counter output pulses are proportional in size to the X-ray photon energy. By applying 'pulse height analysis', only pulses within a certain range of heights are accepted. This range is known as the 'window' and the lower limit as the 'threshold'. Unwanted X-rays, such as high-order Bragg reflections, can thus be rejected (Figure 2.11).

2.4 Energy-dispersive spectrometers

Energy-dispersive spectrometers (EDSs) employ pulse height analysis as just described: a detector giving output pulses proportional in height to the X-ray photon energy is used in conjunction with a pulse height analyser (in this

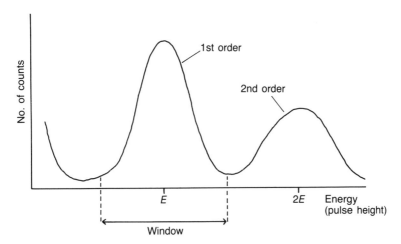

Figure 2.11 Pulse height distribution with pulse height analyser window set to select X-rays of energy E and reject second order reflection of energy $2E$.

case a multichannel type). Rather than a gas-filled counter, however, a solid-state detector is used because of its better energy resolution.

2.4.1 Lithium-drifted silicon detectors

Energy-dispersive (ED) systems attached to electron probes and SEMs nearly always use a 'lithium-drifted silicon', or Si(Li), detector. This device consists of a slice of silicon typically about 4 mm in diameter and 3 mm thick, into which lithium has been introduced by a process known as 'drifting', in order to compensate for residual impurities which would otherwise degrade the performance of the detector. Incident X-ray photons cause ionization in the detector, producing an electrical charge which is amplified by a sensitive preamplifier located close to the detector. Both detector and preamplifier are cooled with liquid nitrogen to minimize electronic noise. A typical detector assembly is shown in Figure 2.12. Detectors can also be made from germanium: hitherto, these have been used mainly for high-energy X-rays (and γ-rays), but are now available in a form suitable for low-energy X-rays and offer somewhat better energy resolution (see below).

2.4.2 Energy resolution

The ED spectrum is displayed in digitized form with the x-axis representing X-ray energy (usually in channels 10 or 20 eV wide) and the y-axis representing the number of counts per channel (Figure 2.13). An X-ray line (consisting of effectively mono-energetic photons) is broadened by the

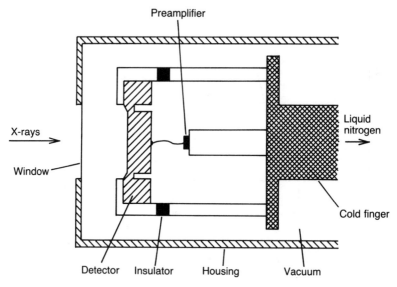

Figure 2.12 Lithium-drifted silicon detector assembly.

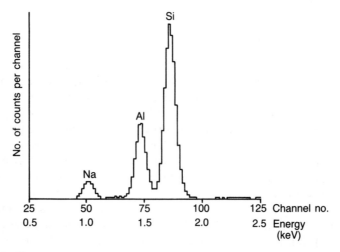

Figure 2.13 ED spectrum of jadeite (part), showing K peaks of Na, Al and Si.

response of the system, producing a Gaussian profile. Energy resolution is defined as the full width of the peak at half maximum height (FWHM). Conventionally, this is specified for the Mn Kα peak at 5.89 keV. For Si(Li) detectors, values of 130–150 eV are typical (Ge detectors can achieve 115 eV). Peak width is a function of energy, as shown in Figure 2.14.

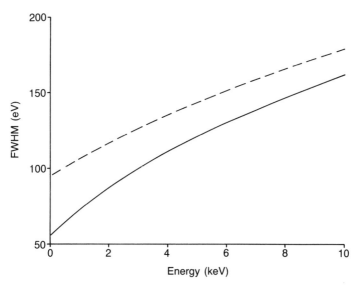

Figure 2.14 Energy resolution of an ED spectrometer (full width at half maximum of peak) as a function of X-ray energy: (—) – 'state of the art' Si(Li) detector (at low count rate), (---) – resolution obtained at high count rate or with lower-grade detector.

The resolution of an EDS is about an order of magnitude worse than that of a WDS, but is good enough to separate the K lines of neighbouring elements (Figure 2.13). EDSs have certain advantages which will be discussed later.

2.4.3 Dead time and throughput

In processing the pulses from a Si(Li) detector prior to pulse-height analysis, it is necessary to use quite a long integrating time to minimize noise. The system consequently has a significant 'dead time', or period after the arrival of an X-ray photon during which the system is unresponsive to further photons. This limits the rate at which pulses can be processed and added to the recorded spectrum. 'Throughput' passes through a maximum above which it decreases with further increases in input count rate (Figure 2.15). The maximum throughput rate is a function of the integration time and the design of the system.

Energy resolution is determined partly by the statistics of the detection process and partly by noise fluctuations in the baseline upon which the pulses are superimposed. The longer the integration time, the more the noise is smoothed out, and the better the energy resolution. There is thus a 'trade-off' between resolution and throughput. Hitherto, maximum throughput rates have been typically in the region of 10 000 counts s^{-1}, but

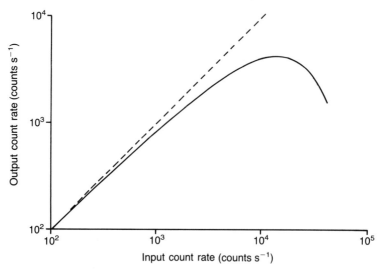

Figure 2.15 Output count rate of an ED system as a function of input count rate, showing the effect of dead time ((---) – no dead time).

recent developments allow rates as high as 30 000 counts s^{-1}. The energy resolution, though significantly degraded at such high rates, is still adequate for many purposes.

2.4.4 Entrance windows

The detector and associated components are normally cooled permanently with liquid nitrogen inside an evacuated housing. X-rays enter via a 'window', usually made of beryllium about 8 μm thick, chosen for its ability to withstand atmospheric pressure while transmitting X-rays down to about 1 keV. The K lines of light elements ($Z < 10$) can be detected only if the beryllium window is either removed completely or replaced by a thin organic film (the adjacent chamber of the instrument being evacuated). New types of window with various compositions (mostly involving B, N or C) have been introduced: these offer better transmission for low-energy X-rays than the standard beryllium type, while still withstanding atmospheric pressure. The detection efficiency at low energies with various types of window is illustrated in Figure 2.16. Even with no window, absorption occurs in the thin layer of gold which forms the front contact of the detector and in the 'dead layer' of silicon which exists at the surface of the detector, reducing the efficiency for low energies.

THE ELECTRON COLUMN

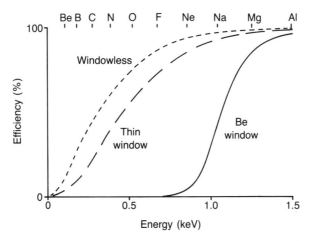

Figure 2.16 Effect of window absorption on detection efficiency of an Si(Li) detector in the low-energy region (element symbols indicate positions of Kα lines).

2.4.5 Collection efficiency

The collection efficiency of a Si(Li) detector is related to the solid angle subtended at the source, given by A/x^2, where A is the area of the detector and x is the distance from the source. Nearly all X-ray photons absorbed in the active region of the detector are detected, though for energies above about 20 keV some pass through and emerge from the back. Absorption in the window reduces the efficiency for low energies as shown in Figure 2.16.

If the detector is reasonably close to the sample (a few centimetres), the detection efficiency is greater than for WDSs – that is, the count rate obtained for a given beam current is higher. However, because of the relatively long dead time, it is necessary either to use a low beam current (e.g. a few nanoamperes) or alternatively to decrease the collection efficiency of the detector. For this purpose some detectors (especially those fitted to SEMs) can be moved away from the specimen by means of a translation mechanism. Another approach is to place a variable aperture in front of the detector. If the efficiency is reduced, a sufficiently high beam current for simultaneous WD analysis can be used without saturating the ED system.

2.5 The electron column

2.5.1 Electron gun

The usual source of electrons in electron beam instruments is a tungsten 'hairpin' filament heated to ~2500°C, from which electrons are emitted

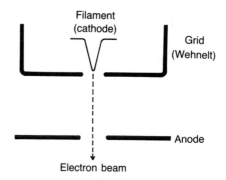

Figure 2.17 Electron gun.

thermionically – that is, they escape from the surface by virtue of their thermal energy. The filament forms the cathode of the electron 'gun' which is configured as a triode, with an anode and a 'grid' or 'wehnelt' (Figure 2.17). A bias voltage applied to the latter restricts electron emission to the tip of the filament. The anode is grounded and the filament is held at a high negative voltage (typically 10–30 kV). Under the influence of the electrostatic field, the electrons are accelerated through the hole in the anode. The emerging beam typically carries a current of around 50 μA.

Lanthanum and cerium hexaboride emitters are available, offering higher currents in the focused beam. Even more intense field emission sources are also available, but the main application of these devices is for high-resolution imaging rather than electron probe analysis, for which the tungsten filament is quite adequate.

2.5.2 Probe-forming system

The electron beam emerging from the gun is focused onto the surface of the specimen by means of magnetic lenses. The effective diameter of the electron source in the gun is about 50 μm. Demagnification by a factor of at least 50 is therefore required to obtain a final beam diameter of less than 1 μm. This can be achieved using the two-lens system shown in Figure 2.18.

In the electron probe, spatial resolution is limited by beam spreading in the sample and there is little advantage in reducing the incident beam diameter much below 1 μm. However, for applications requiring high-resolution scanning electron images, probe diameters as small as 5 nm can be achieved using three lenses (the number commonly incorporated in SEMs). Modern electron microprobes also usually have three lenses, to provide a similar imaging capability, though for X-ray analysis this is unnecessary.

Magnetic lenses suffer from aberrations similar to those occurring in glass lenses used to focus light. In electron lenses, spherical aberration (blurring

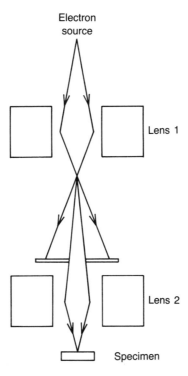

Figure 2.18 Two-lens probe forming system producing a demagnified image of the electron source on the specimen.

of the image caused by the outer regions of the lens focusing more strongly than the centre) is not amenable to correction as in glass lenses. To control this effect, the aperture of the final lens must therefore be limited. This has important implications for the maximum current available in the focused beam (section 2.5.3).

Astigmatism (rotational asymmetry in the focused beam) caused by slight imperfections in the lenses and dirt on the apertures can be corrected with a 'stigmator', consisting of coils designed to distort the beam to an equal and opposite extent to the astigmatism.

2.5.3 Beam diameter and current

The final beam diameter is governed by the diameter of the electron source, the demagnification of the lens system and spherical aberration. To reduce the beam diameter, the demagnification must be increased, entailing stronger lens settings. The result is a more divergent beam, a smaller fraction of which passes through the final aperture. Further, the size of the aperture must be reduced to decrease the effect of spherical aberration. The maximum available beam current thus decreases with decreasing probe diameter.

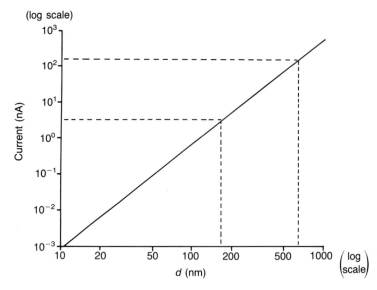

Figure 2.19 Probe current as a function of beam diameter d for a typical system; dashed lines indicate range used for electron probe analysis.

A plot of beam current versus diameter for a typical instrument is shown in Figure 2.19. For X-ray analysis, currents in the range 5–100 nA are normal. Much lower currents are commonly used for producing scanning electron images.

2.5.4 Optical microscope

Electron microprobe instruments incorporate an optical microscope for viewing the specimen, as an aid to the selection of points for analysis. The usual arrangement consists of a reflecting objective mounted coaxially with the electron beam. A hole allows the beam to pass and a 45° mirror deflects the light to the eyepiece. Reflected- and transmitted-light viewing are provided, with polarized light facilities for geological users. The magnification is typically about 300×. Sometimes, a low power microscope with a large field of view is also provided, but this is useful only for initial location of samples.

The coaxial reflecting microscope enables the specimen to be viewed during bombardment with the electron beam, allowing cathodoluminescence (light emission stimulated by electron bombardment) to be observed. This phenomenon is of inherent interest in some geological samples and is also useful for setting-up purposes. The position of the beam relative to the field of the microscope can be located by observing the bright spot produced in a cathodoluminescent material (e.g. MgO). At the same time, the beam focus and astigmatism correction can be adjusted.

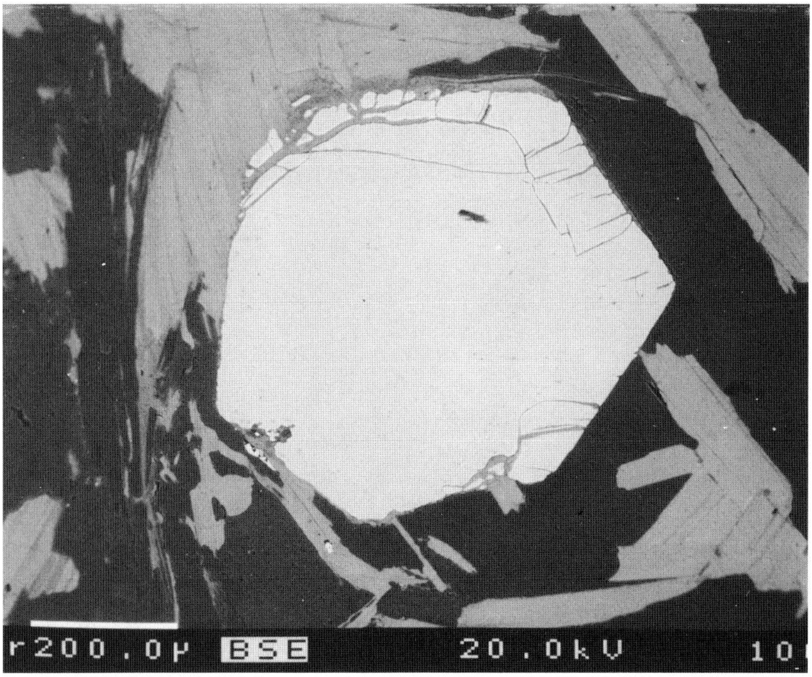

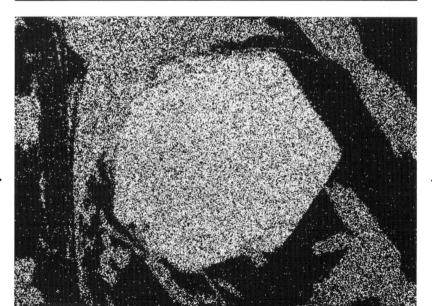

Plate 2.1 Backscattered electron image of schist specimen containing garnet (light), biotite (medium) and other phases, principally quartz (dark).

Plate 2.2 X-ray 'dot-map' showing distribution of iron (same specimen as in Plate 2.1). See Figure 2.21 for line scan along line aa.

2.5.5 Specimen chamber

For electron probe analysis, a specimen holder capable of holding several specimens together with some standards is used. It is important that the front surface of each specimen should be located close to the focal plane of the microscope (and therefore in the correct position relative to the WD spectrometers). Fine adjustment of the height is carried out by mechanical movement of the stage (z-axis), bringing the surface into sharp focus in the microscope. Mechanical x- and y-movements enable areas of interest to be located.

In SEMs somewhat different arrangements are used. The simplest form of mount is the 'stub', onto which the back of the specimen is stuck. In the absence of a high-power optical microscope (which is usually not available in SEMs), uncertainty in the location of the front surface is problematical for WD analysis (section 2.3.3), though it is much less important in ED analysis. The height can be set approximately by obtaining a sharp scanning image for a predetermined 'working distance', as governed by the final lens setting.

2.5.6 Vacuum system

A pressure of less than about 10^{-4} mbar in the column is required to prevent the electrons in the beam being scattered by gas atoms. WDSs must also be evacuated to avoid absorption of long-wavelength X-rays, though the pressure need not be as low.

High vacuum is usually obtained by means of a two-stage pumping system in which a mechanical rotary pump, giving a pressure of about 10^{-1} mbar, is used in series with an oil diffusion pump, producing an ultimate pressure of about 10^{-5} mbar or lower. Sometimes a turbomolecular pump or an ion pump is used in place of the diffusion pump. WDSs need to be pumped only with the rotary pump. Usually there is a specimen airlock chamber which is pumped by the rotary pump only: this avoids the need to vent the whole system when exchanging specimens.

A 'cold trap' cooled with liquid nitrogen is sometimes fitted: this device traps hydrocarbon molecules and minimizes the rate at which carbon contamination is deposited on the specimen, which is especially important in light element analysis. Another way of reducing contamination is to direct a gas jet at the sample via a thin tube, the flow rate being adjusted to avoid an excessive pressure rise in the specimen chamber.

2.5.7 Computer control

In common with other types of analytical instrument, electron probes have become increasingly computerized as digital technology has progressed. To a large extent, instrument functions are now controlled via a keyboard, though manual control is still available on some older instruments.

ED systems are manufactured by different companies from those producing electron beam instruments, and have their own computer systems. They commonly also have facilities for controlling the electron beam (for mapping etc.) via an appropriate interface. Provision for the acquisition of data from WD spectrometers, intended mainly for SEMs fitted with this type of spectrometer, is also an optional feature. There is thus a degree of overlap between the systems. Some steps towards integrating ED systems and electron beam instruments have been made, for example, by operating both through a common user-interface.

2.6 Scanning and mapping

2.6.1 Scanning electron microscopy

SEM images are produced by scanning the beam across the surface of the sample and displaying one of the various types of signal discussed in section 2.6.2 on a simultaneously scanned video display. Advantages of this technique include the potential for high spatial resolution (5 nm), wide magnification range (e.g. 10–100 000×), large depth of field and simple specimen preparation.

The addition of an X-ray spectrometer (usually of the ED type) makes an SEM capable of chemical analysis. Not only can element distribution maps using the X-ray signal be produced, but spot analyses can also be obtained by stopping the scan. Quantitative analysis is possible in principle but is hindered by the lack of features such as beam current regulation, well-defined specimen geometry and a high-power optical microscope (mainly important for WD analysis), which are provided in electron microprobe instruments.

2.6.2 Electron images

Topographic images can be produced by collecting secondary electrons which are dislodged from the surface of the sample as a result of bombardment by the incident beam, and are distinguished by their low energy (<50 eV). The detector usually consists of a scintillator which emits light when bombarded by electrons, and a photomultiplier which picks up the light signal and produces an amplified electrical output. The secondary electrons are attracted to the detector by a positively biased grid. Secondary electron (SE) images are most effective for three-dimensional samples.

Polished samples, as used for X-ray analysis, do not show much contrast in secondary electron images. More valuable information is obtained by using the backscattered electron (BSE) signal, derived from electrons in the incident beam that suffer large deflections in the sample and re-emerge from the surface. Backscattered electrons have higher energies (in the keV range)

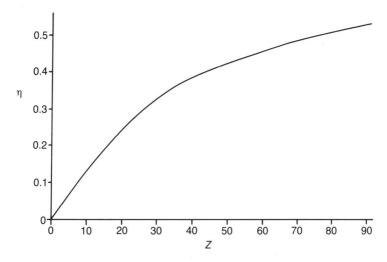

Figure 2.20 Backscattering coefficient (η) as a function of atomic number (Z).

than secondary electrons. They are detected either by solid-state detectors or by scintillators.

The fraction of incident electrons that are backscattered is known as the backscattering coefficient (η) and varies with atomic number, as shown in Figure 2.20. Contrast in BSE images is thus related to atomic number (mean atomic number in the case of a compound). These images therefore contain compositional information which is not specific as to the identity of the elements present, unlike the X-ray images described in section 2.6.3, but have the advantage of better spatial resolution (about $0.1\,\mu m$) and can be produced in a much shorter time. An example is shown in Plate 2.1. This mode of imaging has become a standard technique for petrological applications.

2.6.3 X-ray images

Element distribution images can be formed by a process similar to that just described, except that the signal is derived from an X-ray spectrometer 'tuned' to a characteristic line of the element of interest. In the simplest form of X-ray image (known as a 'dot map') each detected photon produces a bright spot on the screen. When viewed 'live' such images are very 'noisy', but the quality is much better if photographic recording with an exposure of a few minutes is used. Variations in elemental concentration are represented by differences in the density of dots. An example of this type of image is shown in Plate 2.2.

The X-ray signal may be derived from either an EDS or a WDS. In the

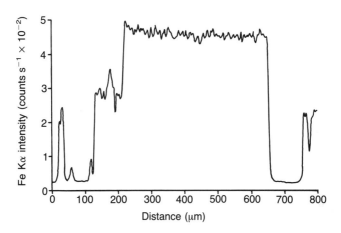

Figure 2.21 Example of a line scan: distribution of Fe along line as in Plate 2.2.

former case, the required characteristic line is selected by means of a 'window' set to the appropriate energy range. The EDS has the advantage of being relatively immune to displacement of the source but is more limited with respect to count rate (section 2.4.3) and is less suitable for low concentrations (<1%), owing to the lower peak-to-background ratio. The sensitivity of WD spectrometers to source displacement can be overcome by moving the specimen stage rather than the beam when scanning.

2.6.4 Line scans

Information on the spatial distribution of an element can also be presented in the form of a line scan, in which the variation in X-ray intensity is displayed as the beam sweeps a line across the specimen (Figure 2.21). Although the information contained in a line scan is limited to one dimension, it has the advantage of requiring much less time to record than a two-dimensional map.

2.6.5 Digital image storage

One result of the increasing 'computerization' of electron probe analysis is that the mode of image acquisition described above has been displaced to a large extent by the storage of digital images in computer memory. Each line is divided into individual 'pixels' and the signal detected at each corresponding point on the specimen is stored. When the image is displayed, the number is converted into an intensity on the display screen. An electron image can be obtained using a fast scan rate, the image being constantly renewed with each successive scan. An X-ray image may be recorded by storing counts obtained in repeated scans and can be viewed while this is in progress. Acquisition can be terminated when enough counts have been accumulated to give an image of satisfactory quality.

An important advantage of digital recording is the ability to carry out manipulations of the kind described in section 2.6.6 separately from the recording process. Images can be stored on hard disc, but the memory required may be as much as 1 Mb per image. It is, therefore, desirable to use some other medium (e.g. optical disc) for long-term storage.

2.6.6 Image display and analysis

Raw X-ray data stored as described above can be manipulated in various ways to produce different types of image. Similar manipulations can also be applied to electron images.

Colour may be used in several ways. One possibility is to represent intensity by a 'false colour' scale which makes compositional variations more visible. Another approach is to assign a different colour to each element present and to combine the images into a single multicoloured image instead of having separate images for each element.

Once an image has been stored in digital form, all the facilities offered by modern image analysis programs become available. Operations such as identifying phases from the recorded intensities, and calculating relative areas and other properties, can easily be carried out, thereby extending considerably the range of information obtainable from the electron probe (or SEM).

2.7 Qualitative analysis

2.7.1 Line identification

The object of qualitative analysis is to find what elements are present in an 'unknown' specimen by identifying the lines in the X-ray spectrum using tables of energies or wavelengths (e.g. Doyle *et al.*, 1979). Ambiguities are rare and can invariably be resolved by taking into account additional lines as well as the main one.

2.7.2 Qualitative ED analysis

The ED spectrometer is especially useful for qualitative analysis because a complete spectrum can be obtained very quickly. Aids to identification are provided, such as facilities for superimposing the positions of the lines of a given element for comparison with the recorded spectrum (Figure 2.22). Owing to the relatively poor resolution, there are cases where identification may not be immediately obvious; e.g. P Kα and Zr Lα (energy difference =

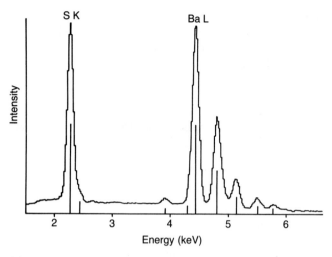

Figure 2.22 Line markers for S K and Ba L lines in the ED spectrum of barite.

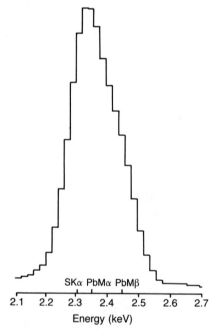

Figure 2.23 Unresolved peaks in an ED spectrum (PbS sample).

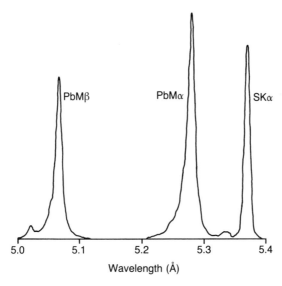

Figure 2.24 WD spectrum of PbS showing well-resolved Pb and S peaks (cf. Figure 2.23).

27 eV), Na Kα and Zn Lα (32 eV), S Kα and Pb Mα (39 eV), and Ti Kα and Ba Lα (42 eV). An example showing unresolved S K and Pb M lines is given in Figure 2.23. Such lines can, however, be resolved easily with a WDS (section 2.7.3).

2.7.3 Qualitative WD analysis

Scanning through the spectrum with a WDS is relatively slow and is necessary only when the resolution of the EDS is inadequate or the concentrations of elements of interest are low. The time required is less if only the relevant parts of the spectrum are scanned.

The S K and Pb M peaks can be resolved easily with a WD spectrometer, as shown in Figure 2.24.

2.8 Quantitative analysis – experimental

2.8.1 Counting statistics

X-ray intensities are measured by counting pulses generated in the detector by X-ray photons which are emitted randomly from the sample. If the mean number of counts recorded in a given time is n, then the numbers recorded in a series of discrete measurements form a Gaussian distribution with a standard deviation (σ) of $n^{1/2}$. A suitable measure of the statistical error in a single measurement is $\pm 2\sigma$ It follows that 4×10^4 counts must be collected to obtain a 2σ precision of $\pm 1\%$ (relative). Such statistical considerations thus dictate the time required to measure intensities for quantitative analysis.

2.8.2 Choice of conditions

The optimum choice of accelerating voltage is determined by the elements present in the specimen. The accelerating voltage (in kV) should be not less than twice the highest excitation energy E_c (in keV) of any element present, in order to obtain adequate intensity. For instance, in silicates, the element with the highest atomic number is commonly Fe, which also has the highest excitation energy (7.11 keV), hence the accelerating voltage should be at least 15 kV. Line intensities increase with accelerating voltage, but so does electron penetration, making spatial resolution worse and increasing the absorption suffered by the emerging X-rays.

The other important variable selected by the user is beam current. The higher the current the higher the X-ray intensity, but there are practical limitations. Some samples are prone to beam damage, which necessitates the use of a low current. In the case of ED analysis, the limited throughput capability of the system has to be considered and a current as low as a few nA may be appropriate. For WD analysis a considerably higher current (up to 100 nA) is normal.

2.8.3 Selection of points

In most geological applications the operator wishes to relate the analytical information to the microstructure of the specimen as seen in the optical microscope. Electron probes usually have a built-in microscope of quite high power, which can be used to identify points for analysis. The carbon coating applied to non-conducting samples has the unfortunate effect of obscuring differences in optical reflectivity, which is sometimes a nuisance. Scanning electron images can be used to locate analytical points but correlation with the structure observed optically may be difficult. It is in any case good practice to examine specimens beforehand and if necessary prepare photographs or diagrams to make areas of interest easier to locate for analysis. In the case of a computer-controlled instrument, the coordinates of points for analysis can be transferred from a digitized mechanical stage mounted on a bench microscope.

2.8.4 WD analysis

A WDS can detect X-rays of only one wavelength at a time: intensities for different elements are therefore measured sequentially (although multiple spectrometers can be used simultaneously). Under typical conditions, count rates of the order of 100 counts s^{-1} per percent concentration are obtained.

QUANTITATIVE ANALYSIS – EXPERIMENTAL

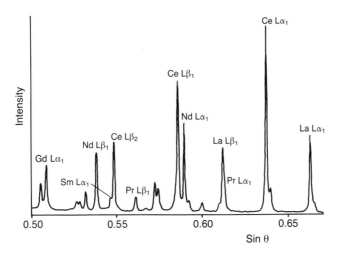

Figure 2.25 WD spectrum of REE in monazite (LiF crystal).

For major elements a counting time of 10 s is therefore often enough to give statistical precision of ±1% relative (section 2.8.1).

The intensity measured with the spectrometer set on a peak includes a contribution from background, mainly originating from the X-ray continuum (section 2.2.4) lying under the peak. Background is usually estimated by interpolating the intensity measured with the spectrometer offset on each side of the peak. When the background is non-linear, or there is interference from a neighbouring peak, more elaborate procedures are required.

For major elements, the peak-to-background ratio is often at least 100:1 and the background correction is almost negligible, but as concentration decreases background becomes more important. For a concentration approaching the detection limit, the intensity at the peak position becomes almost equal to that of the background and both need to be determined with the same precision, hence equal counting times should be used.

In a few applications the spectrum is so crowded that interference between lines becomes a problem, as in the case of rare-earth element L spectra (Figure 2.25). Special measures are required to deal with this situation (e.g. Roeder, 1985).

2.8.5 ED analysis

In ED analysis the whole spectrum is recorded simultaneously. A typical counting time is 100 s, with a count-rate of several thousand counts s^{-1} (for the whole spectrum), giving a relative precision of the order of ±1% for major elements. In an ED spectrum the peaks are relatively broad and the counts are spread over a number of channels. It is therefore preferable to

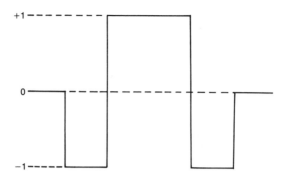

Figure 2.26 'Top hat' function used for filtering ED spectra.

utilize the whole or most of the peak profile in determining the line intensity. One possibility is to sum the contents of the channels containing most of the counts in the peak. Alternatively, a Gaussian function may be fitted to the whole profile.

The approach to background corrections used in WD analysis is generally unsatisfactory for ED analysis, due to the greater width of the peaks. An alternative is to fit a mathematical model for the continuum shape to peak-free regions of the spectrum. Another procedure is to apply a mathematical filter which discriminates against the slowly varying background component of the spectrum. Usually, a 'top hat' filter function is used, consisting of a positive central region with negative lobes on each side (Figure 2.26). The numbers of counts in consecutive channels are multiplied by this function and the results are summed. This process is repeated as the filter function is stepped channel by channel through the spectrum. Any linear component is eliminated and peaks are transformed to the form shown in Figure 2.27. In practice, background non-linearity is negligible within the width of the filter function, so background is effectively removed. Filtered peaks from standards are fitted to the filtered sample spectrum, giving the relative peak intensities, with overlap taken into account.

ED analysis is generally quicker and easier to set up than WD analysis and, due to the higher X-ray collection efficiency, a lower beam current can be used, thereby minimizing damage to sensitive specimens (section 2.8.9). However, WD analysis is essential for low concentrations. The application of ED analysis to silicate minerals has been described by Reed and Ware (1975) and Dunham and Wilkinson (1978). Sometimes, combined ED and WD analysis (ED for major elements and WD for minor elements) is appropriate (Ware, 1991).

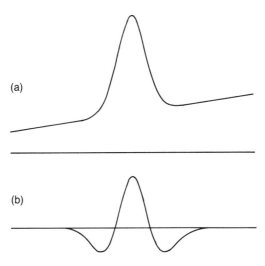

Figure 2.27 Effect of digital filtering with a 'top hat' function, showing the spectrum (a) before and (b) after filtering; note elimination of sloping background.

2.8.6 Light elements

For the most part, quantitative electron probe analysis is concerned with elements above 10 (Ne) in atomic number. However, 'light' elements ($Z < 10$) are sometimes of interest. H, He and Li do not give X-ray peaks, but Be, B, C, N, O and F are detectable. The long wavelengths (low energies) of the K lines of these elements necessitate special experimental techniques. For WD analysis, synthetic multilayer diffracting devices (section 2.3.2) are used. The application of evaporated multilayers to geological light element analysis has been described by Armstrong (1988(b); O, N), McGee, Slack and Herrington (1991; B) and Potts and Tindle (1989; F). For light element ED analysis, the standard beryllium window can either be removed completely ('windowless' operation) or replaced by a thin window (section 2.4.4). Quantitative ED analysis for light elements is somewhat hampered, however, by the poor spectral resolution and by the presence of L and M lines of heavier elements.

The assumption that X-ray spectra are independent of chemical bonding is invalid for light elements. Line positions and shapes vary significantly according to the chemical environment of the emitting atoms. Hence, where the chemical state differs between specimen and standard, a correction must be applied (Bastin and Heijligers, 1990).

Another practical consideration is that long-wavelength X-rays are easily absorbed. The conducting coating applied to insulating specimens must therefore be of strictly controlled thickness and steps should be taken to minimize deposition of carbon from vacuum contaminants during electron bombardment, by the use of a cold trap or a gas jet (section 2.5.6). Further, the correction required for self-absorption in the specimen (section 2.9.5) is critical.

2.8.7 Valence state determination

The chemical effects which are observable for long-wavelength X-ray lines (and are a nuisance in quantitative light element analysis) can in principle be a source of information on the state of the emitting atoms. A possible application is to the determination of the valence state of Fe. Experimental studies (e.g. Albee and Chodos, 1970) have shown that the Fe $L\beta/L\alpha$ intensity ratio varies with Fe^{2+}/Fe^{3+}, but also depends in a complex manner on other factors. Thus, it seems that reliable Fe^{2+}/Fe^{3+} determination is restricted to simple solid-solution series such as Fe-Ti oxides (O'Nions and Smith, 1971).

2.8.8 Standards

The basis of quantitative electron probe analysis is comparison of the intensities of X-ray lines emitted by the specimen with those from reference standards. Usually, a separate standard is required for each element. A standard may be a pure element (when this has suitable properties), though there is some advantage in using standards closer in composition to the analysed material, for example, oxides for silicate analysis. It is difficult, however, to obtain complex mineral standards of accurately known composition which are also homogeneous. Synthetic glasses have possibilities, but available compositions are limited and homogeneity is difficult to achieve in some cases. A list of commonly used mineral standards is given in Table 2.2.

2.8.9 Specimen damage

A focused electron beam can affect the bombarded material by heating or, in the case of insulators, by depositing charge within a small volume. Effects occurring in minerals include loss of CO_2 from carbonates, loss of H_2O from hydrous phases and migration of alkalies (especially Na) within the crystal lattice. If a component such as CO_2 or H_2O is lost, the concentrations of the remaining elements will appear too high. Alkali loss will give low results for the element concerned; typically, there is a progressive decrease in the observed line intensity with time of exposure to the beam, while for other elements there is a correlated increase.

Preventive measures include decreasing the beam current, increasing the beam diameter, scanning the beam and moving the specimen continuously during analysis. Except for the first, these actions entail loss of spatial

Table 2.2 Standards for electron probe analysis

Element	Standard	
	Name	Formula
Na	Jadeite	$NaAlSi_2O_6$
Mg	Periclase	MgO
Al	Corundum	Al_2O_3
Si	Wollastonite	$CaSiO_3$
P	Apatite	$Ca_5P_3O_{12}F$
S	Pyrite	FeS_2
Cl	Halite	$NaCl$
K	Orthoclase	$KAlSi_3O_8$
Ca	Wollastonite	$CaSiO_3$
Ti	Rutile	TiO_2
Cr	Chromite	$FeCr_2O_4$
Mn	Rhodonite	$MnSiO_3$
Fe	Fayalite	Fe_2SiO_4

resolution. In the case of samples prone to electron beam damage, a coating consisting of a good conductor such as aluminium, copper or silver, for example, can be helpful, although this causes increased absorption of emerging X-rays (compared to the usual carbon coating) and the introduction of additional lines in the X-ray spectrum.

2.8.10 Automation

Electron probes have become increasingly 'computerized', with both operation of the instrument and data reduction carried out by a built-in computer. This not only enables instrument functions to be controlled by keyboard commands, but also makes fully automatic analysis possible. The user sets up analysis conditions initially and stores the coordinates of the points to be analysed. The instrument then goes through the analysis procedure on each point in turn, with the computer calculating and storing the results. This mode of operation obviously increases the output of the instrument, since it can be occupied productively unattended, out of normal working hours.

2.9 Quantitative analysis – data reduction

2.9.1 Castaing's approximation

As shown by Castaing (1951), the relative intensity of an X-ray line is approximately proportional to the mass concentration of the element concerned. This relationship is due to the fact that the mass of the sample penetrated by the incident electrons is approximately constant regardless of composition. (The electrons are decelerated by interactions with bound electrons and the number of these per atom is equal to the atomic number which, in turn, is approximately proportional to the atomic weight).

Given this approximation, an 'apparent concentration' (C') can be derived using the following relationship:

$$C' = \left(\frac{I_{sp}}{I_{st}}\right) C_{st} \qquad (2.3)$$

where I_{sp} and I_{st} are the intensities measured for specimen and standard respectively, and C_{st} is the concentration of the element concerned in the standard. To obtain the true concentration, certain corrections are required (section 2.9.2).

2.9.2 Matrix corrections

'Apparent concentrations' require various corrections which are dependent on the composition of the 'matrix' within which the element concerned exists. The first of these corrections arises because the mass penetrated is not strictly constant, but is affected by differences in the 'stopping power' of different elements. This is the 'stopping power correction'. The second allows for the loss of incident electrons which escape from the surface of the sample after being deflected by target nuclei and thus can make no further contribution to X-ray production: this is the 'backscattering correction'. The third takes account of attenuation of the X-rays as they emerge from a finite depth in the sample ('absorption correction'), and the fourth allows for enhancement in X-ray intensity, arising from fluorescence by other X-rays generated in the sample ('fluorescence correction').

The absorption correction is generally the most important. It is dependent on the angle between the surface of the specimen and the X-ray path to the spectrometer, or 'X-ray take-off angle' (Figure 2.28), which is 40° for current models of electron probe, although earlier instruments used different angles. The angle of incidence of the beam is also relevant: standard correction methods assume normal incidence and modifications are required for non-normal incidence (as sometimes applies in SEMs).

2.9.3 Bence–Albee coefficients

A method of applying matrix corrections which is quite often used in geological applications is to lump together the effects described in the previous section and express the total correction as an 'alpha coefficient', the correction factor for element A (F_A) being obtained from the expression:

$$F_A = \Sigma C_i \alpha_i \qquad (2.4)$$

QUANTITATIVE ANALYSIS – DATA REDUCTION

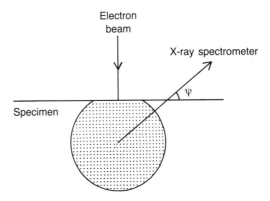

Figure 2.28 X-ray source region, with path of X-rays to the spectrometer (ψ = take-off angle).

The summation is carried out for all elements present.

Alpha coefficients may be derived either empirically or by calculation based on a physical model. Values for oxides given by Bence and Albee (1968) and Albee and Ray (1970) are for an accelerating voltage of 15 kV and an X-ray take off angle of 52.5° (appropriate for the now-discontinued ARL instrument). Armstrong (1988(a)) has given values for the common take-off angle of 40°, including an additional second order term for greater accuracy.

2.9.4 ZAF corrections

The acronym 'ZAF' describes a procedure in which corrections for atomic number effects (Z), absorption (A) and fluorescence (F) are calculated separately from suitable physical models. (The atomic number correction encompasses both the stopping power and backscattering factors described in section 2.9.2). Certain specific methods of calculating these corrections, developed in the 1960s, constitute the standard form of the ZAF correction, which is still used quite successfully. Its main drawback is that the absorption correction is inadequate when the correction is large; alternative absorption correction procedures are therefore preferable, as described in the section 2.9.5.

2.9.5 Phi–rho–z models

To calculate the absorption correction, it is necessary to know the depth distribution of X-ray production in the sample. This is described by the function $\phi(\rho z)$, or phi–rho–z, where z is depth and ρ is density, and takes

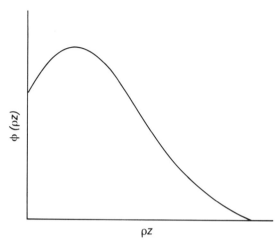

Figure 2.29 'Phi-rho-z' function representing depth distribution of X-ray production (z = depth, ρ = density).

the form shown in Figure 2.29. The exact shape depends on atomic number, and the depth scale is dependent mainly on the accelerating voltage.

In the model used in the original ZAF procedure (Philibert, 1963), the expression for $\phi(\rho z)$ falls to zero at $\rho z = 0$, which is physically incorrect. Several more realistic models have been developed, using parabolic (Pouchou and Pichoir, 1987), Gaussian (Packwood and Brown, 1981; Bastin, van Loo and Heijligers, 1984), exponential (Pouchou and Pichoir, 1988) or quadrilateral (Love, Sewell and Scott, 1984) functions. The term 'phi–rho–z method' is used for correction procedures based on these models. In some cases the absorption and atomic number corrections are combined into a single 'ZA' correction, by making the area of $\phi(\rho z)$ proportional to the generated intensity.

Methods of calculating the other corrections have changed relatively little.

2.9.6 *Absorption coefficients*

X-ray absorption obeys the following law:

$$I = I_0 \exp[-(\mu/\rho)\rho x] \qquad (2.5)$$

where I_0 is the initial intensity and I is the intensity after travelling a distance x through an absorber of density ρ which has a 'mass absorption coefficient' (μ/ρ). Tables such as those of Heinrich (1987) give (μ/ρ) values of pure elements for the principal X-ray lines. The value for a compound is obtained by averaging (using mass concentrations), and such values are used for calculating absorption corrections by the methods mentioned in the preceding sections.

Mass absorption coefficients increase with increasing wavelength (decreasing energy), so corrections are generally larger for elements of low atomic number, which have long-wavelength emission lines. They also increase with the atomic number of the absorber. These trends are, however, modified by the presence of 'absorption edges' related to inner shell excitation. For a given element, absorption increases suddenly at the critical excitation energy as the X-rays then have enough energy to remove an electron from the appropriate shell (for example, μ/ρ of Mg is about eight times that of Si for Al Kα X-rays).

2.9.7 Iteration

Matrix corrections are dependent on composition and in principle cannot be calculated until the true composition is known, which itself requires knowledge of the correction factors. This difficulty is resolved by using an iterative process whereby the composition assumed for calculating the corrections is an estimate derived initially from the apparent (uncorrected) concentrations, and thereafter from the latest corrected values. Usually, convergence occurs after between three and six iterations.

2.9.8 Unanalysed elements

Geological applications often involve analysing silicates and other phases containing oxygen, which must be taken into account when calculating matrix corrections. It is usual to estimate oxygen either from the assumed stoichiometry or as the difference between the sum of the cation concentrations and 100%, rather than by direct determination, which is somewhat difficult and time consuming if worthwhile accuracy is to be achieved.

2.9.9 Accuracy

For major elements, it is usually possible to obtain a statistical precision of ±1% relative (2σ), or better (section 2.8.1). However, various factors apply which limit the accuracy obtainable in the final result. Instrumental instabilities should be less than 1%, and uncertainty in standard compositions may be similarly small (though not always). The largest uncertainties are generally in the matrix corrections, especially the absorption correction when absorption is severe. Generalization is difficult, but ±1% is attainable in favourable cases.

2.9.10 Detection limits

With decreasing concentration, statistical errors and uncertainties in background corrections become dominant. For a concentration in the region of 100 ppm the intensity measured on the peak consists mainly of background. The smallest detectable peak may be defined as three times the standard deviation of the background count.

An order-of-magnitude detection limit estimate can be obtained as follows. If the count rate for a pure element is 10^4 counts s^{-1} and the peak-to-background ratio is 1000:1, the background count rate is 10 counts s^{-1}. In 100 s a total of 10^3 background counts will be accumulated, giving a relative standard deviation of $(10^3)^{1/2}/10^3$, or 0.03. Since the background intensity in this case is equivalent to the peak count rate for a concentration of 1000 ppm, three standard deviations is thus equivalent to a concentration of $0.03 \times 3 \times 1000 = 90$ ppm, which is a typical detection limit for routine analysis.

Reducing the detection limit requires more counts, which can be obtained by increasing the counting time and/or the beam current. By pursuing this approach, detection limits of 10 ppm can be attained, but at this level it is difficult to be sure that errors in background interpolation are not present.

The discussion above refers to WD analysis. In ED analysis, peak-to-background ratios are an order of magnitude lower and count rates are constrained by electronic limitations. Detection limits are typically about 0.1%, although some reduction can be achieved by using long counting times.

Values given here for detection limits refer to samples such as silicates, for which the mean atomic number (which determines continuum intensity) is quite low. Phases containing heavy elements give higher detection limits due to the higher background. Further, detection limits for heavy elements (using L or M lines) tend to be somewhat higher because the peak-to-background ratio is lower than for K lines.

2.9.11 Presentation of results

Quantitative electron probe analysis gives the mass concentrations of the elements present, usually expressed as 'weight percent'. However, atomic concentrations are also useful for some purposes, e.g. calculating mineral formulae. For silicates etc., it is conventional to express results as the weight percentages of oxides, based on an assumed valency for each element. An example showing these forms of output is given in Table 2.3.

Results such as those given in Table 2.3 contain useful internal checks on the quality of the analysis. One of these is the sum of the oxide weight percents, which ideally should be 100%, though for most purposes a range of 99–101% is acceptable. However, the total will be low if water is present

Table 2.3 Results of electron probe analysis of olivine

Element	Element weight %	Element atom %	Atoms per 4 oxygens	Oxide	Oxide weight %
Si	18.72	14.29	0.997	SiO_2	40.04
Mg	27.37	24.13	1.684	MgO	45.38
Ca	0.18	0.10	0.007	CaO	0.25
Mn	0.17	0.07	0.005	MnO	0.22
Fe	10.87	4.17	0.291	FeO	13.98
O[a]	42.69	57.24	4.000	–	–
Total	100.00	100.00	6.984	–	99.87

[a] By difference.

or if some Fe (assumed here to be divalent) is actually trivalent. On the other hand, a high total may be obtained if F or Cl substitutes for O, for example.

Another check on the quality of the analysis is the mineral formula, which in this case is M_2SiO_4, where M represents Mg, Fe, etc. For this purpose it is useful to express the results as number of atoms per n oxygen atoms, where n is appropriate for the mineral concerned ($n = 4$ in this case). Various programs exist for calculating mineral formulae from electron probe data (e.g. Rock and Carroll, 1990).

2.9.12 Estimation of water content

In appropriate cases, the amount of water present can be estimated from the difference between the sum of the oxides and 100%. However, the result is sensitive to errors in the sum arising from random errors in the individual elements (or the omission of an element), and uncertainty as to the valence state of Fe. Water content data obtained in this way should therefore be viewed with caution.

In principle, better results can be obtained if oxygen is determined directly (Armstrong, 1988(b); Nash, 1992). Quantitative oxygen analysis is possible but is subject to the difficulties typical of light elements (section 2.8.6) and the accuracy obtainable by routine procedures is probably insufficient to make it worthwhile in most cases.

References

Albee, A.L. and Chodos, A.A. (1970) Semiquantitative electron microprobe determination of Fe^{2+}/Fe^{3+} and Mn^{2+}/Mn^{3+} in oxides and silicates and its application to petrologic problems. *Am. Mineral.*, **55**, 491–501.

Albee, A.L. and Ray, L.A. (1970) Correction factors for electron probe microanalysis of silicates, oxides, carbonates, phosphates, and sulfates. *Anal. Chem.*, **42**, 1408–14.

Armstrong, J.T. (1988a) Bence–Albee after 20 years: review of the accuracy of α-factor correction procedures for oxide and silicate minerals, in *Microbeam Analysis – 1988* (ed. D.E. Newbury), San Francisco Press, San Francisco, pp. 469–76.

Armstrong, J.T. (1988b) Accurate quantitative analysis of oxygen and nitrogen with a W/Si multilayer crystal, in *Microbeam Analysis – 1988* (ed. D.E. Newbury), San Francisco Press, San Francisco, pp. 301–4.

Bastin, G.F. and Heijligers, H.J.M. (1990) Quantitative electron probe microanalysis of ultralight elements (boron–oxygen). *Scanning*, **12**, 225–36.

Bastin, G.F., van Loo. F.J.J. and Heijligers, H.J.M. (1984) Evaluation and use of Gaussian $\phi(\rho z)$ curves in quantitative electron probe microanalysis: a new optimization. *X-ray Spectrom.*, **13**, 91–7.

Bence, A.E. and Albee, A.L. (1968) Empirical correction factors for the electron microanalysis of silicates and oxides. *J. Geol.*, **76**, 382–403.

Castaing, R. (1951) PhD Thesis, University of Paris.

Doyle, B.L., Chambers, W.F., Christensen, T.M., et al. (1979) $\sin \theta$ settings for X-ray spectrometers. *Atomic Data Nuclear Data Tab.*, **24**, 373–493.

Dunham, A.C. and Wilkinson, F.C.F. (1978) Accuracy, precision and detection limits of energy-dispersive electron-microprobe analyses of silicates. *X-ray Spectrom.*, **7**, 50–5.

Heinrich, K.F.J. (1987) Mass absorption coefficients for electron probe microanalysis, in *Proc. 11th ICXOM* (eds J.D. Brown and R.H. Packwood), University of Western Ontario, London, Ontario, pp. 67–119.

Jenkins, R., Manne, R., Robin, J. and Senemaud, C. (1991) Part VIII. Nomenclature system for X-ray spectroscopy. *Pure and Applied Chemistry*, **63**, 735–46.

Love, G., Sewell, D.A. and Scott, V.D. (1984) An improved absorption correction for quantitative analysis. *J. Physique*, **45** (coll. C2), 21–4.

McGee J.J., Slack, J.F. and Herrington, C.R. (1991) Boron analysis by electron microprobe using MoB_4C layered synthetic crystals. *Am. Mineral.*, **76**, 681–4.

Nash, W.P. (1992) Analysis of oxygen with the electron microprobe: applications to hydrated glasses and minerals. *Amer. Mineral.*, **77**, 453–7.

O'Nions, R.K. and Smith, D.G.W. (1971) Investigations of the $L_{II,III}$ X-ray emission spectra of Fe by electron microprobe. Part 2. The Fe $L_{II,III}$ spectra of Fe and Fe–Ti oxides. *Amer. Mineral.*, **56**, 1452–63.

Packwood, R.H. and Brown, J.D. (1981) A Gaussian expression to describe $\phi(\rho z)$ curves for quantitative electron probe microanalysis, *X-ray Spectrom.*, **10**, 138–46.

Philibert, J. (1963) A method of calculating the absorption correction in electron-probe microanalysis, in *X-ray Optics and X-ray Microanalysis* (eds H.H. Pattee, V.E. Cosslett and A. Engström), Academic Press, New York, pp. 379–92.

Potts, P.J. and Tindle, A.G. (1989) Analytical characteristics of a multilayer dispersion element ($2d = 60 \text{ Å}$) in the determination of fluorine in minerals by electron microprobe. *Mineral. Mag.*, **53**, 357–62.

Pouchou, J.L. and Pichoir, F. (1987) Basic expression of 'PAP' computation for quantitative EPMA, in *Proc. 11th ICXOM* (eds J.D. Brown and R.H. Packwood), University of Western Ontario, London, Ontario, pp. 249–53.

Pouchou, J.L. and Pichoir, F. (1988) A simplified version of the 'PAP' model for matrix corrections in EPMA, in *Microbeam Analysis – 1988* (ed. D.E. Newbury), San Francisco Press, San Francisco, pp. 315–18.

Reed, S.J.B. and Ware, N.G. (1975) Quantitative electron microprobe analysis of silicates using energy-dispersive X-ray spectrometry. *J. Petrol.*, **16**, 499–519.

Rock, N.M.S. and Carroll, G.W. (1990) MINTAB: a general purpose mineral recalculation and tabulation program for Mac computers. *Am. Mineral.*, **75**, 424–30.

Roeder, P.L. (1985) Electron-microprobe analysis of minerals for rare-earth elements: use of calculated peak-overlap corrections, *Can. Mineral.*, **23**, 263–71.

Ware, N.G. (1991) Combined energy-dispersive–wavelength-dispersive quantitative electron probe analysis, *X-ray Spectrom.*, **20**, 73–9.

Further reading

Agarwal, B.K. (1991) *X-ray Spectroscopy*, 2nd edn, Springer-Verlag, Berlin.

Goldstein, J.I., Newbury, D.E., Echlin, P. *et al.* (1992) *Scanning Electron Microscopy and Analysis*, 2nd edn, Plenum Press, New York.

Heinrich, K.F.J. (1981) *Electron Beam X-ray Microanalysis*, Van Nostrand Rheinhold, New York.

Heinrich, K.F.J. and Newbury, D.E. (eds) (1991) *Electron Probe Quantitation*, Plenum Press, New York.

Potts, P.J. (1987) *A Handbook of Silicate Rock Analysis*, Blackie, Glasgow.

Reed, S.J.B. (1993) *Electron Microprobe Analysis*, 2nd edn, Cambridge University Press, Cambridge.

Reimer, L. (1985) *Scanning Electron Microscopy*, Springer-Verlag, Berlin.

Russ, J.C. (1984) *Fundamentals of Energy Dispersive X-ray Analysis*, Butterworths, London.

Scott, V.D. and Love, G. (1994) *Quantitative Electron Probe Microanalysis*, 2nd edn., Ellis Horwood, Chichester.

CHAPTER THREE
Analytical electron microscopy
Pamela E. Champness

3.1 Introduction

There are three phenomena produced by the interaction of a high-energy electron beam (100–1000 keV) with a thin specimen that can be used to provide chemical information in the analytical (transmission) electron microscope (AEM or ATEM). The origin of the most widely used of these, the characteristic X-ray spectrum, is explained in section 2.2.2 and quantification of the X-ray signal in the AEM follows procedures that are modifications of those used in electron probe microanalysis (EPMA).

Not all ionization events caused by bombarding electrons result in X-ray emission. The alternative process is emission of an outer electron that carries the excess energy of the excited atom as kinetic energy. Such Auger electrons are emitted in far larger numbers than X-rays from light elements, while for heavy elements the situation is reversed. However, because they are of very low energy, Auger electrons can only escape from the surface atomic layer. Measurement of the energy of Auger electrons forms the basis of Auger electron spectroscopy (AES), which supplies chemical information about surface layers. At first sight, it would appear to be attractive to detect the Auger signal in the TEM. However, there are a number of practical problems and only a very few laboratories have yet succeeded in performing AES in a TEM. For this reason AES will not be considered further in this chapter.

The third phenomenon that can provide chemical information in the AEM is the inelastic scattering of the primary electron beam. Because most of the primary electrons are transmitted through the thin specimen, they can be collected and any loss in their energy can be measured. This is the basis of electron energy-loss spectroscopy or EELS, which is the subject of the second part of this chapter.

3.1.1 The instrument

The column of a typical AEM is shown schematically in Figure 3.1. The construction of the illumination system is similar to that of the electron probe microanalyser (Chapter 2). However, unlike the EPMA, there are

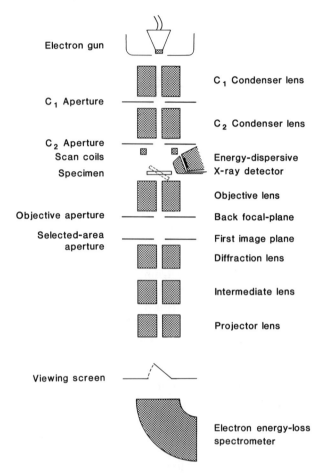

Figure 3.1 Schematic diagram of the design of a typical analytical electron microscope.

lenses below the specimen. The objective lens forms a magnified image of the specimen, which, in the imaging mode of the instrument, is further magnified by the diffraction, intermediate and projector lenses. Images with magnifications of one million times or more and a point resolution of about 2 Å can be obtained in modern TEMs.

In the diffraction mode, the diffraction lens is focused on the back focal-plane of the objective lens and a diffraction pattern is projected onto the viewing screen. The area from which the diffraction pattern comes can be isolated in the image plane of the objective lens by a selected-area aperture (Figure 3.1).

The specimen holder is usually a side-entry rod that holds a 3 mm diameter, thin specimen. It is always possible to tilt the specimen rod about its own axis (by an angle of up to ±70°, depending upon the space available around the objective lens). For crystallographic work, special holders are provided that allow either tilting about a second axis at right angles to the rod axis (usually ±30°) or rotation of the specimen in its own plane.

TEMs have been available since the late 1930s and are mostly used to study the microstructure (second phases, defects, etc.) that are beyond the resolution of the optical microscope and the scanning electron microscope. The first AEM was developed by Duncumb in the mid-1960s and the first commercial instrument became available in 1969. Duncumb fitted a standard 100 keV transmission electron microscope with a mini-lens just above the specimen to form a probe of ~100 nm diameter and with two wavelength-dispersive spectrometers (WDSs) to detect the X-rays (Duncumb, 1966; Cooke and Duncumb, 1969). Because of its higher detection efficiency and its ease of use, the energy-dispersive spectrometer (EDS) has supplanted the WDS for the detection of X-rays in the modern AEM (Figure 3.1). In addition, the electron beam can usually be scanned, as in the EPMA, transforming the instrument into a 'STEM' (scanning transmission electron microscope). 'Dedicated STEMs', in which there are no lenses after the specimen, are also available. These use a field-emission source to provide a probe of high intensity and minimum size. Either type of TEM may be fitted with an EEL spectrometer, usually in addition to an EDS.

3.1.2 The probe-forming system

Because X-ray counts are at a premium in the AEM and spatial resolution for analysis is, to a first approximation, governed by the size of the electron probe (section 3.2), the 'standard' modern instrument is increasingly being fitted with a lanthanum hexaboride (LaB_6) filament which provides a brightness ~10× higher than the tungsten hairpin filament (~10^{10} A m^{-2} sr^{-1} compared with ~10^9 A m^{-2} sr^{-1}). Field-emission guns (FEGs) offer an even higher brightness (~10^{12}–10^{13} A m^{-2} sr^{-1}) but, as they require a vacuum better than 10^{-10} torr, they are an expensive option. FEGs also have the disadvantage that the total current they produce is lower than that produced by LaB_6 and W filaments. On the whole FEGs are restricted to dedicated STEMs, although they are available for the current generation of 'conventional' TEMs.

The size of the electron probe in the AEM depends upon a number of factors, one of which is the state of excitation of the various lenses above the specimen. Thin-film microanalysis is often referred to as 'STEM' microanalysis. However, it is not necessary to operate in the STEM mode to obtain a small probe, as can be seen from Figure 3.2. In some instruments and for certain applications, the focused probe size may be reduced and focused on the specimen in the 'conventional' TEM mode (this is sometimes

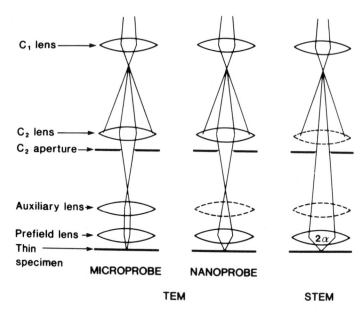

Figure 3.2 The three methods of obtaining a fine probe in the Philips range of transmission electron microscopes. In the 'nanoprobe' mode, the auxiliary lens is effectively switched off, but C_2 is on. In the STEM mode, both the auxiliary and the C_2 lenses are effectively switched off. The convergence angle, 2α, can be varied by changing the size of the C_2 aperture. (From Champness, 1987. Reproduced with permission, Mineralogical Society.)

known as the 'microprobe' mode). When the instrument is operated in the STEM mode, the second condenser has a very small excitation and the auxiliary lens is also effectively switched off. The pre-field of the objective lens is used to focus the electron beam on the specimen. This mode results in smaller spot sizes and larger convergence angles than in the microprobe mode for the same size of C_2 aperture (Figure 3.2).

In certain instruments, notably the Philips range, there is a very useful TEM mode known as the 'nanoprobe' mode, in which, in contrast to the microprobe mode, the auxiliary lens is effectively switched off (Figure 3.2). This results in a range of spot sizes and convergence angles intermediate between those obtained in the STEM and microprobe modes for the same size of C_2 aperture.

3.1.3 Specimen preparation

Specimens for transmission electron microscopy and analysis should, ideally, be less than about 100 nm thick, and even thinner if the highest spatial resolution is required for X-ray analysis or if quantitative EELS is to be performed.

INTRODUCTION

The oldest and simplest method of specimen preparation of brittle materials such as minerals consists of crushing a small quantity of material in absolute alcohol, distilled water or other suitable liquid, and dispersing the resulting suspension onto a stable, structureless film, such as carbon, which is supported on a standard mesh grid 3 mm in diameter. Barber (1993) has described the method in detail. It is ideal for material that is already fine-grained, such as experimental charges. However, crushing is of limited use in cases where spatial relationships between phases or between different regions within the same phase are of interest.

The almost universal method for thinning non-metals for TEM is ion- or atom-beam milling (hereafter called beam milling). Barber (1993) provides a recent review. The starting point for beam milling is usually a $\sim 30\,\mu$m-thick, doubly polished thin section that may have been previously examined in an optical microscope and/or the EPMA. A 3 mm disc is removed from the section and is usually glued to a thick, single-apertured, electron microscope grid, which is normally made of copper, though other materials are available for special applications. The grid serves to support the specimen during thinning and later manipulation. Beam milling involves the bombardment of both specimen surfaces at an angle of about 15° with ~ 5 keV particles, usually argon atoms or ions, while the specimen rotates in its own plane. Thinning is continued until perforation is achieved; the hole(s) in the specimen should then be surrounded by an annulus of material that is thin enough to transmit 100–400 keV electrons. The sample is finally coated with a thin layer of carbon to prevent charging during observation in the TEM. Typical beam thinning rates are $1-20\,\mu$m h^{-1}.

The main drawback of beam milling is the very different rates of thinning shown by different minerals; for instance, garnet and olivine thin much more slowly than feldspar or mica and pyroxenes, and amphiboles are intermediate in sputter resistance. This difference can present real problems in specimens that contain more than one phase: when one phase is suitably thin, the other may be totally opaque to the beam! In such cases it is advisable to start with as thin a thin section as possible and to operate the AEM at as high a voltage as possible (the penetration of the electron beam varies almost linearly with voltage for light elements such those that predominate in silicates).

A further problem of beam milling is that it produces heat and radiation damage in the sample. The radiation damage is normally confined to the production of an amorphous layer up to 50 nm thick on both surfaces of the sample. Although I know of no evidence that the amorphization process leads to a change in the composition of silicates or sulphides, it is known to cause compositional changes in some ceramics and alloys. It is therefore sensible to minimize the thickness of the amorphous layers by decreasing the voltage and the angle of incidence of the ion/atom beam at the end of the thinning process. The use of atom rather than ion thinning also decreases the extent of the damage. Heating effects can be minimized by operating at a low ion/atom current and by the use of a cooling stage.

Beam milling can also be used to prepare specimens from a particulate sample if the powder is first either made into a thin section mounted in epoxy or mechanically embedded in a thin foil of a soft metal such as indium, lead or tin (Boswell, Reece and Gee, 1992). The latter method is the superior of the two because it results in larger thin areas and is more rapid. However, it is only suitable for materials that are harder than the embedding metal.

A third method that can be used to make thin specimens for TEM of non-metals is ultramicrotomy. This has mostly been used to prepare relatively soft materials such as clays and involves sectioning with a diamond knife after encapsulating the sample in a low-viscosity resin (Lee and Jackson, 1975).

3.1.4 Mineralogical applications

The advent of X-ray analysis in the TEM has revolutionized the study of fine-scale microstructures of minerals. It is extremely tedious, if not impossible, to identify a phase from its electron diffraction pattern alone! For the vast majority of cases, the X-ray spectrum provides identification in a matter of seconds (Figure 3.3). If it does not (for instance in the distinction between polymorphs), electron diffraction will usually resolve the matter.

X-ray analysis in the AEM has become almost as routine a technique as EPMA. In general, it is used to obtain quantitative chemical analyses from the same kinds of materials as are examined by EPMA (section 2.1.5), but with a spatial resolution at least two orders of magnitude smaller and an activated volume at least five orders of magnitude smaller. AEM also has the distinct advantage over EPMA that images approaching atomic resolution can be obtained, together with diffraction patterns that provide information about crystallographic orientations etc.

EELS is a much newer technique than X-ray analysis in terms of its routine use in the AEM and, to date, has found few applications in mineralogy for chemical analysis *sensu stricto*. Its main strength lies in the analysis of light elements ($Z \leq 10$), which, apart from oxygen, carbon in carbonates and, to a lesser extent, fluorine and boron, are not of great importance in most minerals.

3.2 The X-ray analysis of thin specimens

3.2.1 Principles of the technique

The main advantage of using a thin specimen for X-ray microanalysis is the improved spatial resolution over that obtainable in a bulk sample. In the

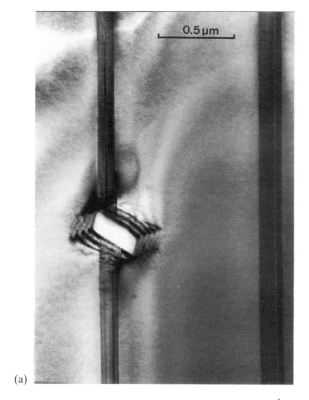

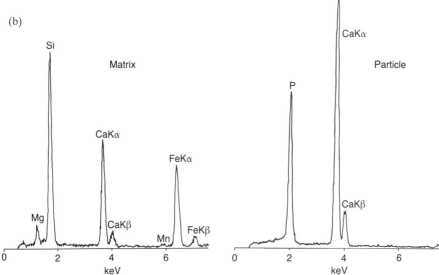

Figure 3.3 (a) Transmission electron micrograph of an apatite particle in a calcic pyroxene from the Bjerkreim–Sokndal lopolith, south-west Norway. The particle has acted as a nucleation site for the orthopyroxene lamellae that run N–S. (From Rietmeijer and Champness, 1982. Reproduced with permission, Mineralogical Society). (b) X-ray spectra from the matrix and particle obtained at 100 kV with a detector that has a standard Be window.

ANALYTICAL ELECTRON MICROSCOPY

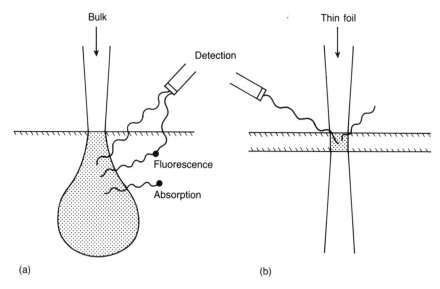

Figure 3.4 Schematic representation of the interaction of a high-energy electron beam with (a) a bulk and (b) a thin specimen. Absorption of and fluorescence by the generated X-rays are significant in the bulk sample, but negligible in the thin foil. (After Lorimer, 1987).

bulk samples normally analysed in EPMA, the electron beam diffuses into the sample to a depth of 1–5 μm. X-rays are produced from a bell-shaped region (Figure 3.4(a)). Thus, the spatial resolution is limited to the micrometre scale. In a thin foil, on the other hand, the electron beam passes through the sample and X-rays are generated in a volume dictated by the size of the focused probe and the extent of the electron scattering, which is a function of both the thickness (t) of the sample and the accelerating voltage (Figure 3.4(b)).

A further advantage of using a thin foil for X-ray microanalysis is that X-ray absorption and secondary fluorescence are minimal and the measured intensity is, to a first approximation, the same as the generated intensity. This fact leads to a much simpler quantification procedure than the ZAF corrections that must be employed for EPMA (section 3.2.4).

(a) Probe size and spatial resolution

It is often assumed that the electron density (number of electrons per unit area) in the electron probe can be represented as a function of radius by a Gaussian function. In this case, 50% of the total current is contained within a disc of diameter equal to the full width at half maximum (FWHM) of the Gaussian and 90% of the total current is contained within the full width at tenth maximum (FWTM). Both of these widths are widely used as definitions of the probe size and spatial resolution for analysis, the former being the normal one quoted by manufacturers. However, Cliff and Kenway (1982) have shown that spherical aberration (the phenomenon by which electrons that are travelling at high angles to the optical axis are brought to focus closer to the lens than those that are at low angles) can produce large 'tails' in the intensity distribution of the highly convergent probes used in AEM (Figure 3.5). Quoting the FWHM of such a probe gives a highly misleading picture of the spatial resolution for analysis. Cliff and Kenway found that, if the spatial resolution is defined as the diameter that contains 90% of the current, the spatial resolution for a STEM probe can be more than 40× the FWHM (Figure 3.5(d)) compared with 1.82× the FWHM for a Gaussian probe. As can be seen from Figure 3.2, it is possible to reduce the convergence of the beam, and hence the tail produced by spherical aberration, by reducing the size of the C_2 aperture, but this is at the expense of the total probe current.

When the analysis probe is a few nanometres in diameter, beam broadening within the specimen plays a critical role in determining the spatial resolution for analysis. Ignoring the effects of diffraction and fast secondary electrons, the average size of the interaction diameter, R, midway through the foil for a probe that is properly apertured, i.e. that is approximately Gaussian, is given by (Michael et al., 1990):

$$R = \frac{[d + (d^2 + b^2)^{1/2}]}{2\sqrt{2}} \qquad (3.1)$$

where d is the incident beam diameter and b is the beam broadening. Using a single-scattering approximation, Reed (1982) showed that:

$$b = 7.21 \times 10^5 \left(\frac{\rho}{A}\right)^{1/2} \left(\frac{Z}{E_0}\right) t^{3/2} \qquad (3.2)$$

where ρ is the density, A is the average atomic weight, Z is the average atomic number, E_0 is the accelerating voltage and t is the thickness of the specimen. To minimize R, one should use as thin a specimen, as small a probe diameter and as high an accelerating voltage, as possible. Unfortunately, the thinner the specimen, the lower is the intensity of the X-ray signal! Nevertheless, a spatial resolution (R) of about 2 nm is attainable for quantitative analysis with current instrumentation using an FEG. With an LaB_6 gun, the equivalent value is about 10 nm.

3.2.2 X-ray detectors

Current AEMs use energy-dispersive spectrometers to detect the X-rays produced by interaction of the specimen with the electron beam. Because of its superior energy resolution and sensitivity to low concentrations (Chapter

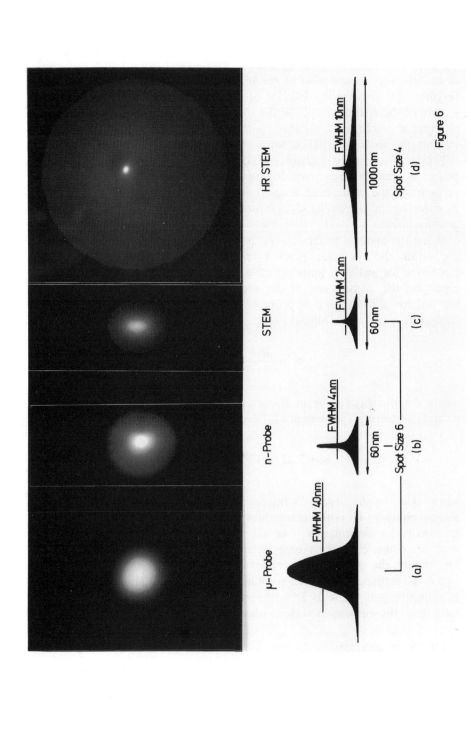

Figure 6

2), there is current interest in fitting a compact WDS within the electron column of the AEM (Goldstein, Lyman and Williams, 1989). However, the efficiency (count rate) of a WDS is much lower than that of an EDS (mostly because of the former's smaller solid angle). Consequently, a WDS system would require a larger probe current and/or a doubly curved diffracting crystal to achieve the same count rate as an EDS.

Most of the EDSs used in AEMs are of the Si(Li) (lithium-drifted silicon) type described in section 2.4.1 and allow detection of elements as light as beryllium ($Z = 4$), depending on the type of window (or lack of one) used. To maximize the flux of X-rays reaching the detector, the solid angle, Ω, (Figure 3.6) subtended by the detector must be as large as possible, current values being $> 0.1\,\text{sr}$. To achieve these values the detector is typically positioned only 15 mm from the sample.

There has been some interest recently in the use of high-purity germanium (HPGe) detectors for AEM. One of the advantages of these detectors is their superior energy resolution (~115 eV for HPGe compared with ~130 eV for Si(Li) at 5.9 keV). This improved resolution is particularly beneficial for the analysis of light elements ($Z < 10$) and has facilitated, for example, the complete separation of $CK\alpha$ and $OK\alpha$ peaks (Griffin and Johnson, 1992). HPGe detectors also have better detection efficiency at high X-ray energies (>20 keV) than Si(Li) detectors. In theory, this improved efficiency allows quantitative analysis using the K lines of even the heaviest elements. However, Zemyan and Williams (1991) have shown that the peak-to-background ratio is so low at these high energies that the detection limit for Pb, for example, is 15 times worse when the $K\alpha$, rather than the lower energy $L\alpha$ peak, is used.

3.2.3 Operational conditions for X-ray analysis of thin specimens

If the X-ray spectrum from a specimen is to be quantified, it must be assumed that (i) none of the X-rays produced in the specimen are absorbed

---

Figure 3.5 Micrographs and schematic drawings of the electron density distributions in the electron probes of the Philips EM400T transmission electron microscope under various operating conditions. A 70 μm C_2 aperture was used in each case and the same excitation of the C_1 lens (spot size 6) was used in (a), (b) and (c). In (d) a higher excitation of C_1 was used (spot size 4). (a) In the conventional or 'microprobe' mode the convergence angle is low and the probe is essentially Gaussian. (b) In the 'nanoprobe' mode the convergence angle is high and there is significant spherical aberration. (c) The STEM mode shows similar characteristics to the nanoprobe mode. (d) In the 'high-resolution' STEM mode (in which C_1 has a somewhat larger excitation than in the normal STEM mode), the convergence angle is even higher and gives rise to a still larger halo. Under the conditions used here, the halo contains an estimated 95% of the total probe current and extends to 1000 nm, despite the probe having a FWHM of only 10 nm. (From Cliff and Kenway, 1982. Reproduced with permission, San Francisco Press).

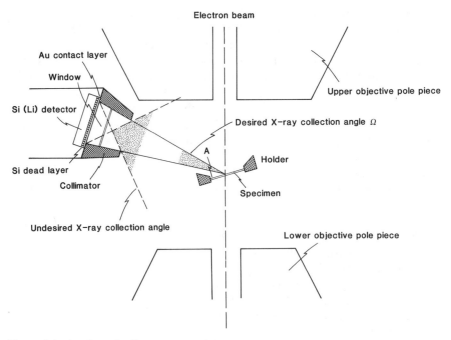

Figure 3.6 A schematic diagram of a Si(Li) solid-state detector and its relationship to the specimen. The collection efficiency is determined by the solid angle, Ω, the nature and thickness of the detector window and the thickness of the Au contact layer and the Si 'dead layer'. Regions of the specimen such as A may not be in line-of-sight of the detector, even with a positive take-off angle, as shown here. In addition to the 'desirable' collection angle, Ω, the total (undesirable) angle of acceptance is shown. This total angle collects X-rays and electrons from all the regions that it can 'see'. (After Williams, 1984).

on their way to the detector and (ii) all the X-rays collected come from the region of interest in the sample and not from other areas of the specimen, the holder, etc. In general, this assumption may not be justified, although modern instruments have modifications that seek to minimize the effects of spurious X-rays contributing to the spectrum. The main reason for these problems is that the AEM uses high-energy electrons and these electrons and the X-rays that are generated are scattered by the specimen in the very constricted region of the microscope stage (Figure 3.6). As is apparent from Figure 3.6, the specimen holder must be in such a position during analysis that no part of it intercepts the cone of X-rays entering the detector. In modern instruments, the detector has a positive 'take-off' angle, normally $\geqslant 20°$. It is also sensible to tilt the specimen towards the detector so as to increase the total take-off angle to about 40°. Tilting the specimen in this way, however, has the disadvantage that the specimen then interacts with its own continuum radiation (section 3.2.3(a)).

Despite a positive take-off angle, there will still be regions such as A in Figure 3.6 from which the X-rays are seriously attenuated. The special holders supplied for analysis are normally made from a material with a low atomic number such as beryllium or carbon to reduce the number of back-scattered electrons and to minimize the effect of spurious characteristic X-ray peaks. One side-effect of this design is that the X-ray spectrum from an area such as A in Figure 3.6 suffers severe absorption of its low-energy peaks, while the higher energy peaks remain unaffected. It is therefore wise to analyse only from the region of the specimen that is on the side furthest from the detector. Usually the detector is mounted at right angles to the axis of the (side-entry) specimen rod, so it is easy to determine the location of the region suitable for analysis at any magnification of the image. For crushed specimens mounted on a mesh grid, attenuation effects will occur in areas of the specimen in the shadow of a grid bar. It is clear from Figure 3.6 that the specimen should be at the correct (eucentric) height if attenuation problems are to be minimized.

(a) Spurious X-rays

Having considered the X-rays that were generated in the specimen but did not reach the detector, I will now consider the even more serious problem of those X-rays that are not generated in the region of interest in the specimen, but do reach the detector. As is apparent from Figure 3.6, the detector 'sees' a large volume of the stage and its environment and collects X-rays (and electrons) from all these regions. There are many sources of spurious X-rays and the ultimate aim is to remove all of them. Those that can, in principle, be eliminated are present even if the electron beam is directed through a hole in the specimen (Figure 3.7) and give rise to the 'hole count'. The sources of the 'hole count' are described in detail by Williams (1984). They include:

- Hard X-rays generated in and transmitted through the condenser aperture. These X-rays can cause fluorescence in the specimen, grid and elsewhere in the area around the specimen.
- Uncollimated electrons and those scattered from the bore of the C_2 aperture can excite X-rays in areas of the specimen remote from the probe and from components below the specimen (e.g. the objective aperture and objective pole piece).

The hole count can be minimized by the following measures, most of which are now standard practice:

- Removal of the objective aperture during analysis.
- The use of an extra-thick, 'top hat' C_2 aperture.
- The insertion of extra ('spray') apertures in the column above the specimen.

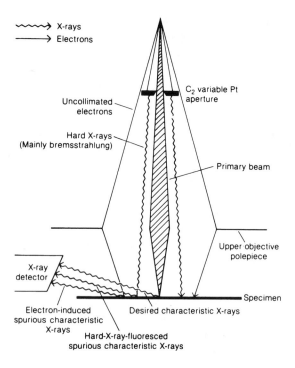

Figure 3.7 Schematic diagram showing the sources of spurious X-rays (the 'hole count') that originate in the specimen and are produced either by stray electrons that travel round the C_2 aperture or by hard X-rays that penetrate the aperture. (From Williams, 1984. Reproduced with permission, Philips Electronic Instruments Inc.).

- The use of materials of low atomic number for the specimen holder and support grid. Alternatively, the specimen grid can be made of a material that is not contained in the specimen; copper is suitable for most silicate analyses. Nickel or gold may be suitable for the analysis of sulphides.

Williams (1986) has suggested that the hole count of an instrument be considered acceptable if, under the experimental conditions that produce ~10 000 counts in the Kα peak of interest in ~100 s live time, the intensity in the same peak in a spectrum recorded in the hole for the same live time is less than the background obtained when the beam was on the sample. If the hole count is not satisfactory, it is possible to subtract the 'in-hole' spectrum from the experimental spectrum before quantification. However, this practice may cause problems if the bulk composition (including the contribution from the specimen grid) is radically different from that of the region of interest.

Another source of spurious X-rays involves the specimen itself and is therefore extremely difficult to eliminate completely. It accounts for the

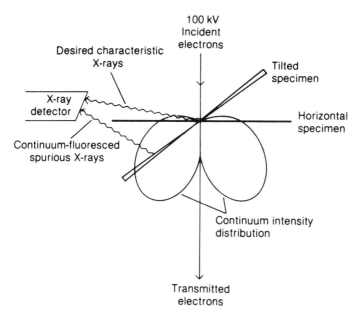

Figure 3.8 Schematic diagram showing the increased interaction of the specimen with its own continuum radiation when it is tilted from the horizontal. This effect produces spurious X-rays from the specimen in regions other than that directly under the probe and it also lowers the peak-to-background ratio of the characteristic peaks. (From Williams, 1984. Reproduced with permission, Philips Electronic Instruments Inc.).

well-known observation that, even if the 'hole count' has been eliminated, Cu X-ray lines invariably appear in the spectrum of a specimen supported on a Cu grid, even if the primary beam is many micrometres from any grid material. The two main effects are:

1. Back scattering of the high-energy incident electrons first from the specimen and then from bulk material such as the objective pole-pieces. These electrons can generate continuous and characteristic X-rays from the specimen, its support grid and the specimen holder.
2. Continuum X-radiation produced in the specimen at the region being analysed will fluoresce distant regions of the specimen, the support grid, etc. (Figure 3.8).

The first problem can be minimized by the use of material of low Z to shield the bulk parts of the specimen area from the back-scattered electrons. The problem of X-ray fluorescence produced by the continuum is one of the factors that lead to positive take-off angles for EDSs. Bishop (1974) pointed out that the continuum intensity generated at ~ 100 kV is peaked in the forward direction (Figure 3.8) and its interaction with the specimen is minimized if the holder is at $0°$ tilt.

(b) Choice of kV

Theoretical treatments (e.g. Joy and Maher, 1977) show that the peak-to-background ratio should increase with kV and therefore the sensitivity of AEM should improve (section 3.2.5(b)). Until recently, it has not been possible to demonstrate this improvement because the spurious X-rays that are generated in the specimen, as described in the section 3.2.3(a), contribute to the background, an effect that increases with kV. However, when the specimen chamber is 'cleaned up', the improvement can be realized (Nicholson *et al.*, 1982; Zemyan and Williams, 1992; Zaluzec, 1992). It is, therefore, preferable to operate at the maximum kV, with the added advantage that the spatial resolution is improved (equation 3.2) and, for minerals and other non-metals, radiation damage is minimized (section 3.2.5(c)). However, it should be emphasized that the 'thin-film criterion' cannot be assumed to hold for analyses at high voltage and correction for absorption, and possibly fluorescence, may have to be made, even for elements $Z \geqslant 11$ (sections 3.2.4(d), 3.2.4(e)).

(c) Contamination of the specimen

Contamination manifests itself as conical carbonaceous deposits that build up on both sides of the thin foil when the electron beam is focused on the specimen during analysis (Figure 3.9). Although the contamination spots have useful applications, such as in measuring the thickness of the foil (section 3.2.4(f)) and in indicating the occurrence of drift of the probe or the specimen during analysis, it has three main undesirable consequences:

1. It will preferentially absorb low-energy X-rays emanating from the specimen. This effect is particularly important if a thin-window or windowless detector is used to detect light elements.
2. It will increase the X-ray background and hence reduce the peak-to-background ratio.
3. Scattering within the cone may cause more spreading within the specimen than would otherwise occur and thus the analysed volume will increase.

Modern AEMs have relatively clean vacuum systems, with residual hydrocarbons in the specimen area being $<10^{-10}$ torr. The source of most of the contamination is the specimen and its holder. Williams (1984) lists the precautions that should be taken to minimize this source of contamination during the handling and loading of the specimen. For beam-milled specimens, it is additionally important that any of the adhesive that is used to cement the support gird to the specimen is removed from the exposed surface before milling. The analyst can 'fix' most of any residual contamination by flooding the specimen for about 15 min before analysis with a completely defocused beam with the C_2 aperture removed. It is always advisable to use the cold trap that is located below the specimen when performing analyses; use of a specimen cooling stage will also restrict diffusion of the hydrocarbons to the site of analysis.

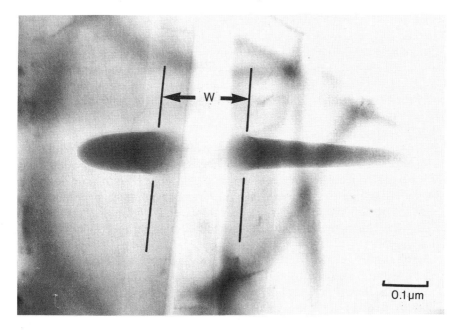

Figure 3.9 Contamination marks produced on the top and bottom of a specimen of stainless steel by a focused probe. The specimen has subsequently been tilted by 35°. A measure of the thickness of the specimen can be obtained from W, the apparent separation of the two contamination marks. This method overestimates the thickness, as can be seen from the projected width of the slip plane (the white band) which was vertical when the contamination marks were formed. (From Rae *et al.*, 1981. Reproduced with permission, The Metals Society).

(d) Diffracting conditions

Anomalously high X-ray intensities are generated when the specimen is close to the Bragg condition for diffraction. (This enhanced emission of X-rays forms the basis of the technique of ALCHEMI (atom location by channelling-enhanced microanalysis; see Buseck and Self, 1992, for a review)). As the problem may not be entirely eliminated by taking the ratio of two elements, it is advisable to avoid Bragg contours in the image when performing analyses. The use of a large convergence angle, as occurs in the STEM mode with a focused probe (Figure 3.2) also minimizes the problem. Similar precautions need to be taken in EELS (section 3.3).

3.2.4 Quantitative X-ray analysis of thin specimens

(a) Background subtraction

The X-ray spectrum recorded by the EDS consists of the characteristic peaks superimposed on the continuum background (section 2.8.5). It is necessary to remove the background in order to obtain the integrated intensities. This is achieved by direct calculation or by mathematical filtering using a 'top hat' function, as described in Chapter 2, or by scaling and subtracting a reference background from a material such as carbon (Livi and Veblen, 1987). Once the background has been removed the peak intensities are obtained either by fitting a Gaussian profile or by using reference spectra that have been acquired previously and stored in the computer.

(b) The ratio technique

In a sample that is sufficiently thin for transmission of 100 keV electrons, the incident beam loses only a small amount of energy and the ionization cross-section (the probability of an electron producing an ionization event) is constant along the electron path. To a first approximation, as noted above, X-ray absorption and secondary X-ray fluorescence within the specimen can be ignored. Under these conditions the 'thin-film' criterion applies (Cliff and Lorimer, 1975).

The absolute X-ray intensity is a function of the thickness of the specimen, as well as of the composition, but the ratio of the measured X-ray intensities, I_A/I_B (with their backgrounds subtracted) for two elements A and B, is independent of thickness. This ratio can be simply related to the corresponding ratio of the weight fractions (or to the atomic ratios) of the elements, C_A/C_B, by the equation:

$$\frac{C_A}{C_B} = k_{AB} \frac{I_A}{I_B} \qquad (3.3)$$

where k_{AB} is a 'sensitivity factor' (the 'Cliff–Lorimer factor') that accounts for the relative efficiency of production and detection of the X-rays. At a given accelerating voltage, k_{AB} is independent of specimen thickness and composition. If $I_A\ldots I_N$ are measured simultaneously, as is usual with an EDS, the measurements are independent of variations in the probe current.

The k_{AB} factor is not a fundamental constant because it depends upon such things as the composition and thickness of the detector window and because, even for a given detector, it will change if contamination builds up on the window. However, k_{AB} values for a particular instrumental arrangement can be stored and used long after they have been measured to obtain concentrations in unknowns. Thus, no standardization is normally necessary at the time of analysis.

As absolute X-ray intensities are not used in the quantification, there is no internal check on the quality of the AEM analysis provided by the analysis total (unlike EPMA; section 2.9.11) and an assumption must be made about normalization, e.g. $\Sigma C_n = 1$ if all the elements can be detected, as in sulphides. For silicates such as olivines or pyroxenes, in which X-rays from all the elements except oxygen can be measured quantitatively, the normalized concentration of an element A as a proportion of the total cations is given by:

$$\frac{C_A/C_{Si}}{C_A/C_{Si} + C_B/C_{Si} + \ldots C_n/C_{Si}} = C_A \qquad (3.4)$$

These concentrations can then be converted to oxide weight percents and totalled to 100% (e.g. Table 3.1). The chemical formula can be calculated in the usual way to a suitable number of oxygens (six in the case of the orthopyroxene in Table 3.1). The resulting number of cations in each site may give an indication of the quality of the analysis (as in the case of the orthopyroxene in Table 3.1).

Problems arise in cases where there are elements other than oxygen that cannot be detected. For hydrated samples, assuming that all the cations can be detected, an oxide analysis total appropriate to the mineral type (e.g. 95% for micas, Shau et al., 1991) can be assumed or the formula can be normalized to an appropriate number of oxygen atoms (11 or 22 for micas; Table 3.1).

In the general case, it is recommended that, where possible, normalization be carried out on the basis of the known number of cations in a particular crystallographic site, e.g., the tetrahedral site in feldspars (Peacor, 1992). Apparent cation deficiencies in another site could indicate either that an undetectable element such as Li was present, or that mass loss had occurred during analysis (section 3.2.5(c)). In those structures that contain Fe, if it can be assumed that all the cations have been detected and that no mass loss has occurred, the charge balance can be obtained by adjusting the Fe^{3+}/Fe^{2+} ratio.

(c) Determination of k_{AB} factors

The k_{AB} factors are usually determined experimentally from well-characterized, homogeneous standards, the reference element, B, in equation 3.3 being Si for silicates (Figure 3.10) and S for sulphides. Mineralogists normally use natural mineral standards, but the glasses produced by NIST (the former US National Bureau of Standards) are extremely useful for the determination of k_{AB} factors for elements that do not occur in large concentrations in minerals (Sheridan, 1989). Because the quality of analyses obtained using equation 3.3 is critically dependent on the accuracy of the k_{AB} values, it is vitally important that these values are measured with care and it is advisable to use several standards for each element. If the ratio C_A/C_B is plotted against I_A/I_B, the slope of the line gives k_{AB}. For elements $Z \geq 12$, k_{AB} factors can now be measured with an error in the range 1–4% relative (Wood, Williams and Goldstein, 1984; Sheridan, 1989). The determination of k_{AB} factors for light elements presents particular problems and has been discussed at length by Westwood, Michael and Notis (1992).

Table 3.1 AEM analyses of an orthopyroxene (columns 1 and 2) and a biotite (columns 3 and 4)

	1		2		3		4
SiO_2	48.01	Si	1.87 ⎫ 2.00	SiO_2	38.13	Si	5.31 ⎫ 8.00
Al_2O_3	4.88	Al^{IV}	0.13 ⎭	Al_2O_3	23.20	Al^{IV}	2.69 ⎭
		Al^{VI}	0.09			Al^{VI}	1.12
TiO_2	0.00	Ti	0.00 ⎫	TiO_2	1.58	Ti	0.16 ⎫
FeO	29.91	Fe^{2+}	0.97	FeO	13.75	Fe^{2+}	1.60 ⎬ 5.71
MnO	1.89	Mn	0.06 ⎬ 2.01	MnO	0.00	Mn	0.00
MgO	15.31	Mg	0.89	MgO	13.58	Mg	2.83 ⎭
Na_2O	0.00	Na	0.00 ⎭	Na_2O	0.62	Na	0.16 ⎫
CaO	0.00	Ca	0.00	CaO	0.00	Ca	0.00 ⎬ 1.78
K_2O	0.00	K	0.00	K_2O	9.14	K	1.62 ⎭
Total	100.00	O	6		100.00	O	22

The oxide weight percents in columns 1 and 3 were derived, assuming a total of 100%, using equation 3.4. The atomic formulae in columns 2 and 4 were calculated assuming a total of 6 and 22 oxygens and a total of 2 and 8 (Si + Al) for the orthopyroxene and the biotite respectively.

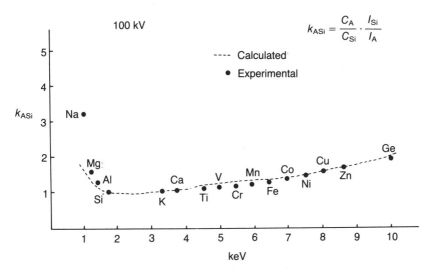

Figure 3.10 Experimentally determined k_{ASi} factors vs. $K\alpha$ X-ray energy at 100 kV. Calculated values are also shown. (From Lorimer, 1987. Reproduced with permission, Mineralogical Society).

McGill and Hubbard (1981) described a method by which k_{AB} factors may be obtained from beam-sensitive materials. They plotted the apparent variation of k_{NaSi} for albite ($NaAlSi_3O_8$) with time (Figure 3.11) and extrapolated to zero time to find the true value.

An alternative method for dealing with beam-sensitive materials has been suggested by van der Pluijm et al. (1988). This method involves analysing the unknown under the same conditions of current density, thickness (as judged by the total count rate) and analysis time as the standard. As the method implicitly assumes that the rate of mass loss is the same in the standard and unknown, these samples should be as similar in composition and structure as possible. Clearly, when it is necessary to use beam-sensitive standards, as large a beam diameter as possible should be used. It is also helpful to cool the specimen (section 3.2.5(c)).

If a suitable standard containing the two elements of interest cannot be found, k_{AB} factors can be obtained from two standards, e.g.

$$k_{ASi} = k_{AB} \times k_{BSi}$$

In cases where no suitable standards are available, the k_{AB} factors must be calculated from:

$$k_{AB} = \frac{Q_B \omega_B a_B A_A \varepsilon_B}{Q_A \omega_A a_A A_B \varepsilon_A} \qquad (3.5)$$

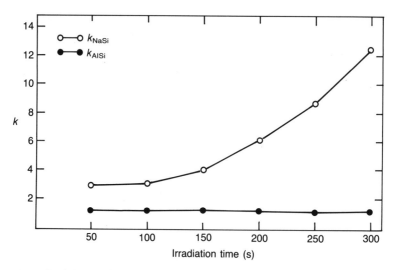

Figure 3.11 Variation in k_{NaSi} and k_{AlSi} with radiation time for albite. (From McGill and Hubbard, 1981. Reproduced with permission, The Metals Society).

Where Q is the ionization cross-section for X-rays, i.e. the probability that an electron will excite an atom, ω is the fluorescence yield (X-rays emitted per ionization), a is the relative transition probability $[K_\alpha/(K_\alpha + K_\beta)$ for K lines], A is the atomic weight and ε is the efficiency of the detector for the X-rays from the particular element.

Of the terms in equation 3.5, ε and Q are the most difficult to calculate. Currently, the calculation of k_{AB} values for K lines above 1.5 eV in energy is in error by ~10–15%, mainly because of the uncertainty in Q (Williams, 1984; Joy, Romig and Goldstein, 1986). The detector efficiency depends upon the thickness of the beryllium window, the gold contact layer, the silicon deadlayer and the active silicon layer and is particularly difficult to determine for X-ray energies <1.5 keV, for which absorption is significant. Calculation of k_{AB} values is therefore not recommended for light elements ($Z < 13$). Calculation of k_{AB} values is not recommended for L lines either.

(d) Breakdown of the thin-film criterion; absorption in the specimen
When the thin-film criterion breaks down, it is usually because the effects of absorption are significant. Goldstein et al. (1977) proposed that, for any set of two elements A and B in equation 3.3, an absorption correction is necessary if:

$$[(\mu/\rho)^B_{spec} - (\mu/\rho)^A_{spec}] \cdot \rho \cdot (t/2) \cdot \operatorname{cosec} \alpha > 0.1 \quad (3.6)$$

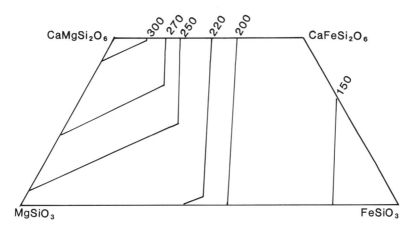

Figure 3.12 The maximum thickness (in nm) of Ca-Mg-Fe pyroxenes for which absorption corrections can be ignored. (After Nord, 1982).

where ρ is the density, t is the thickness and $(\mu/\rho)_{\text{spec}}^{\text{B}}$ is the absorption coefficient of element B in the specimen and α is the take-off angle for the detector (assuming zero tilt of the specimen). The maximum thickness for which an absorption correction is unnecessary is thus:

$$t_{\max} = \frac{0.2}{[(\mu/\rho)_{\text{spec}}^{\text{B}} - (\mu/\rho)_{\text{spec}}^{\text{A}}] \cdot \rho \cdot \operatorname{cosec} \alpha} \quad (3.7)$$

Notice that it is the difference in the absorption coefficients for the two elements that is the important factor. If two elements are adjacent in the periodic table, their values of μ/ρ will be very similar (except near an absorption edge) and the ratio of the intensities of their X-ray lines will be little affected by absorption, whatever the thickness of the specimen.

Nord (1982) calculated the value of $t_{\max}$ for Mg/Si, Ca/Si and Fe/Si for members of the pyroxene quadrilateral. Figure 3.12 shows a compilation of the minimum value of $t_{\max}$ for all three elemental ratios and indicates that analysis must be carried out in areas that are less than 130–300 nm thick, depending on the bulk composition, if absorption effects are to be insignificant. Similar results would be obtained for other silicates. As it happens, the maximum thickness for which microstructures in silicates can be observed using 100 kV electrons is about 200 nm, so if microscopy can be carried out in an area of the foil at ~100 kV, it can be assumed that the foil fulfills the thin-film criterion for elements $Z \geq 11$. For higher voltages or lighter elements, this rule of thumb cannot be used and care must be taken to work in suitably thin areas or, alternatively, to correct for absorption using one of the methods described below.

Goldstein *et al.* (1977) have shown that the effects of absorption can be included in equation 3.3 if the observed intensity ratio I_A/I_B is replaced by I_{A0}/I_{B0}.

$$\frac{I_{A0}}{I_{B0}} = \frac{I_A}{I_B} \times \frac{[(\mu/\rho)_{spec}^A] [1 - \exp - \{(\mu/\rho)_{spec}^B \cdot \rho t \cdot \operatorname{cosec} \alpha\}]}{[(\mu/\rho)_{spec}^B] [1 - \exp - \{(\mu/\rho)_{spec}^A \cdot \rho t \cdot \operatorname{cosec} \alpha\}]} \quad (3.8)$$

This correction is usually available within the software supplied with the AEM; only the appropriate k_{AB} value and the thickness of the specimen need to be input. Methods by which the thickness of the sample can be measured are described in section 3.2.4(f).

The need to measure the thickness in order to take account of absorption in the specimen can be circumvented by a method described by Van Cappellen et al. (1984), Van Cappellen (1990), Horita, Takeshi and Nemoto (1986) and Horita, Sano and Nemoto (1987). If it is assumed that all X-rays are generated at one half of the foil thickness, instead of throughout the foil, equation 3.8 may be simplified (Goldstein et al., 1977):

$$\frac{I_A}{I_B} = \frac{I_{A0}}{I_{B0}} \exp \{-[(\mu/\rho)_{spec}^A - (\mu/\rho)_{spec}^B] \cdot \rho \left(\frac{t}{2}\right) \operatorname{cosec} \alpha\} \quad (3.9)$$

which can be rewritten as:

$$\ln (I_A/I_B) = \ln (I_{A0}/I_{B0}) + \Delta_{BA} t \quad (3.10)$$

Where $\Delta_{BA} = 0.5 \ [(\mu/\rho)_{spec}^B - (\mu/\rho)_{spec}^A] \cdot \rho \operatorname{cosec} \alpha$. Thus $\ln (I_A/I_B)$ is linearly related to t and its extrapolation to $t = 0$ gives I_{A0}/I_{B0}, the ratio that would be obtained with no absorption.

There are a number of parameters that are proportional to t and can be measured much more easily than the thickness itself. Van Cappellen et al. (1984) and Van Cappellen (1990) evaluated two indirect measures of the mass thickness; the use of a window in the continuum and the net integral of a peak that is not significantly absorbed. They found that there is a systematic difference between the results obtained from the two methods, which they attributed to the fact that not all the continuum radiation originates from the specimen (section 3.2.3(a)). They therefore recommended the use of a net peak integral or the sum of a number of net peak integrals.

It should be pointed out that, as equation 3.10 was derived from an approximate form of the absorption equation, the linearity may not hold if either the thickness or the difference in the absorption coefficients of the two elements is very large (Westwood, Michael and Notis, 1992). Also, the beam current and irradiation time should be constant for each calibration point. It is advisable for each data point to carry the same weighting. The same number of counts should therefore be collected for each point and then the counts should be normalized for the same irradiation time.

Obviously, for the extrapolation method to be applied, regions with different thicknesses need to be available in the regions of the specimen to be analysed. This may not be possible for the analysis of an 'unknown', but is usually possible when a standard is being used to derive k_{AB} factors. Indeed, it is advisable always to derive k_{AB} factors in this way so as to be certain that the values are free from the effects of absorption. This procedure is particularly important at voltages >100 kV and for the lighter elements that can be detected with thin-window and windowless detectors. Figures 3.13(a) and (b) show plots for a k_{AB} factor and for a ratio of two intensities, respectively, derived by the extrapolation method.

(e) Fluorescence in the specimen

In thicker specimens, the characteristic X-ray intensity emitted by an element A may be enhanced by secondary X-ray fluorescence from the characteristic X-rays emitted by a second element B. This phenomenon leads to an apparent increase in the concentration of A, but is rarely a problem in practice, particularly in silicates, because fluorescence efficiencies are low for $Z < 20$ and tend to be negligible, except for heavier elements of almost adjacent atomic number (e.g. Cr excited by Fe). Nockolds *et al.* (1980) derived a correction factor for fluorescence in thin foils that is proportional to $t \ln t$. This correction factor is available in most software packages supplied with AEMs, but the effects of absorption are almost always much more serious than those of fluorescence in specimens of similar thickness. Alternatively, Van Cappellen (1990) has shown that the extrapolation method corrects for fluorescence, although generally with less accuracy than for absorption.

(f) Measurement of the sample thickness

As pointed out in section 3.2.4(d), it is not always possible to use the extrapolation method to correct for absorption and fluorescence in the specimen. In these cases the thickness must be measured directly. There are various methods by which this can be done. The two easiest are the parallax and contamination-spot methods. The former method relies on the specimen containing a planar feature, such as a stacking fault. If the feature is initially perpendicular to the foil surface and is tilted through an angle θ about an axis parallel to its interface, the projected width W is related to the thickness t as $t = W \operatorname{cosec} \theta$. This method will underestimate the thickness of beam-milled minerals by twice the thickness of the amorphous layer on the surface of the specimen.

When a probe a few nanometres in diameter is focused on the specimen, a contamination spot forms on the top and bottom surfaces under all but the most stringent conditions of specimen cleanliness and vacuum (Figure 3.9). If the foil is then tilted through an angle θ, the separation of the contamination spots can be used to calculate the thickness as described

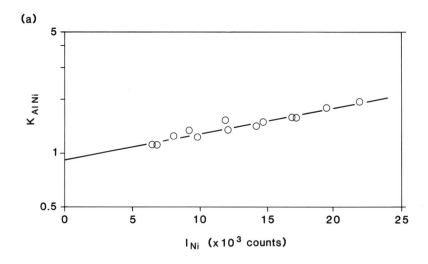

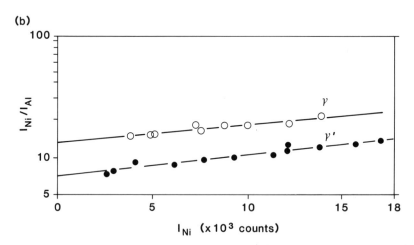

Figure 3.13 (a) k_{AlNi} plotted against the intensity of the Ni Kα peak for the same irradiation time and beam current at different thicknesses in a Ni-Al-Mo alloy standard. The value at $I_{\text{Ni}} = 0$ corresponds to that for an infinitely thin specimen. (b) Measured ratios of intensities of Ni and Al characteristic radiation plotted against I_{Ni} for γ and γ' phases in a Ni-Al-Mo alloy. The values of the ratios at $I_{\text{Ni}} = 0$ correspond to those required for conversion to weight ratios using equation 3.3. Both experiments were performed at 200 kV. (After Horita *et al.*, 1989).

above. The method is quick and easy to use, but it is difficult to decide at which position the contamination spots should be measured (Figure 3.9). The method can seriously overestimate t, the error being greater the thinner the specimen (Rae, Scott and Love, 1981). This overestimate is the result of a number of factors, the most important of which is the fact that the contamination spots sit on an almost invisible 'foothill' of contamination, with the result that the true thickness is smaller than it appears to be.

One of the most accurate, if time-consuming, methods for the measurement of thickness in crystalline materials is convergent-beam electron diffraction (CBED). Under 'two-beam' diffraction conditions (i.e. when only one set of lattice planes is diffracting strongly), 'Kössel–Mollenstedt' fringes are observed in the transmitted and diffracted discs in CBED, the spacing of which can be used to measure the thickness of the specimen. The technique was first described by Kelly *et al.* (1975) and the application and limitations of the technique have been discussed by Allen (1981). With care, the thickness can be measured with an accuracy of ±5% in metallic systems. However, the technique probably has a very limited application in mineralogy (Champness, 1987). The large lattice parameters of minerals, and hence their small reciprocal lattice dimensions, make it difficult to obtain diffraction discs which contain sufficient fringes (at least three). Also, it is difficult to obtain true 'two-beam' conditions and hence the fringe spacing is difficult to interpret. Because, like the parallax method, the CBED method measures the 'crystalline' thickness, it will underestimate the thickness to the value of the thickness of the two amorphous layers.

Another possible indirect measure of the thickness is the ratio of the X-ray intensities from two elements whose stoichiometry is constant, one of which is strongly absorbed, for instance Si/O in certain silicates (Van Cappellen, personal communication). The thickness in equation 3.8 can be adjusted until the 'correct' value of the Si/O ratio is obtained.

Morris, Ball and Statham (1980) described two methods by which the effective mass path-length in equation 3.9 (i.e. the factor $\rho(t/2)\operatorname{cosec}\alpha$) can be derived directly from the X-ray data. The first uses the ratio of the K and L lines for a major element in the sample and is an iterative process requiring knowledge of the appropriate k_{AB} factors and absorption coefficients. This method clearly requires a high concentration of an element with $Z \geq 27$, so is not generally applicable to silicates.

The second method requires several spectra to be recorded from the same area of the specimen at different tilt angles, thus producing a series of spectra of known relative path lengths. Each of the spectra is processed and concentrations are normalized to 100%. A value of the mass path length at zero tilt is then chosen (by an iterative process) that gives the same corrected value of the analyses at the different tilts. Although this method is more generally applicable than the first one, it is time consuming and more prone to errors.

Porter and Westengen (1981) suggested using a thin standard similar in composition to the specimen of interest to calibrate X-ray intensity in terms of thickness. As the intensity of the EDS X-ray spectrum is a linear function of the foil thickness until the effects of absorption become significant, the variation of the X-ray intensity versus thickness in the standard can be used to find the thickness in the 'unknown'. However, the method also requires the measurement of the beam current on-line, and the accurate measurement of the thickness in the standard.

If you are fortunate enough to have an EELS system, the most promising universal method for measuring the specimen thickness is the EELS log-ratio method (Malis, Cheng and Egerton, 1988). There is a very simple relationship between thickness and the ratio between the number of zero energy-loss electrons and the total incident electrons (section 3.3.2(b)). Once the calibration has been carried out, the method can be used 'on-line', is relatively rapid and is accurate to within ±20%.

(g) Analysis of small particles

Cliff *et al.* (1984) have described a simple method for obtaining quantitative analyses of particles that are smaller than the beam size or enclosed by the matrix. In either case, the apparent compositions will have some contribution from the matrix. If absorption of the X-rays in the specimen is not significant, the X-ray intensity I_A from an element A is proportional to the weight fraction, C_A, of that element and the total X-ray intensity, I_A^T, from a thin sample which contains second-phase particles is:

$$I_A^T \propto C_A^M L_M + C_A^P L_P \tag{3.11}$$

Where C_A^M and C_A^P are the weight fractions of element A in the matrix and in the particle respectively, and L_M and L_P are the respective electron path lengths. The path lengths may be 100% in the matrix (Figure 3.14(a)), 100% in the particle (Figure 3.14(b)) or partially in both (Figure 3.14(c)). If element A is concentrated in the matrix and element B is concentrated in the precipitate, and the particle and the matrix both have a fixed composition, a plot of C_A against C_B will be a straight line (Figure 3.15). The effect of changing the relative values of L_M and L_P will be to move point C along the line (Cliff *et al.*, 1984). If the precipitate is very small, it may be impossible to obtain the point B, but the data will extrapolate through the composition of the particle and this fact may be used, along with other information, to identify it.

Lorimer *et al.* (1984) used the extrapolation method to determine the composition of particles and 'dendrites' in sphalerite. The analytical data from the two phases are shown in Figure 3.16, with all the other elements plotted against Zn. If it is assumed that neither phase contains Zn, the data extrapolate to pyrite (FeS_2) for the particles and to chalcocite (Cu_2S) with a small amount of Fe, for the dendrites. These identifications were confirmed by electron diffraction.

Cliff *et al.* (1984) identified small (20 nm) Ca-rich particles in a metamorphosed olivine using the extrapolation technique. SiO_2, FeO and MgO

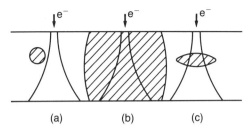

Figure 3.14 Three possible electron beam paths in a sample containing second-phase particles (shown hatched). (From Lorimer *et al.*, 1984. Reproduced with permission, San Francisco Press).

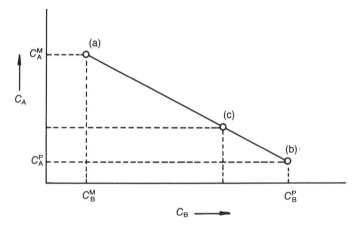

Figure 3.15 Schematic plot of the apparent concentration of element A vs. the apparent concentration of element B for the three situations in the sample of Figure 3.13. (From Lorimer *et al.*, 1984. Reproduced with permission, San Francisco Press).

compositions were plotted against CaO (Figure 3.17). The compositions of diopside, D, and tremolite, T, are indicated in the figure. All but one of the 20 analyses plot on the line that extends to diopside, rather than to tremolite. The wide scatter in the data for the particles suggests that the diopside has a variable composition, but its identification is unambiguous. Although it was clearly not necessary to do so in the case of the sulphides or the diopside, it is possible to apply an absorption correction to the data (Horita, Sano and Nemoto, 1989).

(h) X-ray mapping

Although X-ray images similar to those obtained in the EPMA can be obtained in the AEM if the X-ray signal from the EDS is used to modulate the cathode ray tube of the STEM while the beam is scanned in a raster, such images have rarely been produced because of the long times currently

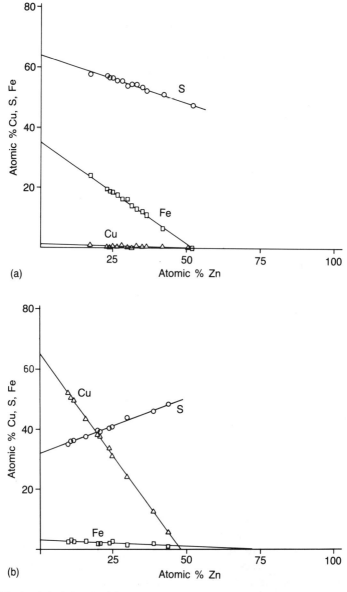

Figure 3.16 Analytical data for (a) small, second-phase particles of FeS$_2$ and (b) 'dendrites' of Cu$_2$S in ZnS. (From Lorimer et al., 1984. Reproduced with permission, San Francisco Press).

needed (1 to 10 or more hours) and the inadequate stability of the instrument over such times. However, the advent of AEMs with FEGs, EDSs with high collection angles, digital image storage and drift-correction software will make the acquisition of X-ray images more feasible in the near future.

THE X-RAY ANALYSIS OF THIN SPECIMENS

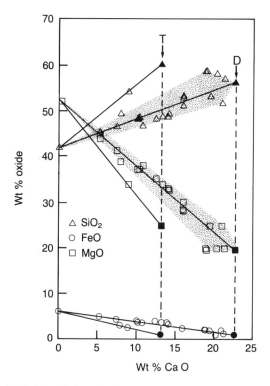

Figure 3.17 Variation in SiO_2, FeO and MgO concentration vs. CaO for second-phase particles in olivine. The point labelled T represents the composition of tremolite and D represents the composition of diopside. (After Cliff *et al.*, 1984).

(Drift-correction software can be used in the STEM mode. The acquisition of the X-ray spectrum (whether in point or in scan mode) is temporarily suspended while a STEM image is acquired. The software then measures any drift since the last image was acquired and predicts and corrects for any drift that occurs in the next acquisition period).

3.2.5 Limits of X-ray microanalysis in thin specimens

(a) Accuracy of quantification

The accuracy of the quantitative data obtained by application of equation 3.3 is ultimately limited by the counting statistics for the collection of X-rays, both in the determination of the k_{AB} factors and in the acquisition of the intensities $I_{A,B}$. Assuming Gaussian counting statistics, the standard deviation, σ, is equal to $I^{1/2}$, where I is the number of counts (section 2.8.1). The percentage error in the number of counts for 95% confidence (2σ) is therefore $2I^{1/2}/I \times 100$. In equation 3.3, the total percentage error in C_A/C_B

for a single measurement is equal to the square root of the sum of the squares of the percentage errors in I_A, I_B and k_{AB}. Clearly the total error can never be less than the error in k_{AB} alone. If the error in k_{AB} is 2% and 10 000 counts are collected for I_A and I_B ($\sigma = 1\%$), the total error is 3.5%. If the error in k_{AB} is reduced to 1% and if 100 000 counts could be collected for I_A and I_B ($\sigma = 0.32\%$), the total error would be 1.3%. However, such large intensities are not normally achievable because of the inefficiency of the EDS spectrometer geometry and the instability of the electron probe and the specimen stage, which restrict the counting times that can be used. The recent introduction of drift-correction software will make longer counting times feasible.

The error in I_A/I_B can clearly be further reduced if several readings are taken. Currently, this is more practical than using long counting times. The absolute error at the 95% confidence level in n readings of I_A/I_B is (Romig and Goldstein, 1980):

$$\frac{1}{\sqrt{n}} \frac{t_{95}^{n-1}}{\overline{(I_A/I_B)}} \times 100 \qquad (3.12)$$

where t_{95}^{n-1} is the Student t value, σ is the standard deviation for n readings and $\overline{I_A/I_B}$ is the average value of I_A/I_B.

It should be emphasized that the above treatment is based on the random errors from counting statistics alone. Many other factors, such as the accuracy of the chemical composition of the standard, the presence of spurious X-rays, the lack of precise knowledge of the thickness of the specimen (for absorption and fluorescence corrections), the background-fitting routine and the deconvolution of overlapping peaks will contribute systematic errors. Taking all the above factors into account, the accuracy for routine AEM of major elements in minerals is currently probably no better than ±5% relative to the amount present.

(b) Minimum detectable concentration and mass

The smallest concentration of an element, C_{min}, that can be detected corresponds to the smallest significant X-ray signal that can be measured above the background.

$$C_{min} \approx (I_P \times [I_P/I_B] \times \tau)^{-1/2} \times 100 \, wt\% \qquad (3.13)$$

Where I_P and I_B are the net integrated intensities (in counts per second) in the peak and background respectively for a pure element standard and τ is the analysis time in seconds (Ziebold, 1967).

As the peak-to-background ratio theoretically increases with voltage, there should also be an increase in sensitivity. This improvement is now being achieved experimentally (Zaluzec, 1992). Currently, values of

0.1–0.2 wt% can be obtained using an EDS for elements with $Z > 10$. A $C_{min} < 0.1$ wt% should become routine with a FEG at 300–400 kV (Goldstein and Williams, 1992). The increases in counting times that will be made possible by better stability of the specimen stage and drift-correction software, improvements in instrumentation that will lead to better peak-to-background ratios (section 3.2.3(a)) and the increase in the solid angle of the EDS to ~0.3 sr that will become possible with redesign of the EDS–AEM interface should improve the sensitivity of X-ray analysis in the AEM even further.

Another measure of the sensitivity of AEM is the minimum detectable number of atoms, N_{min}, or the minimum detectable mass (MDM). For thermionic electron sources, values of MDM are 10^{-19} to 10^{-20} g, equivalent to 100–1000 atoms of Fe. With FEGs and the instrumental improvements listed above, it should soon be possible to detect a single atom (Cliff and Lorimer, 1992; Goldstein and Williams, 1992)!

The spatial resolution, R, as given by equations 3.1 and 3.2, C_{min} and N_{min}, are not independent of one another:

$$N_{min} \sim C_{min} n t R^2 \qquad (3.14)$$

where n is the average number of atoms per unit volume and t is the specimen thickness. In order to improve R, one can decrease the probe size, d, (equation 3.1) and use a thinner specimen to minimize b, (equation 3.1) but both of these steps will increase (degrade) C_{min} and N_{min} because I_P (equation 3.13) will be lower (decreasing d and using a thinner specimen will also reduce the accuracy of the analysis for the same reason; d can only be decreased by using a smaller condenser aperture or by increasing the excitation of the C_1 lens; both procedures result in a decrease in the probe current). Goldstein and Williams (1992) have shown that C_{min} is optimized at about two to four times the minimum spatial resolution. However, the detection of single atoms will only be possible for $R \leq 1.5$ nm because of the R^2 dependence of N_{min} in equation 3.14.

(c) Specimen damage during analysis

The discussion in previous sections does not take into account the fact that the current densities required for high spatial resolution, X-ray mapping and maximum analytical sensitivity can induce selective loss of elements from the region being analysed. It is well known that loss of volatiles such as CO_2, H_2O and alkalis is a problem in EPMA (Chapter 2). Current densities in a focused probe in AEM are typically $\sim 2 \times 10^6$ A m^{-2} for a LaB$_6$ gun and $\sim 10^9$ A m^{-2} for an FEG, one to four orders of magnitude higher than is used in EPMA. Champness and Devenish (1990) and Devenish and Champness (1993) have shown that all silicates suffer some mass loss at the highest current densities used in AEM, but that there is a threshold of the current density for each element in a particular structure for which no loss occurs. For instance it was found that the threshold values of the current density for which no loss occurs for any element is $\approx 10^5\,\text{A}\,\text{m}^{-2}$ for diopside and $\sim 3 \times 10^4\,\text{A}\,\text{m}^{-2}$ for margarite mica. Notice that both these values are lower than the current density in a focused beam from a LaB_6 gun.

It is important that, where possible, the analyst operates below the current density at which damage occurs if quantitative results are required. Clearly, defocusing the electron beam is more effective in minimizing mass loss than rastering a focused beam over the same area. The effect of mass loss may also be minimized by using the highest voltage available, by using a cooling stage and by analysing as thick an area as is possible (van der Pluijm, Lee and Peacor, 1988). However, the advantage of the last strategy must be balanced against the isadvantage of the increase in absorption effects that will result.

3.3 Electron energy-loss spectroscopy

3.3.1 Principles of the technique

The energies of the electrons transmitted through a thin specimen fall into three ranges (Figure 3.18). Those that experience negligible loss in energy produce the narrow and intense 'zero-loss peak'. The energy range approximately 5–50 eV is known as the 'low-loss region' and is largely produced by the collective vibration of the entire distribution of valence electrons – so-called 'plasmon scattering'. For metals such as Al or Mg that contain a large number of free electrons, the plasmon peaks are well defined and Gaussian in shape and their positions can give compositional information. However, for materials such as minerals, in which free electrons are not abundant, the plasmon peaks are broad and the compositional information that they contain is not accessible. However, the intensity of the plasmon peaks can be used to determine the thickness of mineral specimens (section 3.3.2(b)).

The 'characteristic edges' in the 'high-loss' region are the result of the inner-shell excitations that give rise to characteristic X-rays and Auger electrons. It is these edges that are normally used for chemical analysis. The energies of the EEL edges are slightly above those of the corresponding X-ray fluorescence peaks. At each edge there is a rapid increase in the intensity of the spectrum followed by a gradual decrease to background levels (Figure 3.18). Qualitative analysis is carried out by noting the energies of the edges that are present and comparing them with tabulated values. However, quantification is much less straightforward than in the case of EDS because of the asymmetric shape of the edge, the low peak-to-background ratio (Figure 3.18) and the fact that there is often significant overlap of the tail of an edge from one element with the edge of another. Because the intensity in

ELECTRON ENERGY-LOSS SPECTROSCOPY

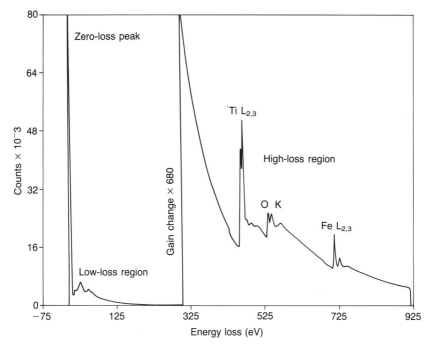

Figure 3.18 Electron energy-loss spectrum of ilmenite showing the zero-loss, low-loss and high-loss regions of the spectrum. Notice the large change in gain that is needed at about 300 eV in order to make the characteristic Ti, O and Fe edges visible. Also, notice how much weaker the Fe edge is, compared with the Ti edge, even though these elements occur in equal atomic proportions in the ilmenite.

the characteristic edges is so low, the spectrum is usually plotted with a large change in 'gain' at about 50 eV with the vertical scale magnified 100–200 times (Figure 3.18).

As the ionization cross-section decreases with atomic number, one great advantage of EELS is that it is possible to detect the edges from elements such as hydrogen, helium and lithium that cannot be detected by X-ray analysis. Above an energy loss of about 2 keV, the number of losses is so small that it is impractical to measure them. Hence, for elements He to Si, K excitations are used for analysis, for Al to Sr, $L_{2,3}$ excitations are used and for Rb to Os, $M_{4,5}$ excitations are used. (See section 2.2 for an explanation of the nomenclature of electron excitations. The $L_{2,3}$ edge contains contributions from the L_2 and L_3 edges which cannot be resolved by the spectrometer; similarly, the $M_{4,5}$ edge contains contributions from the M_4 and M_5 edges). However EELS is not normally used for the analysis of heavier elements, X-ray analysis being preferred.

The EEL spectrum can be used to determine more information about the specimen than its chemical composition. In particular, 'fine structure' may be visible before and after the edges. Energy loss near-edge structure (ELNES) is found within about 30 eV of the edge and is determined by the chemical binding state of the element, while extended energy-loss fine structure (EXELFS) provides information on the number and type of neighbours surrounding the element. EXELFS is analogous to the oscillations seen in X-ray absorption edge fine structure (EXAFS) in synchrotron X-ray spectra. These techniques have been the main application of EELS to mineralogy to date, but are beyond the scope of this chapter. The interested reader is referred to Spence (1988) or Buseck and Self (1992).

(a) The spectrometer

The spectrometer is normally mounted after the specimen and is the final component in the AEM (Figure 3.1). Thus, no modifications are needed to the microscope itself. In this experimental arrangement, a magnetic prism spectrometer is used to deflect all the electrons through about 90°. The exact deflection will depend on the energy of the electrons and these will therefore be dispersed into a spectrum of energies (Figure 3.19). The magnetic prism also acts as a lens and, as the object plane of the spectrometer is normally the crossover of the projector lens, electrons with the same energy are focused to a point in the image plane of the spectrometer.

There are two possible ways in which the data may be collected. In the 'serial' system, electrons with a given narrow range of energy loss pass through a slit in the image plane of the spectrometer. Electrons of different energies are scanned sequentially across the slit by varying the magnetic field of the spectrometer, leaving the spectrum for a fixed dwell time at each energy loss (Figure 3.19(a)). The electrons are counted by a scintillation counter and photomultiplier behind the slit and the total signal for each energy loss is assigned to a channel in a multichannel analyser (MCA). A complete spectrum can be collected in this manner, which is analogous to the collection of an X-ray spectrum by WDS.

The drawback of the serial collection system is that only about 0.1% of the spectrum is being recorded at any one time. The recent introduction of the 'parallel' EELS system ('PEELS') has overcome this problem and is expected to result in EELS becoming more widely used for chemical analysis in the AEM. In PEELS, a large fraction of the whole energy spectrum is collected simultaneously by a YAG scintillator and semiconductor photodiode array in the image plane of the spectrometer (Figure 3.19(b)). A variable magnifying quadrupole lens is placed before the photodiode array so as to further disperse the spectrum. The PEELS system is about 300 times more efficient than serial EELS and useful spectra can be collected in a fraction of a second. This improved efficiency leads to advantages such as reduction of both radiation damage and carbon contamination (the presence

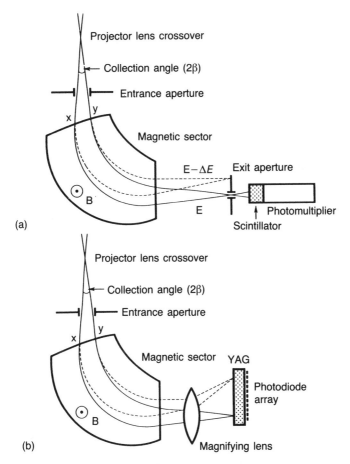

Figure 3.19 Schematic diagram of a magnetic-prism spectrometer. (a) Serial data collection system consisting of a slit, scintillator and photomultiplier detector. (b) Parallel data collection system consisting of a YAG scintillator, magnifying quadrupole and photodiode array. Both (a) and (b) show the trajectories of electrons with two energies, E and $E - \Delta E$. (From Buseck and Self, 1992. Reproduced with permission from the Mineralogical Society of America).

of carbon can be a problem because its K edge may overlap with edges of interest as well as increasing the background under the edges) and makes EELS mapping possible on a reasonable time scale (section 3.3.6).

(b) Coupling of the spectrometer to the microscope

There are two optical arrangements that can be used in EELS, each allowing a different setting of the collection semi-angle β into the spectrometer. In the 'image-coupling' mode, a diffraction pattern is seen on the viewing screen of the microscope and there is an image of the specimen at the object

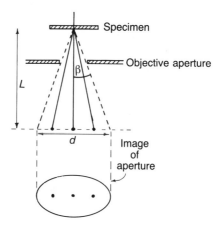

Figure 3.20 Determination of the collection angle β for diffraction coupling of the spectrometer to the microscope. For a camera length L, the image of the objective aperture has a diameter of d, allowing β to the calculated by simple geometry. (From Budd and Goodhew, 1988. Reproduced with permission, Royal Microscopical Society).

plane of the spectrometer. β is then determined by the camera length, L, of the diffraction pattern and the diameter, d, of the entrance aperture of the spectrometer; hence $\beta = d/2L$, radians (Figure 3.20).

In the 'diffraction-coupling' mode, an image is seen on the viewing screen and the diffraction pattern occurs at the object plane of the spectrometer. In this mode β is determined by the size of the objective aperture and is normally measured from the diffraction pattern of a standard specimen (e.g. Al or Au), with and without the objective aperture in place (Budd and Goodhew, 1988). The 'diffraction-coupling' mode is not recommended if simultaneous EDS is required because of the resultant backscattering of electrons from the objective aperture.

Usually, one wants to acquire a spectrum from a specific area. One way of defining the area is to select a portion of the image with an aperture; either the selected-area aperture in the image-coupling mode or the spectrometer-entrance aperture in the diffraction-coupling mode. However, because of the large angles of scattering and the large energy losses involved in EELS, which introduce spherical and chromatic aberrations respectively, contributions to the spectrum can arise from areas well outside the region that has been selected. It is therefore recommended that the area for analysis be selected by focusing the illumination, as is the procedure for X-ray analysis in the EPMA or the AEM.

3.3.2 Quantification of the EELS spectrum

(a) Background subtraction

As in the case of X-ray analysis, the background must first be removed before the intensity under the edge can be determined. In EDS, background subtraction is a relatively straightforward procedure, but in EELS it presents considerable problems because of the much lower signal-to-background ratio and the asymmetric shape of the edges. The background has been found to fit a power law of the form $I = AE^{-r}$, where A and r are empirical constants specific to the energy range of interest and I is the intensity at energy E. In the standard method of background subtraction, a region of width 50–100 eV just before the edge is selected, the above function is fitted, and it is then extrapolated beneath the edge. It is not easy to judge whether the fit is good (e.g. Figure 3.21)! The only simple check is to examine the fit beyond the edge, where it should asymptotically approach the measured data, but not go above it. If the fit is not good, the only remedy is to change the width and/or the position of the fitting region, but this may not be possible if there are other edges nearby.

With parallel EELS, it is possible to produce a 'first-difference spectrum' by recording two spectra separated by a small energy shift and then subtracting one from the other (Figure 3.22). This process removes slowly varying parts of the spectrum, such as the background, leaving the edges

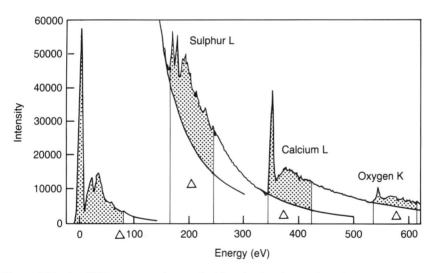

Figure 3.21 An EEL spectrum from anhydrite showing integration windows suitable for quantitative analysis, Δ, and background fitting under the S, Ca and O peaks using the power-law approach. The integrated intensities are shown shaded. (From Buseck and Self, 1992. Reproduced with permission from the Mineralogical Society of America).

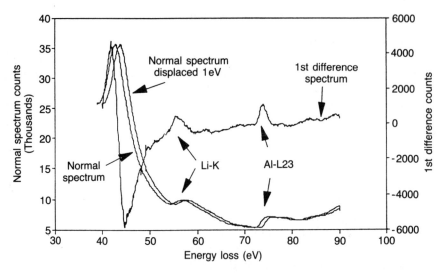

Figure 3.22 Two energy-loss spectra from an Al–Li alloy recorded with an energy displacement of 1 eV and subtracted to produce the first difference spectrum. (From Goldstein and Williams, 1992. Reproduced with permission, San Francisco Press and J.A. Hunt).

essentially unchanged (Goldstein and Williams, 1992). Because of its simplicity and the fact that it is less susceptible to spectral artefacts than the power-law approach, the difference-spectrum method is becoming more popular as parallel EELS becomes more widely available.

(b) Processing the spectra
Once the background has been removed, the intensities under the edges and in the zero-loss and low-loss peaks can be obtained in one of two ways. An integration of the remaining intensity can be carried out over an energy window, Δ (Figure 3.22), which should be the same width as the window that is used to fit the background in the power-law method. The optimum value of Δ is ~0.15 E_K, where E_K is the intensity of the ionization edge, to give the best signal-to-background value, but may have to be smaller if there are other edges close by. Alternatively, multiple least-squares (MLS) methods can be used to determine the intensity under the edge. These methods require the comparison of standard reference spectra to the unknown. MLS routines are not yet commercially available.

Chemical analysis using EELS can be obtained as the number of atoms per unit area, N. For a K edge:

$$N = \frac{I_K(\beta, \Delta)}{I_0(\beta, \Delta)} \times \frac{1}{\sigma_K(\beta, \Delta)} \qquad (3.15)$$

where $I_K(\beta, \Delta)$ is the measured intensity of the edge for a collection semi-angle β and energy window Δ, $I_0(\beta, \Delta)$ is the intensity of the incident beam, estimated by measuring the integrated intensity of the zero-loss and low-energy regions up to the energy Δ (Figure 3.22) and $\sigma_K(\beta, \Delta)$ is the appropriate partial ionization cross-section which may be determined either by calculation (e.g. Egerton, 1986) or from tabulated values (Egerton, 1981). (A partial cross-section is used because the intensity in the edge is integrated over only part of its total energy range and over part of the total angular range, i.e. β. Note that the symbol σ is used for the cross-section in EELS, whereas Q is used in X-ray spectroscopy). Although equation 3.15 gives an absolute value of N without the need to resort to a standard, the thickness, t, has to be measured if a concentration is required.

The need to measure t can be avoided if the atomic ratio of two elements A and B is obtained from the expression:

$$\frac{N_A}{N_B} = \frac{I_{KA}(\beta, \Delta) \times \sigma_{KB}(\beta, \Delta)}{I_{KB}(\beta, \Delta) \times \sigma_{KA}(\beta, \Delta)} \qquad (3.16)$$

(note: the intensities for the two elements may be measured from different atomic shells). Equation 3.16 eliminates the need to scan across the zero-loss peak and low-loss region (other than to check that the specimen is suitably thin; see below). The cross-sections may be determined as described above or their ratios may be determined experimentally from suitable standards and treated in a similar way to the k_{AB} factors in EDS analysis (sections 3.2.4(b) and 3.2.4(c)). If this method is adopted, it is clearly essential that the operating conditions for the microscope and the spectrometer are identical for the standard and the unknown. An advantage of using standards is that β does not have to be measured.

The thickness of the specimen can be obtained from the expression:

$$t = \lambda \ln(I_T/I_0) \qquad (3.17)$$

where λ is the total inelastic mean free path, I_0 is the intensity in the zero-loss peak and I_T is the integrated intensity in the whole spectrum. Malis, Cheng and Egerton (1988) have measured λ for a number of pure metals, alloys and oxides and derived an equation relating λ to the mean atomic number of the specimen, the incident energy of the electron beam and β. Equation 3.17 gives t to an accuracy of $\sim \pm 20\%$.

3.3.3 Operational conditions

(a) Spectrometer parameters
The spectrometer parameters that must be set are the entrance aperture, the slit width (for a serial detector) and the energy range to be collected. The

slit width affects both the energy resolution and the number of counts recorded. It is important to decide the resolution required for the application so that counts are not lost unnecessarily by setting the width of the slit too narrow. The ultimate resolution attainable is normally limited by the energy spread of the electron source. Thus, resolutions of between 1 and 1.5 eV can be obtained with LaB_6 guns and FEGs routinely produce resolutions less than 1 eV (0.27 eV has been attained).

(b) Other considerations

As for the case of X-ray analysis, it is essential to ensure that the recorded EEL spectrum does not contain contributions other than from the region of interest in the specimen. Fortunately, it is difficult for radiation from elsewhere to reach the spectrometer and such spurious effects are therefore rare. Williams (1984) described how the spurious effects that have their origin in the spectrometer and the processing system may be recognized and eliminated.

The electron channelling that gives rise to anomalously high X-ray counts when the specimen is close to the Bragg condition also affects the EEL spectrum. Thus the precautions described in section 3.2.3(d) for EDS should also be applied to EELS.

Although the number of ionization events increases with the depth of penetration of the electron beam, the number of multiple-scattering events also increases with depth. These latter events remove intensity from the zero- and low-loss regions of the spectrum and transfer them to the high-loss region, hence enhancing the background below the characteristic edges. As the background intensity grows faster with thickness than the edge intensities, it is more difficult to measure the intensity associated with the edge and even to detect the presence of an edge in samples that are too thick. In addition, equations 3.15 and 3.16 are only applicable to samples that are thinner than one plasmon mean free path.

The usual test for acceptable sample thickness is determination of the ratio of the counts in the zero-loss peak, I_0 to those in the first plasmon peak, I_P. The maximum acceptable value of I_P/I_0 depends on the specimen and needs to be determined experimentally, but thicknesses of between 15 and 60 nm have been found to be suitable for many minerals at 100 kV (Buseck and Self, 1992). The maximum usable thickness could be increased by using a higher accelerating voltage: the total mean free path increases by a factor of about two at 300 kV compared with the value at 100 kV. However, electron spectrometers with high energy resolution are not yet available to deal with 300–400 keV electrons.

The maximum usable thickness can also be extended if multiple-scattering events are removed from the spectrum by deconvolution (Egerton, 1986). However, as deconvolution does not improve the signal-to-noise ratio and can introduce artefacts, it is better to use specimens that are thin enough to be free of multiple scattering. This imposes a severe practical limitation on the maximum usable thickness of the specimen.

The compositions derived from equations 3.15 and 3.16 may need correction if the EEL spectrum has been collected using an electron beam whose convergence angle, α, is equal to or larger than β. The equation for this correction is given by Joy, Romig and Goldstein (1986).

3.3.4 Spatial resolution

In contrast to EDS, beam spreading is not important in EELS because the entrance aperture of the spectrometer collects only those electrons that emanate from the specimen in a narrow cone of semi-angle β. Thus, EELS has a better spatial resolution than EDS under the same conditions of kV and thickness. However, experiments suggest that the improvement is small for an FEG because the high current densities that can be achieved in very small probes mean that very thin foils can be used and beam spreading is minimized (Titchmarsh, 1989).

3.3.5 Detection limits and accuracy

Because the EEL spectrometer collects a large fraction of the transmitted electrons (even a well-designed energy-dispersive X-ray spectrometer collects only about 1.3% of the characteristic X-rays), EELS has an inherently higher efficiency than EDS, particularly at low energy. However, EELS has the poorer signal-to-background ratio. The best sensitivity will be achieved with parallel collection and an FEG. Leapman and Newbury (1992) have demonstrated a sensitivity of <10 ppm in some standard glasses prepared by NIST and Mory and Colliex (1989) have shown that it should be possible to detect a single atom of thorium on a carbon film. Unfortunately, these levels of sensitivity are unlikely to be achieved in minerals because of the problem of specimen damage (section 3.2.5(c)).

Because of the high signal-to-background ratio and the difficulties with background subtraction, EELS is inherently less accurate than EDS. However, in favourable cases, an accuracy of better than 5% relative can be achieved with the ratio method, compared with about 10% for the determination of N, the number of atoms per unit area, from equation 3.15.

3.3.6 EEL spectrum imaging

Because of its higher efficiency, EELS is a more promising candidate for elemental mapping of thin specimens than EDS. In addition, any of the other information in the EEL spectrum, e.g. chemical bonding, dielectric constant or thickness, can be mapped. With a parallel collection system and advanced software for processing the spectra, a complete spectrum can be

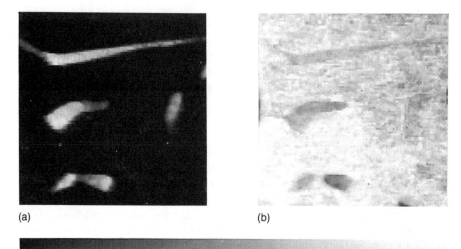

Figure 3.23 Digital EELS maps showing atomic % Be (a) and Cu (b) in a Cu-Be-Co alloy. The instrument was a dedicated STEM (VG HB501). 1024 EELS channels were collected and stored for each pixel in an 128 × 128 pixel spectrum-image in 25 min. The width of the images is 136 nm. The upper grey scale at the bottom of the figure varies from 0 (left) to 61 atomic % for Be (right) and from 0 (left) to 100 (right) atomic % for Cu. The Be K characteristic edge was quantified by linear least-squares fitting of a standard power law to the background. Because the Co and Cu M characteristic edges overlap, the Cu M edge was quantified by suppression of the background by the first difference method and the fitting of a reference spectrum using the multiple least-squares method. The relative thickness of the foil was calculated using the EELS log-ratio method. A Co image indicated that there is no Co in the particles. The images of the portions of the particles that run through the foil (the brightest areas in the Be image) indicate that their composition is approximately $CuBe_2$. (From Hunt, unpublished).

stored at each pixel in a STEM image in about 1 h for a 128 × 128 pixel map (Hunt and Williams, 1991). Clearly drift-correction software is essential, as is the measurement of and the correction for variations in the beam current. Storage of the complete spectrum has the advantage that the same spectral data can be processed by different routines and compared. The presence of unsuspected elements can also be verified long after the data have been acquired.

Figure 3.23 comprises two digital EELS images from an Cu-1.85 wt% Be-0.2 wt% Co alloy (Hunt, unpubl.). The images show the concentration of Be (left) and Cu (right) in atomic percent, the top grey scale at the bottom of the figure varying from 0 (left) to 61 atomic percent (right) for Be and from 0 to 100 atomic percent for Cu.

An (analog) energy-loss image can be obtained directly on the viewing screen of the TEM if an energy filter is inserted into the imaging system of the TEM. At the present time, the only commercial systems are the Zeiss energy-filtering electron microscopes which, until recently, had a maximum operating voltage of 80 kV and did not allow the use of a goniometer stage. Thus applications have mainly been confined to the biological sciences. The new 120 kV version of the instrument incorporates a goniometer stage and is therefore expected to find more materials (and mineralogical) applications.

3.4 EDS versus EELS

Ideally, all analytical electron microscopes would have a windowless EDS and a parallel EELS system. There is no doubt that, if finance were limited and it were only possible to purchase one of the two systems, the choice of the mineralogist for routine chemical analysis would be for (windowless) EDS. EDS is able to detect and quantify all but the lightest elements much more easily than EELS and the restriction on usable thickness is not nearly so severe. The large number of elements that the average mineral tends to contain will give problems for background subtraction in EELS (because of the close proximity of other edges). However, PEELS and the difference-spectrum method show great potential here (Figure 3.23).

In its favour, EELS has better energy and spatial resolution than EDS and can detect all elements, even H and He. Parallel EELS will surely find applications in the detection and quantification of, for instance, Li in micas, clays and Mn ores, Be in bauxite minerals and B in shales. EELS mapping also shows great promise because of its versatility and the inherent efficiency of the parallel detector. However, quantification of EEL spectra is far more complex than for X-ray spectra and there are many pitfalls to await the inexperienced user. There is a need for the development of manageable, user-friendly software for which the operator input is minimal, as is now the case for X-ray analysis of thin films.

3.5 Future directions of AEM instrumentation

Many of the recent or imminent developments in AEM have already been mentioned in the text. Some of the important ones are drift-correction software, HPGe X-ray detectors, beam-current measurement, a compact WDS system, and parallel EELS and EELS imaging. Other desirable improvements would be: (i) improved source stability, (ii) a multi-specimen holder, (iii) higher detector take-off and collection angles, (iv) improved stage stability, (v) ultra-high vacuum in the specimen area and (vi) the development of an electron spectrometer with a high resolution that can handle 300–400 kV electrons.

Dedicated STEMs already have many of the attributes outlined above. However, their use in mineralogy is unlikely to become widespread because of their cost and their lack of flexibility for routine imaging and diffraction.

Acknowledgements

I should like to thank G.W. Lorimer, D.B. Williams, J.A. Hunt, P.B. Kenway and G. Cliff for invaluable discussions. The editors, J.F.W. Bowles, P.J. Potts and S.J.B. Reed made a number of suggestions that have improved the text.

References

Allen, S.M. (1981) Foil thickness measurement from convergent-beam diffraction patterns. *Phil. Mag.*, **A43**, 325–35.

Barber, D.J. (1993) Preparation of rock, mineral, ceramic and glassy materials, in *Procedures in Electron Microscopy* (eds A.W. Robards and A.J. Wilson), Wiley, Chichester, 1140 pp.
Bishop, H.E. (1974) Recent instrumental developments in microanalysis, in *Advances in Analysis of Microstructural Features by Electron Beam Techniques*, Metals Soc., London, pp. 1–18.
Boswell, E., Reece, M.J. and Gee, M.G. (1992) Preparation of hard particle powders for examination in the transmission electron microscope. *J. Microsc.*, **167**, 123–6.
Budd, P.M. and Goodhew, P.J. (1988) *Light Element Analysis in the Transmission Electron Microscope: WEDS and EELS*, Oxford University Press/Royal Microscopical Society, 73 pp.
Buseck, P.R. and Self, P. (1992) Electron energy-loss spectrometry (EELS) and electron channelling (ALCHEMI), in *Minerals and Reactions at the Atomic Scale: Transmission Electron Microscopy* (ed. P.R. Buseck), Mineralogical Society of America, Washington, pp. 141–80.

Champness, P.E. (1987) Convergent beam electron diffraction. *Mineral. Mag.*, **51**, 33–48.
Champness, P.E. and Devenish (1990) Elemental mass loss in silicate minerals during X-ray analysis. *Trans. Roy. Microsc. Soc.*, **1**, 177–80.
Cliff, G. and Kenway, P.B. (1982) The effects of spherical aberration in probe-forming lenses on probe size, image resolution and X-ray spatial resolution in scanning transmission electron microscopy, in *Microbeam Analysis – 1982* (ed. K.F.J. Heinrich), pp. 107–10.
Cliff, G. and Lorimer, G.W. (1975) The quantitative analysis of thin specimens. *J. Microsc.*, **103**, 203–7.
Cliff, G. and Lorimer, G.W. (1992) AEM: from microns to atoms, *Proc. 50th Ann. EMSA Meeting*, pp. 1464–5.
Cliff, G., Powell, D.J., Pilkington, R. *et al.* (1984) X-ray microanalysis of second phase particles in thin foils, in *Electron Microscopy and Analysis, 1983* (ed. P. Doig), Inst. Phys., Bristol, pp. 63–6.
Cooke, C.J. and Duncumb, P. (1969) Performance analysis of a combined electron microscope and microprobe analyser 'EMMA', in *Proc. 5th Intl. Congr. on X-ray optics and Microanalysis* (eds G. Mollenstedt and K.H. Gaukler), pp. 245–7.

Devenish, R.W. and Champness, P.E. (1993) The rate of mass loss in silicate minerals during X-ray analysis, *Proc. 13th Intl. Congr. on X-ray optics and microanalysis*, Manchester, 1992, Inst. Physics, London and Bristol, pp. 233–6.
Duncumb, P. (1966) Precipitation studies with EMMA-4 – A combined electron microscope and X-ray analyser, in *The Electron Microprobe* (eds T.D. McKinley, K.F.J. Heinrich and D.B. Wittry), Wiley, New York.

Egerton, R.F. (1981) Values of K-shell partial cross-sections for electron energy-loss spectrometry. *J. Microsc.*, **123**, 333–7.

REFERENCES

Egerton, R.F. (1986) *Electron Energy Loss Spectroscopy*, Plenum Press, New York, 410 pp.

Goldstein, J.I., Costley, J.L., Lorimer, G.W. and Reed, S.J.B. (1977) Quantitative X-ray analysis in the electron microscope, in *SEM/77* (ed. O. Johari), IITRI, Chicago, pp. 315–24.
Goldstein, J.I., Lyman, C.E. and Williams, D.B. (1989) The wavelength-dispersive spectrometer and its proposed use in the analytical electron microscope. *Ultramicroscopy*, **28**, 162–4.
Goldstein, J.I. and Williams, D.B. (1992) X-ray microanalysis and electron energy loss spectrometry in the analytical electron microscope: review and future directions. *Microbeam Anal.*, **1**, 29–53.
Griffin, B.J. and Johnson, A.W.S. (1992) Experiences with HPGe EDS detectors on a Philips EM430 and a JEOL 6300 FESEM, *Proc 50th Ann. Meeting EMSA*, pp. 1232–3.

Horita, Z., Sano, T. and Nemoto, M. (1987) Simplification of X-ray absorption correction in thin-sample quantitative microanalysis. *Ultramicroscopy*, **21**, 271–6.
Horita, Z., Sano, T. and Nemoto, M. (1989) Energy dispersive X-ray microanalysis in the analytical electron microscope. *ISIJ International*, **29**, 179–90.
Horita, Z., Takeshi, S. and Nemoto, M. (1986) An extrapolation method for the determination of Cliff–Lorimer k_{AB} factors at zero thickness. *J. Microsc.*, **143**, 215–31.
Hunt, J.A. and Williams, D.B. (1991) Electron energy-loss spectrum-imaging. *Ultramicroscopy*, **38**, 47–73.

Joy, D.C. and Maher, D.M. (1977) Sensitivity limits for thin specimen X-ray analysis. *Scanning Electron Microscopy*, **1**, 325–33.
Joy, D.C., Romig, A.D. and Goldstein, J.I. (eds) (1986) *Principles of Analytical Microscopy*, Plenum, New York.

Kelly, P.M., Jostens, A., Blake, R.G. and Napier, J.G. (1975) Determination of foil thickness by scanning transmission electron microscopy. *Phys. Stat. Sol.*, **A31**, 771–9.

Leapman, R.D. and Newbury, D.E. (1992) Trace element analysis of transition elements and rare earths by parallel EELS. *Proc. 50th Ann. Meeting EMSA*, pp. 1250–1.
Lee, S.Y. and Jackson, M.L. (1975) Micaceous occlusions in kaolinite observed by ultra-microtomy and high resolution electron-microscopy. *Clays and Clay Minerals*, **34**, 125–9.
Livi, K.J.T. and Veblen, D.R. (1987) 'Eastonite' from Easton, Pennsylvania: a mixture of phlogopite and a new form of serpentine. *Amer. Mineral.*, **72**, 113–25.
Lorimer, G.W. (1987) Quantitative X-ray microanalysis of thin specimens in the transmission electron microscope; a review. *Mineral. Mag.*, **51**, 49–60.
Lorimer, G.W., Cliff, G., Champness, P.E. *et al.* (1984) In situ X-ray microanalysis of second phase particles in thin foils, in *Analytical Electron Microscopy – 1984* (eds D.B. Williams and D.C. Joy), San Francisco Press, San Francisco, pp. 153–6.

Malis, T., Cheng, S.C. and Egerton, R.F. (1988) EELS log-ratio technique for specimen-thickness measurement in the TEM. *J. Electron Microsc. Technique*, **8**, 193–200.
McGill, R.J. and Hubbard, F.H. (1981) Thin film k-value calibration for low atomic number elements using silicate standards, in *Quantitative Microanalysis with High Spatial Resolution* (eds G.W. Lorimer, M.H. Jacobs and P. Doig), Metals Society, London, pp. 30–4.
Michael, J.R., Williams, D.B., Klein, C.F. and Ayer, R. (1990) The measurement and calculation of the X-ray spatial resolution obtained in the analytical electron microscope. *J. Microsc.*, **147**, 289–303.
Morris, P.L., Ball, M.D. and Statham, P.J. (1980) The correction of thin foil microanalysis data for X-ray absorption effects, in *Electron Microscopy and Analysis 1979* (ed. T. Mulvey), Inst. Phys., Bristol, pp. 413–6.

Mory, C. and Colliex, C. (1989) Elemental analysis near the single-atom detection level by processing sequences of energy-filtered images. *Ultramicroscopy*, **28**, 2339–46.

Nicholson, W.A.P., Gray, C.C., Chapman, J.N. and Robertson, B.W. (1982) Optimising thin film X-ray spectra for quantitative analysis. *J. Microsc.*, **125**, 25–40.

Nockolds, C., Nasir, M.J., Cliff, G. and Lorimer, G.W. (1980) X-ray fluorescence correction in thin foil analysis and direct methods for foil thickness measurement, in *Electron Microscopy and Analysis 1979* (ed. T. Mulvey), Institute of Physics, Bristol, pp. 417–20.

Nord, G.L. Jr (1982) Analytical electron microscopy in mineralogy; exsolved phases in pyroxenes. *Ultramicroscopy*, **8**, 109–20.

Peacor, D.R. (1992) Analytical electron microscopy: X-ray analysis, in *Minerals and Reactions at the Atomic Scale: Transmission Electron Microscopy* (ed. P.R. Buseck), Mineralogical Society of America, Washington, pp. 113–40.

Porter, D.A. and Westengen, H. (1981) STEM microanalysis of intermetallic phases in an Al–Fe–Si alloy, in *Quantitative Microanalysis with High Spatial Resolution* (eds G.W. Lorimer, M.H. Jacobs and P. Doig), Metals Society, London, pp. 94–100.

Rae, D.A., Scott, V.D. and Love, G. (1981) Errors in foil thickness measurement using contamination spot method, in *Quantitative Microanalysis with High Spatial Resolution* (eds G.W. Lorimer, M.H. Jacobs and P. Doig), Metals Society, London, pp. 57–62.

Reed, S.J.B. (1982) The single scattering model and spatial resolution in X-ray analysis of thin foils. *Ultramicroscopy*, **7**, 405–9.

Rietmeijer, F.J.M. and Champness, P.E. (1982) Exsolution structures in calcic pyroxenes from the Bjerkreim–Sokndal lopolith, SW Norway. *Mineral. Mag.*, **45**, 11–24.

Romig, A.D. and Goldstein, J.L. (1980) Determination of the Fe–Ni and Fe–Ni–P phase diagrams at low temperatures (700 to 300°C). *Metall. Trans.*, **11A**, 1151–9.

Shau, Y.-H., Feather, M.E., Essene, E.J. and Peacor, D.R. (1991) Genesis and solvus relations of submicroscopically intergrown paragonite and phengite in a blueschist from northern California. *Contrib. Mineral. Petrol.*, **106**, 367–78.

Sheridan, P.J. (1989) Determination of experimental and theoretical k_{ASi} factors for a 200-kV analytical electron microscope. *J. Electron Microsc. Technique*, **11**, 41–61.

Spence, J.C.H. (1988) Techniques closely related to high-resolution electron microscopy, in *High Resolution Transmission Electron Microscopy* (eds P.R. Buseck, J.M. Cowley and L. Eyling), Oxford University Press, pp. 190–243.

Titchmarsh, J.M. (1989) Comparison of high spatial resolution in EDX and EELS analysis. *Ultramicroscopy*, **28**, 347–51.

Van Cappellen, E. (1990) The parameterless correction method in X-ray microanalysis. *Micro. Microanal. Microstruct.*, **1**, 1–22.

Van Cappellen, E., Van Dyck, D., Van Landuyt, J. and Adams, F. (1984) A parameterless method to correct for X-ray absorption and fluorescence in thin film microanalysis. *J. Phys. Colloq. Fr.*, **45**, C2, 411–14.

van der Pluijm, B.A., Lee, J.H. and Peacor, D.R. (1988) Analytical electron microscopy and the problem of potassium diffusion. *Clays and Clay Minerals*, **36**, 498–504.

Westwood, A.D., Michael, J.R. and Notis, M.R. (1992) Experimental determination of light-element k-factors using the extrapolation technique: oxygen segregation in aluminium nitride. *J. Microsc.*, **167**, 287–302.

Williams, D.B. (1984) *Practical Analytical Electron Microscopy in Materials Science*, Philips Electronic Instruments Inc., Mahwah, New Jersey, 153 pp.

REFERENCES

Williams, D.B. (1986) Standardised definitions of X-ray analysis performance criteria in the AEM, in *Microbeam Analysis – 1986* (eds A.D. Romig and W.F. Chambers), pp. 443–8.

Wood, J.E., Williams, D.B. and Goldstein, J.I. (1984) Experimental and theoretical determination of k_{AFe} factors for quantitative X-ray microanalysis in the analytical electron microscope. *J. Microsc.*, **133**, 255–74.

Zaluzec, N.J. (1992) Current performance limits for XEDS in the AEM. *Proc. 50th Ann. Meeting EMSA*, pp. 1466–7.

Zemyan, S.M. and Williams, D.B. (1991) X-ray analysis of heavy elements by use of L and K series lines, in *Microbeam Analysis – 1991* (ed. D.G. Howitt), pp. 134–6.

Zemyan, S.M. and Williams, D.B. (1992) Peak-to-background measurements on a 300 kV TEM/STEM. *Proc. 50th Ann. Meeting EMSA*, pp. 1236–7.

Ziebold, T.O. (1967) Minimum detectability limits in electron probe microanalysis. *Anal. Chem.*, **39**, 858–63.

CHAPTER FOUR
The nuclear microprobe – PIXE, PIGE, RBS, NRA and ERDA

Donald G. Fraser

4.1 Introduction

Like the electron microprobe (Chapter 2), the scanning nuclear microprobe uses a focused beam of charged particles to excite samples. The focused beam usually consists of protons, but can include heavier particles such as alpha particles, oxygen ions or sulphur ions. In modern instruments, the beam can be scanned and rastered to produce line scans and two-dimensional elemental maps in addition to performing spot analyses. This instrumentation is often referred to as the scanning proton microprobe (or SPM). The incident proton beam is most commonly used to generate characteristic secondary X-rays from a sample in an analogous way to the electrons in the electron microprobe. In the nuclear microprobe literature this phenomenon is known as particle-induced X-ray emission or PIXE.

One of the great attractions of the nuclear microprobe is that the same beam technology can be used to generate other analytical signals based on quite different physical principles. The techniques involved include methods based on the energies of particle recoil, such as Rutherford backscattering (RBS) and elastic recoil detection analysis (ERDA), together with a variety of methods based on nuclear reactions induced in the sample. These are grouped together as nuclear reaction analysis (NRA).

Because of the high mass of the proton in comparison with electrons, incident protons lose a smaller fraction of their energy during each interaction with electrons in the atoms of the sample. This property gives the proton microprobe two major analytical advantages when compared with the electron microprobe: (i) higher spatial resolution because the beam is scattered less within the sample and (ii) a lower X-ray background (bremsstrahlung) caused by deceleration of the incident beam. The low X-ray background makes it possible to use the proton microprobe to measure trace element concentrations down to levels of about 1 ppm on a 1 μm beam spot, depending on the excitation characteristics of the element being analysed and of other elements present in the sample matrix. These advantages are partly offset by the greater technical difficulties and expense inherent in producing and maintaining a well-focused proton beam and by the greater depth of penetration of protons as compared with electrons in a sample. For a 3.5 MeV proton beam in a typical geological specimen, penetration (and hence generation of X-rays) occurs to a depth of about 100 μm. Thus, not only must care be taken to identify material lying below the surface of a specimen, but sophisticated software is required to quantify the X-ray generating process at different depths within a sample as the primary proton beam decelerates and loses its energy.

So far, the commonest use of the scanning proton microprobe is for PIXE analyses and most of this chapter will be concerned with this technique.

The PIXE technique was developed after the use of the electron microprobe had already become routine. Following the demonstration that much lower bremsstrahlung is obtained by using a proton beam as compared with electrons (Sterk, 1965; Khan, Potter and Worley, 1966), it was shown that quantitative analyses could be carried out by using an energy-dispersive Si(Li) spectrometer to detect the X-ray spectrum (Johansson, Akselsson and Johansson, 1970). In these early studies, an unfocused proton beam was used to excite the sample. Before true proton microprobes could be developed, further technical advances had to be made in the design and construction of suitable electromagnetic lenses. Since the field strength required for focusing varies according to the product (particle mass × particle energy), a 3 MeV proton microprobe lens requires a field strength of some 400 times that used in a 30 keV electron microprobe. Cookson and Pilling (1970) used four standard quadrupole accelerator beam lenses to achieve focusing to a spot of about 3.6 μm in horizontal and vertical dimensions. This instrument was successfully used to measure elemental maps using PIXE and RBS (Cookson, Ferguson and Pilling, 1972) and profiles of light elements were determined by NRA (Pierce, 1972). Further improvements in quadrupole-lens design led to the first 1 μm proton microprobe in 1981 (Watt *et al.*, 1982) and sub-micrometre beam spots are now attainable by the best instruments. An overview of applications to geological materials is given by Fraser (1990), and detailed reviews of PIXE are given by Johansson and Campbell (1988) and of PIXE and other nuclear microbeam analytical methods by authors in Watt and Grime (1987) and Petit, Dran and Della Mea (1990). The reader is referred especially to the excellent reviews of these developments and of the instrument itself by Cookson (1987(a), 1987(b)).

4.2 The proton microprobe

A typical scanning proton microprobe requires an ion source, a focusing arrangement and various detectors; a typical instrument is shown in Figure 4.1. Negatively charged hydrogen ions generated in a gas discharge are first accelerated to a potential of +1.5–2.0 MeV in the first stage of a tandem

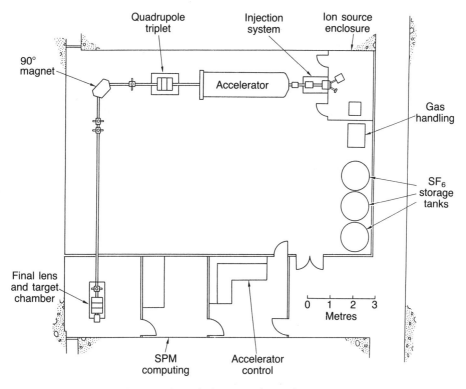

Figure 4.1 Schematic diagram of a typical proton microprobe.

van de Graaff accelerator. They are then stripped of their electrons as they pass through a stripper gas and the resulting H$^+$ ions are accelerated away by the same +1.5–2.0 MeV potential in the second stage to achieve their final energy of 3.0–4.0 MeV. This high-energy proton beam is collimated, can be bent through 90° to save laboratory space and is then focused using a set of quadrupole magnets to achieve a final beam spot on the target, which can be 1 μm or less in diameter.

This highly focused proton beam can be used to perform point analyses or may be scanned over the target to build up a line scan or a two-dimensional image. In the case of geological samples, the target is usually a 100 μm thick electron-microprobe section or a polished and mounted mineral grain and can be viewed using an optical microscope. The position of the impinging proton beam can often be seen as a faint optical anodoluminescence. This anodoluminescence could undoubtedly be used for analysis in a similar way to cathodoluminescence, but little work has so far been carried out on this possibility. In addition to producing optical luminescence, the incident protons generate (i) secondary X-rays in the sample, (ii) scattered protons with energies which vary with the mass of the scattering nucleus (Rutherford backscattering) and (iii) at high energies, nuclear reactions which can be used for nuclear reaction analysis. On the Earth Sciences beamline of the Oxford scanning proton microprobe, the sample chamber is designed to allow the sample to be moved by programmable stepper motors in three dimensions. X-rays are collected and analysed using a Si(Li) detector and the energies of backscattered protons are simultaneously detected using a silicon surface barrier detector.

4.3 The PIXE technique

The high-resolution scanning proton microprobe (SPM) in PIXE mode functions like an electron microprobe but, instead of an electron beam, it uses a highly focused beam of protons to excite the sample. As in the case of electrons, the incident proton beam rapidly dissipates much of its energy on entering a sample by means of inelastic collisions with the electrons of the sample atoms. Since a proton is 1836 times more massive than individual electrons in the target atoms, the incident protons lose only a small fraction of their energy per collision with an electron. This phenomenon gives the proton microprobe two major advantages when compared with the electron microprobe: (i) higher spatial resolution because the beam is scattered less within the sample and (ii) a lower X-ray background (bremsstrahlung) caused by deceleration of the incident beam. The low X-ray background makes it possible to use the proton microprobe to measure trace element concentrations down to levels of around 1 ppm on a 1 μm beam spot, depending on the particular element and matrix (section 3.3.7). This detection limit is far below the best levels attainable using the electron microprobe and approaches the best limits currently available using the ion probe and X-ray microprobes, but at higher spatial resolution.

4.3.1 X-ray background

When a charged particle decelerates, it radiates its excess energy as braking radiation (bremsstrahlung). In the electron microprobe, the primary electron beam decelerates very rapidly when it meets the electrons of the target atoms and the intensity of primary electron bremsstrahlung is high, thus limiting the lower detection limit of characteristic X-rays from elements in the sample. The X-ray and proton microprobes minimize this problem in different ways by using beams of photons and protons respectively. The X-ray microprobe (Chapter 5) uses a synchrotron as the source of an intense X-ray beam which is then collimated and focused using modern X-ray mirrors (e.g. Frantz et al., 1988). The reason for the success of the X-ray microprobe in achieving low detection limits is that X-ray scatter, which contributes to the detected background, is low during the X-ray fluorescence process. This technique currently has a best spatial resolution of about $5\,\mu m$ and requires access to an expensive synchrotron source. However, there will undoubtedly be major advances in this technology in the near future.

The scanning proton microprobe minimizes the primary beam bremsstrahlung by using a heavier particle, accelerated using a relatively simple van de Graaff accelerator. The intensity of bremsstrahlung (I) emitted by a decelerating particle can be approximated by:

$$I \propto a^2 \propto (f/m)^2 \qquad (4.1)$$

where a is the acceleration, f is the force exerted on the decelerating particle and m is its mass. Since the mass of the proton is 1836 times greater than that of an electron, as noted above, and it experiences similar coulombic repulsive forces (f), the primary proton bremsstrahlung should be a factor of $1/(1836)^2$, or approximately 3×10^6, lower in intensity than that in the electron microprobe. In practice, this factor is lowered because the proton undergoes more collisions than the electron during its passage through the target. A plot of proton bremsstrahlung as a function of X-ray energy is shown in Figure 4.2.

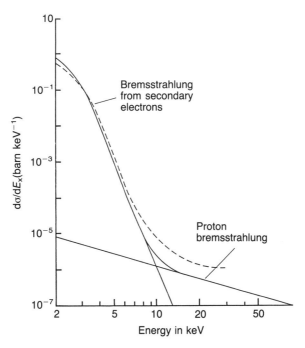

Figure 4.2 Comparison of proton and secondary electron bremsstrahlung at 90° to the probe for a thin carbon film at 2 MeV. The full curves are calculated and the dashed curve is measured. (Folkmann et al., 1974).

However, the detection limits of the proton microprobe are, unfortunately, not 10^6 times lower than those of the electron microprobe. To generate characteristic X-rays, an inner orbital electron must be ejected from a sample atom by interaction with the incident beam. The resulting secondary electrons, and other electrons emitted from the sample during this excitation process, also slow down as they interact with other atoms in the sample thus generating a secondary electron bremsstrahlung contribution to the background X-ray spectrum. A plot of secondary electron bremsstrahlung as a function of X-ray energy is also shown in Figure 4.2, and it can be clearly seen that for energies up to about 11 keV (corresponding to about Se K_α), the background is dominated by the secondary electron bremsstrahlung. A comparison of the X-ray spectra of a dolomite sample excited both by the electron microprobe and by the proton microprobe is given in Figure 4.3, which indicates the improvement in peak-to-background ratio which can be achieved in practice by using the proton microprobe.

4.3.2 Sample thickness

The efficiency of generation of X-rays from a given element decreases with decreasing energy of the exciting particle. Thus, since the protons in the proton beam progressively lose energy with depth in a sample, PIXE analysis is simplest in either of two extreme cases. In the first, the sample is kept very thin so that the energy of the proton beam does not decrease significantly during its passage through the specimen. The 'stopping power', $S(E)$, of a sample for particles of energy E, varies with the sample density and can be expressed as:

$$S(E) = 1/\rho \cdot \frac{dE}{dx} \qquad (4.2)$$

where ρ is the sample density and dE/dx the rate of energy loss in the sample.

For a 2.5 MeV beam, the stopping powers of Si and Fe, for example, are 98.2 and 74.9 keV mg^{-1} cm^{-2} respectively (Anderson and Ziegler, 1977; Ziegler et al., 1985). The total depth of penetration, D, may thus be calculated by integrating:

$$D = \int_0^D dx = \int_{E_0}^0 dE/(dE/dx) \qquad (4.3)$$

This expression gives penetration depths for samples made of Si and Fe respectively of 68 and 27 μm for 2.5 MeV protons (Johansson and Campbell, 1988). Protons of energy 3.0 and 3.5 MeV will penetrate correspondingly deeper. Using equation 4.2 and the above data, the thickness required to

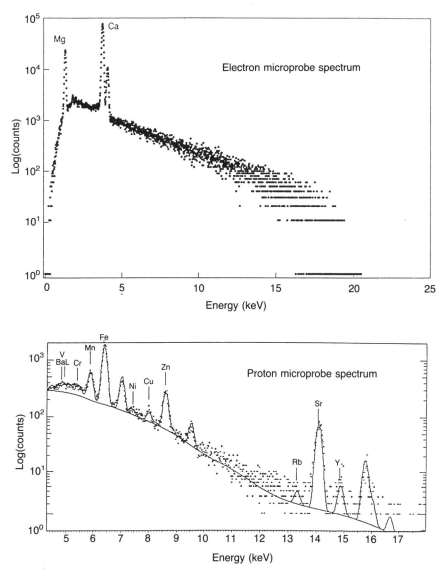

Figure 4.3 Comparison of the X-ray spectra observed from a dolomite sample in both the electron microprobe and the proton microprobe. The curve for the proton microprobe was severely attenuated below about 4.5 keV using a perspex filter to reduce the X-ray counts from the major elements present.

reduce the energy of such a beam by a factor of, say, 2% is thus 2.2 μm for Si and 0.85 μm for Fe. Using such thin films, detection limits of 0.1–1 ppm can be achieved for some elements supported on a low atomic number membrane, e.g. a carbon film.

Although such thin-film criteria can be achieved for biological specimens and in some other special cases, for most geological specimens it is impractical to attempt to work with such extremely thin materials. The usual approach is to use the other end-member case, in which the beam is completely stopped within the target. For the analysis of geological samples using a 3.5 MeV beam, this thickness is less than about 100 μm, depending on the composition of the sample.

4.3.3 Analysis of 'thick' geological specimens

Most geological samples have mean atomic numbers in the range 14–26 (Si to Fe). A polished thin section 100 μm thick will thus be thick enough to completely arrest protons with incident energies of up to 3.5 MeV. This technique, in which the proton is completely stopped by the target is sometimes referred to as thick target PIXE, or TTPIXE. The calibration of TTPIXE analyses is usually undertaken either by comparison with standards of similar composition which have been analysed by other techniques (e.g. Fraser *et al.*, 1984), or by means of sophisticated peak-stripping and modelling software which may allow standardless analysis (e.g. Ryan *et al.*, 1988). This standardless approach may also make use of knowledge of the major element composition of the sample obtained by prior analysis using the electron microprobe, for example. This approach minimizes any systematic errors which may occur, such as errors in the measurement of total beam charge deposited in the sample. Using this method, determinations of trace element concentrations down to 5 ppm in rock samples have recently been reported (Griffin *et al.*, 1988, 1989) and their results for reference materials agree well with the published values. A full description of this technique is given in Ryan *et al.* (1988).

4.3.4 Detectors and filters

Although wavelength-dispersive (crystal) spectrometers have been widely used on the electron microprobe, most PIXE work has been carried out using energy dispersive Si(Li) detectors. A good crystal spectrometer can give energy resolution of <10 eV, depending on the X-ray energy, which is far better than the 130–140 eV typical of a good Si(Li) detector. However, this resolution is achieved at the cost of a greatly reduced solid angle of detection and detector efficiency. In addition, only one element per spectrometer can be analysed simultaneously. Since the main advantage of the proton microprobe lies in the analysis of trace-element concentrations, the rate of data acquisition for a given beam current and counting time are critical, and so a large Si(Li) detector, which is capable of accepting a larger solid angle of detection from the sample, is usually preferred. A further problem encountered in the determination of trace element concentrations in geological samples lies in the high X-ray yields from major elements in the sample. The resultant high X-ray intensities lead to increased dead time and spectral distortions caused by pulse pile-up in the detector counting system which increase the time and difficulty in analysing for trace elements. For these reasons, X-ray filters are usually interposed between the sample and the detector to reduce the intensity of low-energy X-rays. A wide variety of filters has been used, including plastics of different thicknesses, metal foils, compound materials, films containing a critical absorber and 'magic' or 'funny' filters containing one or more holes, which may be used to manicure the X-ray background. The absorption characteristics of all of these filters must be calibrated experimentally, but their use is almost essential for the analysis of trace elements in geological materials. Further details of the proton microprobe technique are given in recent reviews elsewhere (Johansson and Campbell, 1988; Watt and Grime, 1987).

4.3.5 Beam calibration

Accurate measurement of the incident beam current is essential if high-quality quantitative analyses are to be performed. Because the proton source is a gas discharge and not a relatively stable filament (as in the electron microprobe), the proton current is unstable and usually fluctuates in an unpredictable way during an analysis. Thus, the proton current cannot be determined at the beginning and end of a run, for example, but must be monitored continuously. This can be done by measuring the charge deposited on the sample. However, care must be taken to allow for the reduction in the apparent measured charge caused by secondary electrons emitted by the sample. As an alternative, a rotating beam chopper or vane coated with an element whose characteristic X-rays can be monitored (e.g. Volkov *et al.*, 1983) can be used to monitor charge. This vane intercepts the beam and reduces its intensity by the fraction of time for which the chopper is in the beam path. However, a disadvantage is that beam fluctuations may not be adequately averaged out. For these reasons, it can be helpful to use an internal standard procedure by which the instrument is calibrated with reference to the concentration of major elements analysed at marked spots on the specimen. In the case of a line scan, for example, this can often be achieved by marking the specimen with the beam itself at the end of an analysis, by traversing the beam briefly in a direction normal to the measured line scan. In this way, a faint mark or cross is usually observable on the sample which can be used as a marker for analysis by both proton and electron microprobes. Similar procedures can be applied for point analyses and two-dimensional maps. The X-ray count data obtained for the rest of the line scan or two-dimensional map can then be conveniently converted into concentration if the major element used for calibration is also measured along the scan and used as an internal standard. This method, as applied to some of our recent work, is described in detail by Wogelius *et al.* (1992). An alternative to the above procedure is to combine PIXE with RBS measurements to determine the beam current from major elements in the sample; this procedure is discussed below.

4.3.6 Detection limits

Using a knowledge of X-ray production cross-sections and of sources of X-ray background, the signal-to-background ratio, and hence limits of detection of different elements in different matrices, can be estimated. Thus, detection limits for the particular element in question vary with the energy of the proton beam and with the nature and composition of the major element matrix. A plot of detection limit *versus* atomic number for trace elements in thin targets is shown in Figure 4.4. Organic specimens are composed principally of elements of low atomic number and so bremsstrahlung and interference effects from major element X-rays are low, thus enabling sub-part per-million detection limits to be achieved.

Detection limits for trace elements in geological materials are significantly higher. The presence of major elements of similar atomic number to the trace elements of interest causes high characteristic X-ray backgrounds which cannot always be effectively filtered out; the X-ray background arising from bremsstrahlung is also much higher. A comparison has been made of PIXE using 2.5 MeV protons and bulk XRF analysis and the results are reproduced in Figure 4.5. For most elements heavier than S, PIXE detection limits lie in the range 2–40 ppm.

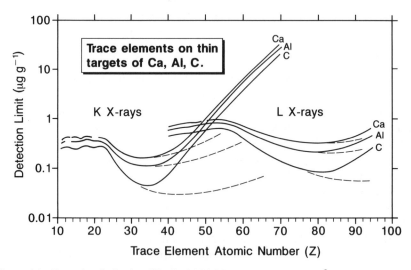

Figure 4.4 Detection limits for $100\,\mu C$ of 3 MeV protons on $1\,mg\,cm^{-2}$ targets of C, Al and Ca. Compare with the proton microprobe spectrum in Figure 4.3. (After Folkmann, 1976).

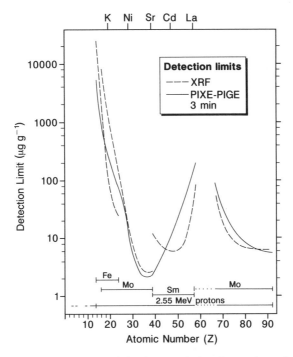

Figure 4.5 Comparison of detection limits for a typical rock sample analyzed by 2.5 MeV protons (PIXE) and by XRF. (After Carlsson and Akselsson, 1981).

4.4 Geological applications of PIXE

The chief advantage of micro-PIXE is the ability to carry out trace element analyses with high spatial resolution. In addition, if a beam scanning device is used, line scan elemental profiles and two-dimensional trace element maps can be obtained. However, the production of two-dimensional trace-elemental maps is time consuming since between four and eight hours of beam time are usually required to build up a trace element map with reasonable counting statistics. The capabilities of the technique can be illustrated by reference to some recent examples.

4.4.1 Trace element zoning in secondary dolomite

Although there have been numerous analyses of geological materials using the PIXE technique, very few have used the high resolution which is now available using focused beams. A good example of the resolution now possible is given in a recent paper in which a dolomite crystal, which showed well-developed cathodoluminescent zoning, was shown to have zoning of

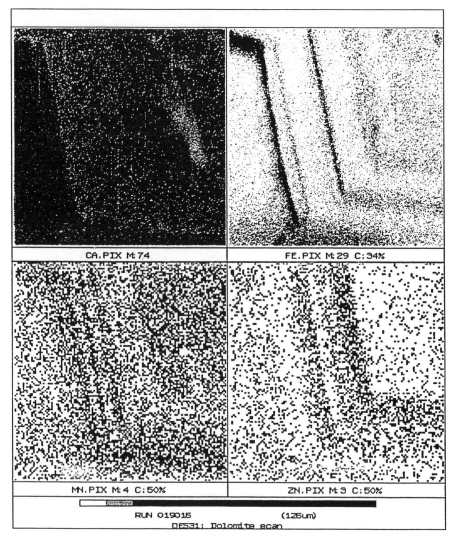

Figure 4.6 Elemental zoning of Ca, Fe, Mn and Sr in a 125 × 125 µm area of a dolomite crystal. The original image is colour coded to distinguish concentration differences as shown by the scale bar at the lower edge.

Fe, Mn and Zn on the micrometre scale (Fraser, Feltham and Whiteman, 1989). The Mn zonation coincided with the cathodoluminescence. A two-dimensional map of the concentrations of these elements is shown in Figure 4.6 and line profiles obtained by scanning across the zoning pattern are shown in Figure 4.7.

The complex pattern of zoning shown in Figure 4.6 indicates that the

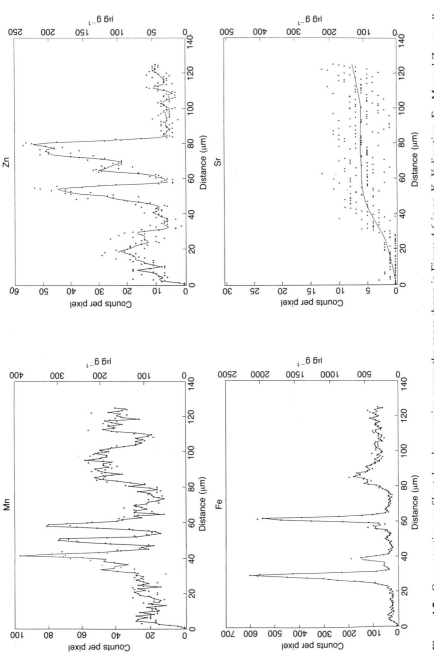

Figure 4.7 Concentration profiles taken by scanning across the zones shown in Figure 4.6 in an X–Y direction. Fe, Mn and Zn are all strongly zoned and are uncorrelated. In contrast, the distribution of Sr seems more homogeneous, although the statistical errors on the count rates are high.

chemical potentials of the Fe, Mn and Zn components varied abruptly and independently in the pore water solution. Comparison with the concentration profile of Sr suggests that these abrupt variations were probably caused by changing redox conditions. Further quantitative analysis of this sample was carried out by calibrating spot analyses by PIXE against corresponding electron microprobe analyses (Wogelius *et al.*, 1992).

4.4.2 Studies of mantle materials

Proton microprobe studies of elemental distribution in mantle xenoliths have been made by Fraser *et al.* (1984) and by Griffin *et al.* (1988, 1989). In the first study, Fraser *et al.* used the high resolution and scanning capabilities of the Oxford proton microprobe to study the distribution of Sr in garnet lherzolite xenoliths. No software was available at that time for standardless analysis and so the Sr X-ray yields were calibrated with reference to clinopyroxene samples in which Sr had been previously determined by isotope-dilution mass spectrometry. The results showed clearly that in a garnet lherzolite xenolith from which radiogenic ^{87}Sr could be readily acid leached, the Sr was located predominantly along grain boundaries and cracks in the specimen. These data demonstrate clearly that in the specimen studied, Sr had been introduced at a late stage, after the primary texture of the rock had formed (Figure 4.8).

More recently, Griffin *et al.* have followed up this work by using the proton microprobe to examine trace element distributions in garnets from sheared mantle xenoliths (Griffin *et al.*, 1989) and in silicate inclusions in diamonds (Griffin *et al.*, 1988). The trace element data presented in these studies showed marked trace element zoning in garnets from high-temperature garnet lherzolite xenoliths and that, as in the case of Sr noted above, the concentrations of several elements including Ti, Zr, Y and Cr were increased by late-stage processes. These data indicate that considerable care should be taken in using the chemical compositions of garnets from mantle xenoliths to infer compositions of primary mantle material.

4.4.3 PIXE analysis of fluid inclusions

Geochemical studies of fluid inclusions have been handicapped by the difficulty of analysing their inorganic constituents *in situ*. The laser Raman microprobe is non-destructive and allows the contents of most polyatomic molecular species to be determined (Chapter 10). However, analysis of the important inorganic solutes is difficult. Crush-leach analyses suffer from an inability to allow detailed investigations of individual inclusions or changes in fluid composition within a population. Moreover, there are the added problems of the destructive nature of the technique, and difficulties related

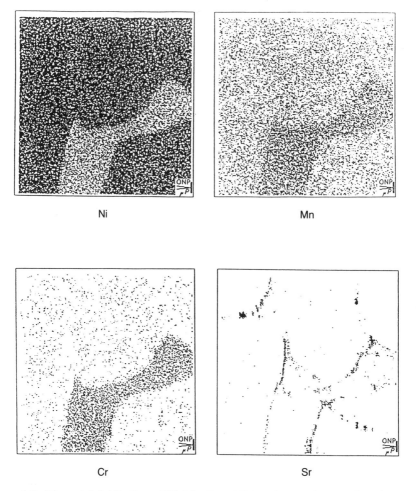

Figure 4.8 Elemental distributions of Ni, Mn, Cr and Sr in a 4 mm × 4 mm section of a garnet lherzolite xenolith. Ni, Cr and Mn are distributed uniformly into either the olivine (large area) or angled orthopyroxene grains. Dark colours represent high concentration in this black and white image. In contrast, Sr is located along cracks and grain boundaries and was clearly introduced at a late stage, after the texture of the rock had formed.

to physical adsorption and differential leaching of daughter minerals. Ion probe erosion and laser ablation have also been used to perform semiquantitative analyses (e.g. Shepherd, Rankin and Alderton, 1985).

The relatively deep penetration of the proton beam in the scanning proton microprobe makes it very well suited to the *in situ* analysis of fluid inclusions and several successful studies have been reported (Horn and Traxel, 1987; Ryan *et al.*, 1991). In the most recent study (Ryan *et al.*, 1991), the mineral studied was considered as a layered structure of which the fluid inclusion was the only layer of unknown composition. Careful use of software, which was adopted to consider X-ray production and emission from layered targets, allowed successful standardless analyses of a fluid inclusion in a quartz-topaz-tourmaline rock from the St Austell granite. Results for most elements were in good agreement with those obtained by crush-leach analyses of parts of the same specimen.

The analysis of layered specimens by proton microprobe in this way is not easy and requires careful application of quite sophisticated software. However, it does allow successful and non-destructive analyses to be obtained from minerals which are not themselves chemically zoned.

4.5 Nuclear reaction analysis (NRA) and PIGE

Although most ions in a nuclear microprobe operated in the MeV-range interact coulombically with atoms in a target to produce X-ray emissions, a small proportion causes changes in the nuclear structure of sample atoms. The products of such changes can be detected and used to analyse for the presence of particular elements or isotopes. The group of techniques which make use of beam-induced nuclear reactions is referred to as nuclear reaction analysis (NRA). When such nuclear reactions cause the emission of gamma rays, these can be analysed using suitable detectors, e.g. Ge(Li), in a way analogous to the analysis of X-rays in PIXE and the technique is referred to as particle induced gamma emission (PIGE).

For particles to cause changes in nuclear structure, they must overcome the 'Coulomb barrier' of the nucleus. This barrier increases with increasing atomic number of the target nucleus and so for yields to be useful, the technique is restricted to light elements. However, since the elements which can be detected in this way cannot be readily analysed by PIXE, NRA and PIGE are useful complementary techniques to PIXE if light element analyses are important. A large number of NRA investigations have been made and many are referred to in Cookson (1987(c)) and Petit, Dran and Della Mea (1990). Elements analysed by NRA and PIGE include Li, Be, B, C, N, O, F and Na. Some nuclear reactions have sharp resonances in their cross-sections – i.e. the nuclear reactions have a high yield only at very sharply defined incident beam energies. Since the energy of the incident particle decreases predictably with depth, the yields of such reactions can be used for non-destructive depth profiling (see also RBS below). An example of this technique is the use of the sharp ^{23}Na(p,α)^{20}Ne reaction, which has a resonance at 592 keV with a width of 0.6 keV, to measure the distribution of Na in silicates with a depth resolution of about 10 nm (Della Mea *et al.*, 1983). Similarly, hydrogen profiling can be accomplished both by NRA using the H(^{15}N,γ)^{12}C reaction or by ERDA (see below). Detection of the emitted γ-rays allows depth profiling with a resolution which varies from 3 nm at the surface to 10 nm at a depth of 100 nm.

4.6 Rutherford backscattering (RBS)

4.6.1 The RBS technique

RBS is based on measuring the energies of backscattered particles from the beam which have collided elastically with atoms in the sample. The collision process depends essentially on coulombic repulsions between the nuclei of the incident ions and the nuclei of target atoms and is described by classical mechanics. Detailed descriptions of the RBS technique are available elsewhere (e.g. Chu, Mayer and Nicolet, 1978; Cookson, 1987(b)).

The energies of the small proportion of ions from the beam which are backscattered by this process are measured using a silicon surface-barrier detector. For a thin target containing atoms of different mass, well-resolved peaks are observed, each corresponding to beam particles of different energy which have collided with the various atoms of different mass in the target (Figure 4.9(a)). Scattering from thicker samples of the same target material causes each peak to be broadened on its low-energy side (Figure 4.9(b)), because beam particles which penetrate to greater depth in the sample in general lose part of their energy in low-angle collisions with other target

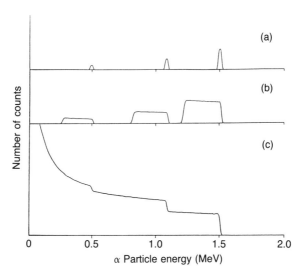

Figure 4.9 Schematic diagram of RBS Spectra for 2 MeV alpha particles backscattered from (a) a thin target, (b) an intermediate thickness target and (c) a thick target containing equal proportions of Co, Al and C.

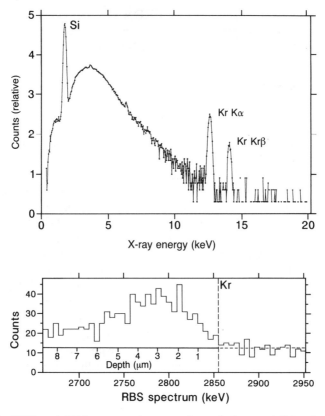

Figure 4.10 PIXE and RBS spectra of an experimental charge of SiO_2 glass previously exposed to Kr at high P, T (sample prepared by Dr M. Carroll). The spectra were obtained for a total beam charge of $0.645\,\mu C$.

atoms before Rutherford backscattering can take place. However, the high energy edge of each peak remains fixed and corresponds to elastic scattering off atoms on the surface. The low-energy tail carries information about the depth distribution of the target atoms (Figure 4.9(b)).

Most geological targets are thicker than the target represented in Figure 4.9(b) and are typically presented as polished sections $30-100\,\mu m$ thick. In these circumstances, scattering takes place with gradual loss in particle energy over most of the depth of the target. Thus, the RBS spectrum retains the high-energy edges corresponding to RBS from atoms of different mass lying on the sample surface, but the peaks which are observed in very thin targets become plateaux stretched to yield a low-energy tail for each mass. The plateau peaks overlap one after the other and their effect is summed by the detector so that the spectrum resembles a series of steps (Figure 4.9(c)).

The information contained in an RBS spectrum can thus be used to obtain

CONCLUSIONS

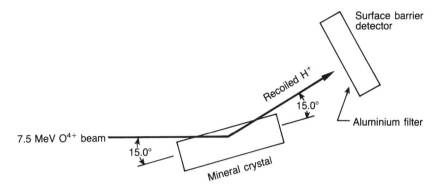

Figure 4.11 Experimental geometry for elastic recoil detection analysis (ERDA).

information about the depths at which different elements occur in a sample. An example of PIXE and RBS applied to an experimental charge is shown in Figure 4.10. This diagram represents a SiO_2 glass specimen exposed to Kr at high P, T to investigate noble gas diffusion in the glass. The PIXE spectrum shows the Kr Kα and Kβ peaks for Kr while the RBS spectrum collected simultaneously shows the distribution of Kr in the top 6 μm of sample. The depth resolution which can be obtained for silicates varies from about 20 nm near the surface to approximately 50 nm at depths of around 1 μm. RBS is likely to have significant future application in the Earth Sciences, especially for studies of diffusion and leaching.

4.7 Elastic recoil detection analysis (ERDA)

In ERDA, the forward elastic recoil of atoms out of the sample is measured. To achieve useful results, the mass of the incident beam particle should exceed that of the target atom. In other respects ERDA is similar in practice to RBS. Because of the kinematics of the collision process, ERDA is most useful for investigating surface concentrations and a typical experimental geometry is shown in Figure 4.11. Although heavy ion beams (and large accelerators) are required to perform ERDA on atoms heavier than H, measurements of H distribution can be carried out using He beams with a small accelerator and are of considerable interest in studies of leaching and low-temperature mineral reaction kinetics.

4.8 Conclusions

The high-resolution nuclear microprobe is now capable of providing a wide variety of techniques which allow trace element analyses and depth profiling at concentration levels down to a few parts per million from specimen areas of 1 μm in diameter or even below, although in most silicates the best detection limits for many elements are about 10 ppm. The difficulties in performing quantitative analyses on thick geological samples now lie primarily in the selection of suitable filters to reduce the X-ray intensities detected from major elements in the sample. In some cases, for example measurements of lanthanide concentrations using their L lines, the extra resolution of a crystal spectrometer could be attractive to allow separation the K lines of major elements present, but this would be at the expense of longer analysis times.

In comparison with the synchrotron X-ray microprobe, the proton or nuclear microprobe has advantages in terms of spatial resolution, cost, variety of analytical methods and ease of beam scanning. However, for those with access to a synchrotron, the X-ray microprobe should offer much shorter counting times.

Acknowledgements

Part of the work reviewed was funded by the NERC, under grant GR3/7342 which the author gratefully acknowledges. It is also a pleasure to acknowledge discussions with Roy Wogelius, David Feltham and Mark Whiteman and the assistance of the Oxford proton microprobe group, whose expertise was essential for the performance of some of the work described here. I also thank Dr N. Charnley for his assistance in obtaining the electron microprobe X-ray spectrum shown in Figure 4.3. The SiO_2 glass represented in Figure 4.10 was prepared by Dr M. Carroll.

References

Anderson, H.H. and Ziegler, J.F. (1977) *The Stopping and Ranges of Ions in Matter*, Vol. 2, Pergamon Press, New York.

Carlsson, L.-E. and Akselsson, R.K. (1981) *Adv. X-Ray Anal.* **24**, 313.

Chu, W.-K., Mayer, J.W. and Nicolet, M.-A. (1978) *Back-scattering Spectrometry*, Academic Press, New York.

Cookson, J.A. (1987a) Historical background, in *Principles and Applications of High-Energy Ion Microbeams* (eds F. Watt and G.W. Grime), Adam Hilger, Bristol, pp. 1–20.

Cookson, J.A. (1987b) Analytical techniques, in *Principles and Applications of High-Energy Ion Microbeams* (eds F. Watt and G.W. Grime), Adam Hilger, Bristol, pp. 21–78.

Cookson, J.A. (1987c) Microbeam applications in metallurgy and industry, in *Principles and Applications of High-Energy Ion Microbeams* (eds F. Watt and G.W. Grime), Adam Hilger, Bristol, pp. 273–98.

Cookson, J.A. and Pilling, F.D. (1973) Use of focused ion beams for analysis *Thin Solid Films*, **19**, 381–5.

Cookson, J.A., Ferguson, A.T.G. and Pilling, F.D. (1972) Proton beams, their production and use. *J. Radioanal. Chem*, **12**, 39–52.

REFERENCES

Della Mea, G., Dran, J.-C., Petit, J.-C. *et al.* (1983) Use of ion beam techniques for studying the leaching properties of lead-implanted silicates. *Nucl. Instr. Meth. Phys. Res.*, **218**, 493–9.

Folkmann, F., Gaarde, C., Huus, T. and Kemp, K. (1974) Proton induced X-ray emission as a tool for trace element analysis. *Nucl. Instr. Methods*, **116**, 487.

Frantz, J.D., Wu, Y., Thompson, A.C. *et al.* (1988) Analysis of fluid inclusions by X-ray fluorescence with synchrotron radiation. *Ann. Rep. Geophys. Lab*, 62–9.

Fraser, D.G. (1990) Applications of the high resolution scanning proton microprobe in the Earth sciences: an overview. *Chem. Geol.*, **83**, 27–37.

Fraser, D.G., Feltham, D. and Whiteman, M. (1989) High resolution scanning proton microprobe studies of micron-scale zoning in a secondary dolomite: implications for studies of redox behaviour in dolomites. *Sediment. Geol.*, **65**, 223–32.

Fraser, D.G., Watt, F., Grime, G.W. and Takacs, J. (1984) Direct determination of strontium enrichment on grain boundaries in a garnet lherzolite xenolith by proton microprobe analysis. *Nature*, **312**, 352–4.

Griffin, W.L., Jaques, A.L., Sie, S.H. *et al.* (1988) Conditions of diamond growth: a proton microprobe study of inclusions in West Australian diamonds. *Contrib. Mineral. Petrol.*, **99**, 143–58.

Griffin, W.L., Smith, D., Boyd, F.R. *et al.* (1989) Trace element zoning in garnets from sheared mantle xenoliths. *Geochim. Cosmochim. Acta*, **53**, 561–7.

Horn, E.E. and Traxel, K. (1987) Investigations of individual fluid inclusions with the Heidelberg proton microprobe. A non-destructive analytical method. *Chem. Geol.*, **61**, 29–35.

Johansson, S.A.E. and Campbell, J.L. (1988) *P.I.X.E.: A Novel Technique for Elemental Analysis*, John Wiley, 347 pp.

Johansson, T.B., Akselsson, K.R. and Johansson, S.A.E. (1970) X-ray analysis: elemental trace analysis at the 10^{-12} g level. *Nucl. Instr. Meth.*, **84**, 141–3.

Khan, J.M., Potter, D.L. and Worley, R.D. (1966) Proposed method for microgram suface-density measurements by observation of proton-produced X-rays. *J. Appl. Phys.*, **37**, 564–7.

Petit, J.-C., Dran, J.-C. and Della Mea, G. (1990) Energetic ion beam analysis in the Earth Sciences. *Nature*, **344**, 621–6.

Pierce, T.B. (1972) Charged particle activation analysis. *J. Radioanal. Chem.*, **12**, 23–38.

Ryan, C.G., Clayton, E.J., Griffin, W.L. *et al.* (1988) SNIP, a statistics-sensistive background treatment for the quantitative analysis of PIXE spectra in geoscience applications. *Nucl. Inst. Meth., Phys. Res.*, **B34**, 396–402.

Ryan, C.G., Cousens, D.R., Heinrich, C.A. *et al.* (1991) Quantitative PIXE microanalysis of fluid inclusions based on a layered yield model. *Nucl. Instr. Meth. Phys. Res.*, **B54**, 292–7.

Shepherd, T.J., Rankin, A.H. and Alderton, D.M.H. (1985) *A Practical Guide to Fluid Inclusion Studies*, Blackie, London.

Sterk, A.A. (1965) X-ray generation by proton bombardment. *Adv. X-ray Anal.*, **8**, 189–97.

Volkov, V.N., Vykhodets, V.B., Golubkov, I.K. *et al.* (1983) Accurate light ion beam monitoring by backscattering. *Nucl. Instr. Methods*, **B205**, 73–7.

Watt, F. and Grime, G.W. (1987) *Principles and Applications of High Energy Microbeams*, Adam Hilger, Bristol, 399 pp.

Watt, F., Grime, G.W., Blower, G.D. *et al.* (1982) The Oxford 1 μm proton microprobe *Nucl. Instr. Meth. Phy. Res.*, **B197**, 65–77.

Wogelius, R.A., Fraser, D.G., and Feltham, D.J. (1992) Trace element zoning in dolomite: proton microprobe data and thermodynamic constraints on fluid compositions. *Geochim. Cosmochim. Acta*, **56**, 319–34.

Ziegler, J.F., Biersack, J.P. and Littmark, U. (1985) *The Stopping and Range of Ions in Solids*, Vol. 1, Pergamon Press, New York.

CHAPTER FIVE
Synchrotron X-ray microanalysis
Joseph V. Smith and Mark L. Rivers

5.1 Introduction

5.1.1 Advantages and limitations of conventional X-ray fluorescence analysis

X-ray fluorescence analysis (XRFA) is a standard technique that is used routinely for the analysis of rocks and minerals. Potts (1987; Chapter 8) covers the theory and practice. When an X-ray photon interacts with an atom, there is a well-defined probability that the photon energy will drive out (i.e., ionize) an electron from a bound state. The excited atom regains equilibrium by emission of either an X-ray photon or an Auger electron, both with characteristic energy determined by the quantum numbers of the initial and final electronic states and the atomic number of the atom. Measurement of the energy of the emitted X-ray photon identifies the type of atom (except for accidental overlap within the experimental resolution). Determination of the flux of this particular type of photon with respect to the total flux from all X-ray photons from all atoms in a sample provides a measure of the atomic concentration in the bulk sample. There are many details involving the relative probabilities of the various processes, the competition with coherent diffraction and other scattering processes, the perturbation of electronic states of an atom from changes in chemical bonding, and the actual experimental details of the X-ray source, the detectors and the algorithms for the mathematical analysis. Nevertheless, the technique is fundamentally straightforward and potentially accurate to the per cent level of the amount present, when suitable standards are available. Classical X-ray fluorescence analysis is technically easiest for those elements that can be determined from K line fluorescent photons with energies in the 1–20 keV range. The spectrum becomes progressively more complex from the K to the L and M line series. Analytical complications from energy shifts related to chemical bonding and electronic state only become significant at lower energy, but these perturbations can be valuable to a structural crystallographer/geochemist.

Whereas the determination of the major elements in bulk specimens is not limited by the number of photons from a conventional laboratory source,

Microprobe Techniques in the Earth Sciences. Edited by P.J. Potts, J.F.W. Bowles, S.J.B. Reed and M.R. Cave. Published in 1995 by Chapman & Hall, London. ISBN 0 412 55100 4

the analysis of trace elements becomes a trade-off between sample size, atomic concentration and time chosen for data collection. The 'white' beam from a conventional hot-cathode X-ray source provides a flux high enough for the analysis of many trace elements of geological significance down to the part-per-million range for samples several millimetres across (the thickness is also important). However, trace elements cannot be measured on a micrometre sample with a conventional hot-cathode source, even with a focused electron beam on a cooled rotating anode. To reach combined partper-million/micrometre sensitivity, the X-ray source must become more brilliant by several factors of ten. Synchrotrons provide X-ray beams with that extra brilliance.

5.1.2 Comparison of the various microanalytical techniques

Synchrotron X-ray microanalysis permits the coupling of fluorescence measurements with diffraction and absorption spectroscopy to obtain not only the concentration of an element, but also its electronic state and topochemistry. The physical laws involving interactions of X-rays with matter are 'well behaved' and interpretation of the observed data is reasonably straightforward. X-rays deposit little energy in a sample during analysis, and many samples can be studied outside the vacuum system. The inherent polarization of a synchrotron beam coupled with tuning through an absorption edge allows better detection limits than for conventional XRF. In general, the higher the atomic number and the more refractory the element, the greater is the advantage of synchrotron-based techniques over competitors. Samples consisting of multi-component particles can be 'flooded' with the X-ray beam to obtain a bulk analysis. Tomography can be used to produce a three-dimensional image. Because little damage is done by a synchrotron beam except in extreme conditions, a synchrotron study can be followed by destructive techniques. Specimen preparation is of critical importance, and doubly polished thin sections with near-parallel surfaces are desirable. It is important to emphasize that a beam of hard X-rays penetrates right through most samples being analysed. Hence, the depth resolution is often governed by the sample itself, even when the incident beam is very narrow.

Electron probe microanalysis will remain the workhorse for the analysis of major and minor elements in thin sections of stable minerals to the ~2–10 μm level, but the bremsstrahlung radiation associated with the excitation process restricts the detection limit of trace elements. Moreover, the sensitivity of electron probe microanalysis (EPMA) falls off seriously for $Z > 30$ (Zn), while synchrotron X-ray fluorescence analysis (SXRFA) is very easy and highly effective for the K spectra of Z up to ~60 (Nd) and can be extended up to U. Glasses, liquids and minerals containing volatile elements and molecules are difficult or impossible to analyze by EPMA. Whenever possible, a routine electron probe microanalysis should be made before SXRF analysis to obtain accurate values for major and minor elements that can then serve as internal standards for SXRF of trace elements.

Secondary ion mass spectroscopy (SIMS), also known as the ion probe, has uniquely valuable analytical features including a capability for determining isotopic ratios for certain elements, depth profiling with nanometer resolution downwards and $\sim 10\,\mu$m sideways, and very high sensitivity for easily ionized elements. Disadvantages are the lack of a simple algorithm for modelling the sputtering process, high sensitivity to variation of electrical conductivity across a multi-grain sample, and the migration of impurities – especially ones containing H – into the sputtered area. For certain light elements, including Li and Na, the sensitivity is very high.

Particle-based techniques, commonly using proton or deuteron beam excitation, also have unique analytical features. Nuclear reaction analysis (NRA) is very effective for the determination of certain light elements. Gamma-ray emission (PIGE) is very effective for F above the $\sim 10\,\mu$g g^{-1} level for samples thick enough to stop a megavolt proton beam: SIMS is sensitive to lower levels, but more difficult to quantify. Particle-excited X-ray emission (PIXE) is easy to quantify in refractory samples thick enough to stop the beam ($\sim >20\,\mu$m), but the sensitivity drops off severely for $Z >$ 30 (Zn). The heating from the heavy particles severely limits the combination of small spot and trace concentration. The billiard-ball technique of Rutherford backscattering (RBA) is useful for depth profiling at the sub-micrometre level, but has inferior detection limits. Synchrotron techniques have a substantial advantage for unstable samples, and this advantage increases for the heavier elements.

Electron microscopy has special features, including diffraction and imaging approaching the atomic level. Analytical electron microscopy has excellent spatial resolution, but is limited to the determination of the major elements in most samples; however, progress is being made towards the identification of small atomic clusters. Scanning electron microscopy is a workhorse for the determination of topographic features, but quantification of the X-ray signal from a rough surface is difficult. Heating and electrical damage represent a limitation in the SEM analysis of some samples, and synchrotron soft X-ray microscopy should carve a niche here to the sub-micrometre level.

To conclude, the tendency to attempt everything on one instrument is doomed to failure. As high-speed data transmission becomes routine, and off-site analysis becomes psychologically accepted, we can expect that a geoscientist will use a combination of analytical techniques at several sites to solve problems of increasing complexity. Let us turn now from these general matters to the details of synchrotron microanalysis.

5.1.3 Advantages/disadvantages of synchrotron sources: general aspects

A synchrotron storage ring consists of a near-circular vacuum tube in which 'bunches' of high-energy electrons or positrons are stored as they circulate near the velocity of light under the control of synchronized electromagnetic fields. The electrons are guided through the ring by steering, focusing and bending magnets, and the energy loss is compensated by input from radio-frequency cavities. Each bending magnet perturbs the electron path from a straight line into a circular arc by an inward centripetal force. As each bunch of charged particles is accelerated inward, an instantaneous jet of 'white' X-rays (Figure 5.1) is emitted along the forward tangent in the form of a fan emanating from each bending magnet.

Each 'flash' of X-rays lasts a fraction of a nanosecond. The circumference of a high-energy ring is a fraction of a kilometre. The electrons, which travel at almost the velocity of light, take several microseconds for each circuit. Usually, there are several dozen bunches separated by tens to hundreds of nanoseconds. The time structure is not used for XRFA and the most obvious advantage of a synchrotron storage ring is simply brute force: the high brightness and brilliance allow the measurement of trace elements on micrometre samples. In section 5.2.2, three figures of merit for the X-ray beams from a synchrotron source are defined and compared: flux, brightness and brilliance. Brightness measures the number of X-ray photons passing through a simple collimator. Brilliance is relevant when the entire X-ray beam is focused onto a small sample. The total flux of X-ray photons is appropriate for a large sample. For microanalysis, brightness and especially

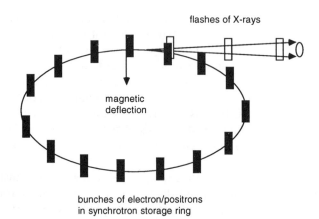

Figure 5.1 Concept of a synchrotron storage ring. Bunches of electrons or positrons circulate just under the velocity of light in a near-circular vacuum tube. The centrifugal force is compensated by magnetic deflection. At any instant, the magnetic deflection generates a narrow flash of X-rays along the forward tangent.

brilliance are the key figures of merit. These simple statements are sufficient here, but must be carefully qualified in the detailed technical description.

A second advantage is that the X-ray beam is polarized in the plane of the storage ring and collection of fluorescence photons parallel to the acceleration vector (i.e. radius at the bending magnet) almost eliminates the background from Rayleigh and Compton scattering.

The provision of a multiple array of magnets in a straight section of the storage ring between two bending magnets is a more subtle reason for a synchrotron to be advantageous for microanalysis. Two devices are used. In a 'wiggler' source with N magnets, the intensities from the left and right wiggles add up to give N times the flux obtained from a simple bending magnet. The undulator source is more sophisticated: here, the magnetic displacement is small enough for the X-ray jets from each magnet to undergo phase interference to yield a harmonic spectrum. The energy of each harmonic can be tuned merely by changing the magnetic field, and a sweep can be made across the absorption edge of a chosen element. Furthermore, the spatial squeezing of the X-ray jet from an undulator harmonic is highly advantageous for focusing the beam onto a tiny spot.

Figure 5.2 shows schematically the history of the increase of brilliance from the conventional hot-cathode tube in the 1940s, via the early bending-magnet sources (1970s), to an undulator in a third-generation synchrotron source (1990s).

This substantial increase in the brilliance of X-ray sources has had the important effect of promoting the development of a number of emerging techniques which permit analysis of surface layers, the characterization of the chemical bonding of the element yielding the fluorescent signal, three-dimensional tomography and so on. Absorption spectroscopy and diffraction are being used to characterize the electronic state and chemical coordination of an element whose concentration is being measured by X-ray fluorescence. This chapter concentrates on synchrotron X-ray fluorescence microanalysis because of the relationship to the analytical techniques considered in other chapters. However, the essentials of some other synchrotron-based techiques are also covered.

Synchrotron-related science is just emerging from the realm of the technical wizards to become part of the standard scientific disciplines. The basic principles and practice have been established, but several decades of technical developments of even greater wizardry are still to come. Most current applications are carefully selected to trigger off new areas of research using samples that are just technically accessible on the prototype instruments. Many desirable experiments are still beyond reach. Systematic exploitation of research areas is just beginning as the pioneering experiments become standardized and accessible to the entire research community.

The acronym SRIXE was used by Jones (1992) for synchrotron radiation-induced X-ray emission. This term embraces synchrotron X-ray fluorescence

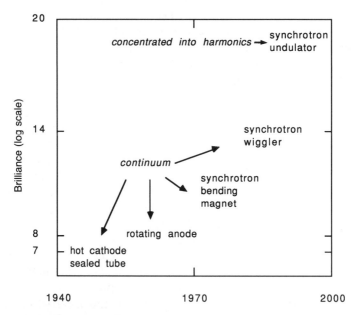

Figure 5.2 Increasing brilliance of X-ray sources. The brilliance is defined in terms of number of photons per second per source area (mm^2) per solid angle (milliradian^2) per energy band width (0.1%). It is pertinent to microanalysis. Note that the four younger sources are distributing their total energy into a continuum ('white light' in synchrotron jargon), whereas the undulators at the third-generation sources will concentrate the photons into narrow harmonic peaks. A plot of total flux would not show such a large increase with time. Caution: details vary tenfold among sources of a particular type – hence the schematic nature of the drawing.

microanalysis (SXFMA), synchrotron X-ray absorption spectroscopy (SXAS) and synchrotron X-ray diffraction (SXRD). As these techniques come into regular (but not routine!) use, they are opening up new applications in mineralogy, petrology and geochemistry. Because they are also used in solid-state physics, structural and physical chemistry and in the biological sciences, synchrotron-related science and engineering is breaking down the artificial barriers between the traditional disciplines.

Each synchrotron storage ring is very expensive, and involves the collaboration of some hundreds of scientists and technicians. Hence, a complex psychology is needed for success – not to mention the physical stamina to work at a remote site at erratic hours. It is much easier to have sole control of an in-house instrument, no matter how limited its analytical range, than to participate in the politicized atmosphere of a large operation. However, by the year 2000, remote operation of instruments should become so routine that 'standard' experiments on a synchrotron beamline will be run by an instructor teaching one or more classes of students at any number of 'wired' locations. An on-site technician would load samples into an instrument on a synchrotron beamline and two-way video and sound would be used for control of the analyses. In a sense, this is a natural development of the current use of electronic mail, and indeed is the way in which large telescopes are operated today. Frontier experiments will continue to challenge the wizards and their discipline-oriented colleagues, and some may never become standard.

It is extremely difficult to become both a technical wizard and a discipline-oriented expert, fully versed in the geological provenance of samples worthy of sophisticated study. Hence, mutually respectful collaboration between scientists and engineers with complementary abilities is the basis of successful SXFMA. The synchrotron-based expert will design a better instrument after ascertaining the analytical needs of the field-based expert. Conversely, the field-based expert will select better problems if the basic parameters of the instrument design are understood and if the experimental protocol is optimized. The physical and psychological pressures on beamline scientists are intense, especially because most beamlines are understaffed. A beamline serving a large community, such as geoscientists performing XRF microanalysis, needs at least three scientist/engineers to maintain and upgrade the beamline, help the 'customers' with standard analyses and pioneer new experiments.

Inadequate funding for all aspects of the materials sciences has hampered the design, construction and operation of synchrotron X-ray sources. Most beamlines do not reach their potential in spite of the extraordinary dedication of the beamline scientists and visitors. In general, as much capital investment is needed for the beamlines as for the storage ring. Perhaps the most efficient and productive beamlines have been those dedicated to a specific mission for a strong user group, with systematic upgrading as the technology became understood.

[JVS asked the editors to print the following personal statement. The future success of scientific discoveries based on instrument development cannot be divorced from social and political matters. Taxpayers, represented by elected politicians, will no longer finance all the new ideas of scientists and engineers. New instruments cost more money and must compete against other instruments. Geoscientists face increasingly difficult problems that require the combined data from several instruments. Hence, an intense competition for research funds has arisen. 'Big science' might be perceived to conflict with 'little science'. A third-generation synchrotron source is definitely big science when considered as a single unit. The total cost of a high-energy ring and the instruments on several dozen beamlines is almost US$1 000 000 000 (~£500 000 000). However, the synchrotron source should serve at least 1000 scientists for at least 30 years. Hence the initial capital cost per user-year is about £10 000, to which must be added the cost of maintenance and upgrades. The initial capital cost and maintenance/upgrading of other advanced analytical instruments in the geosciences are smaller per instrument, but not per user-year. We must, therefore, balance all the instrumental needs in the geosciences to avoid a neurotic competition. Attaining a balance with other scientific areas, particularly high-energy physics, is a severe challenge. I favour strong military force for defence against possible attacks, but just a minor shift of funds from weapons budgets would easily accommodate the cost of advanced analytical instruments. Furthermore, development of new instruments automatically maintains many of the advanced technical skills needed for survival in a competitive world.]

At SRS, Daresbury, UK, most experimental stations are designed and operated by the synchrotron staff with help from outside experts. Access is gained by submitting proposals. At NSLS, Upton, NY, USA, most beamlines were constructed and are operated by participating research teams (PRTs) who were successful in a peer-reviewed competition. The geoscience community has done most of the XRFA at bending-magnet beamline X26A. Test experiments at high energy have been undertaken at CHESS, Cornell, NY and at the NSLS-X17 wiggler line.

5.1.4 Advantages/disadvantages of synchrotron sources: technical aspects

In addition to the inherent high brilliance of a synchrotron X-ray beam, there are other less obvious advantages over a conventional source.

- The jet of X-rays is highly polarized with the electric vector in the plane of the ring. As a consequence, the background spectrum derived from Rayleigh and Compton scattering of the synchrotron beam can be almost eliminated by collecting fluorescent X-rays in the direction of the magnetic deflection (scatter phenomena cannot occur in the plane at right angles to the plane of polarization). Because the electrons roll slightly around the ideal orbit, the polarization is not perfect. In the centre of the jet of X-rays from a bending magnet, the polarization reaches ~99% for synchrotron sources with tight electron bunches, and falls off to 90% at a vertical displacement of about a millimetre at the sample position.
- The intrinsic collimation of the near-parallel beam (having a divergence of a few milliradians) permits more effective focusing and monochromatization than for the highly divergent beam from a conventional source.
- An undulator acts like a diffraction grating to concentrate the X-ray photons into narrow energy bands which can be tuned merely by varying the magnetic field. A particular chemical element can be 'picked off' at a detection limit lower than the part-per-million level by adjusting the energy band to just above its absorption edge.
- Tomography of millimetre-sized specimens at a resolution approaching the micrometre range is feasible using either total absorption or fluorescence signals. The spatial resolution is much better than for conventional sources, and tuning the synchrotron beam above and below an absorption edge allows tomography for a particular element.
- The special features of a synchrotron source permit fluorescence analysis to be combined with absorption spectroscopy and diffraction at the micrometre level. Hence bulk chemical information from the first technique can be 'broken down' to the level of local chemical bonding.

Although a primary X-ray beam generally deposits much lower energy in a sample than an electron, proton or ion beam, synchrotron beams are becoming so intense that sample damage from heating must be considered.

Furthermore, electronic changes may become significant because of radiation damage.

An important factor is that SXRFA permits a sample to be outside the vacuum system for experiments using hard X-rays. This is particularly relevant for hydrated and other 'soft' samples.

5.1.5 Other X-ray techniques: diffraction and absorption spectroscopy

SXRD is not normally considered as a microanalytical technique, although it deserves to be. The high brilliance of synchrotron sources is allowing the determination in 1994 of the electron density distribution of crystals as small as $20\,\mu$m across, and technical advances should reduce the size to below $5\,\mu$m and perhaps $1\,\mu$m within 5 years. The electron density, of course, is not 'labelled' with atomic number information in conventional diffraction. However, tuning the synchrotron X-ray beam through a particular absorption edge permits identification of that part of the electron distribution belonging to the chosen element. Thus, Fe and Mn occupying the same crystallographic site can be separately identified by measuring the anomalous scattering at their particular K-absorption edges.

SXAS allows determination of the electronic state and chemical coordination of a particular element. The simple concept of a sharp absorption edge (Figure 5.3) must be modified to take into account the interference between the wave function of the ejected electron and those of the electrons in the excited parent atom and its neighbours. The electronic state and general chemical coordination influence the near-edge region of the complex absorption 'edge': X-ray absorption near-edge structure (XANES, often pronounced 'zanes'). The outgoing photoelectron just 'creeps' out from the quantum level of the parent excited atom, and its wave interacts with electronic wave functions of the outer bound electrons of the parent atom. These bound electrons are coupled to all the bound electrons of all the parent atoms in the crystalline or amorphous unit. Hence, XANES is sensitive to the electronic state of the parent atom with respect to the general electronic state of the entire crystal structure, and particularly to the stereogeometry of adjacent atoms. In the extended X-ray absorption fine structure (EXAFS) high-energy region of the absorption spectrum, the outgoing photoelectron 'zooms' out and interacts mainly with the combined electronic wave functions of the electrons in the nearer atoms. The 'ripples' in the high-energy tail of the absorption spectrum are the Fourier transform of the radial distribution function of the electrons of the adjacent atoms, giving a measure of both distance and number. This interpretation is qualitatively simple, but quantitatively difficult, because of the complexity of the phase relationships between the electronic wave functions. Hence, a combination of experimentation on model compounds and quantum-mechanical theory is used in the interpretation.

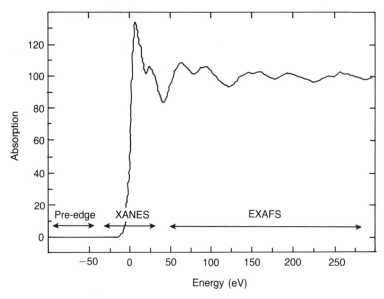

Figure 5.3 Modification of the classical absorption edge by XANES and EXAFS. This example is for the Cu edge of La_2CuO_4. The zero energy is placed at the classical absorption edge ($E = 8979\,eV$). (From Alp, Mini and Ramanathan, 1990).

The sensitivity of SXAS is much lower than that of SXRF, and a larger sample of higher concentration is needed. The increasing brilliance of the present synchrotron sources is allowing the detectable concentration of some chosen elements in XAS to reach the level of 100 ppm for an X-ray beam 0.1 mm across. Several additional factors of ten in the combined sensitivity of concentration and spatial resolution are expected for new synchrotron instruments, and micro-XAS is becoming attainable.

Most XAS is performed on a randomly oriented crystal powder or an amorphous sample. Because the synchrotron X-ray beam is highly polarized, rotation of a single crystal permits three-dimensional evaluation of the electronic state and chemical coordination from XAS. At its simplest, the three-dimensional variation can be just the average of the properties of all crystallographic types of atoms of the chosen element. The coupling of diffraction and absorption spectroscopy in diffraction absorption fine structure (DAFS) is more sophisticated. Collection of each diffraction intensity across the XANES and EXAFS regions of an absorption spectrum can provide three-dimensional information on the electronic state and chemical coordination for each crystallographic site of the chosen element.

A micro-sized sample need not be equidimensional. Synchrotron X-ray sources may be so narrow and near-parallel that analytical data can be obtained from a thin film only a few nanometers across. If the thin film is crystallographically regular in two dimensions, the diffraction pattern will consist of rods in reciprocal space. It can be analysed in a manner similar to that for surface electron diffraction. If the film is chemically distinct from a substrate, the energy of the synchrotron X-ray beam can be tuned to a particular element in the thin film. A near-parallel X-ray beam may be rotated through the angle for total reflection to set up a standing wave in the surface layer. Special variants of XAS and XRD are pertinent to these surface analytical techniques. This chapter will not go into the complex details of these surface techniques. Although the ultimate potential is conceptually known, the scientific interpretation and technological development are still in the domain of the technical wizards.

5.2 Synchrotron storage rings

5.2.1 Theory and practice of X-ray production

(a) Bremsstrahlung

Acceleration of a charged particle generates electromagnetic radiation (bremsstrahlung). Thus, an oscillating electric current in a conductor generates electromagnetic radiation whose directional amplitude, at any instant of time, is determined by a torus-shaped dipole pattern (doughnut without a hole). The instantaneous radiation is 'white', and the continuous spectrum of energy (ε) appears like a whale's back. The *critical energy*, ε_c, expresses the mean energy of the white spectrum of photons. Increasing the acceleration does not alter the general shape but moves the white spectrum to higher energy, with ε_c passing through the infrared, optical, ultraviolet and soft X-ray regions to reach the hard X-ray band needed for XRFA. According to the theory of wave/particle duality, the energy flux (intensity) at a particular point in the white spectrum is represented by a stream of photons, each of energy $\varepsilon = h\nu$, where h is Planck's constant and ν is the frequency.

5.2.2 Ring and bending magnet

(a) Acceleration in synchrotron rings

Radiation was observed half a century ago from betatron and synchrotron accelerators with electrons circulating in closed orbits prior to collision with experimental targets in high-energy physics. To maintain the closed orbit, the electrons were accelerated inward by interaction with a magnetic field which compensates for the centrifugal force (Figure 5.4). At any instant, the inward acceleration generates a torus-shaped dipole pattern in the relativistic reference frame of the electron (i.e. the Cartesian coordinates of the electron

SYNCHROTRON X-RAY MICROANALYSIS

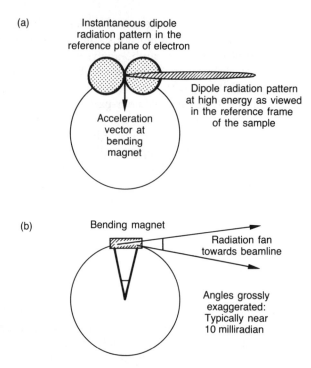

Figure 5.4 Angular controls on the divergence of the X-ray beam from a bending magnet source. (a) Relativistic focusing. The torus-shaped figure-of-revolution about the radial acceleration vector is converted into a narrow jet about the direction of travel (tangent). The relativistic divergences are equal in and perpendicular to the plane of the ring. (b) The horizontal divergence for X-rays from a bending magnet is increased by the subtended angle.

itself, not of the observer in the laboratory). This is exactly the same phenomenon as for an electron being accelerated instantaneously in the wire of a radio antenna. However, the electron in the radio antenna is travelling very slowly, whereas an electron in a high-energy accelerator is travelling at near the velocity of light.

(b) Relativistic transformation

Any photon used in an experiment in the reference frame of the laboratory undergoes a transformation governed by special relativity. The torus is being carried forward by the electron and is transformed into a cigar-shaped jet approximating to an ellipsoid of revolution. Because the Cartesian coordiantes are so similar for both the electron and the observer in a radio experiment, the transformation is trivial. However, the high-speed electron in a synchrotron emits a very narrow jet which becomes even more brilliant with increasing energy. This is the first reason why a synchrotron can give high brilliance. A much higher fraction of X-rays reaches a micro-sample than for a conventional X-ray tube. The normalized energy parameter γ (= kinetic energy of the electron divided by its rest mass, m_0c^2) controls the relativistic transformation. For an electron, $\gamma = 1957E$, for E in GeV. At the critical energy, the opening angle in radians of the relativistically transformed jet is $1/\gamma$. It decreases from 500 μrad at $E = 1$ GeV to 73 μrad at 7 GeV.

(c) Bending-magnet spectra

The instantaneous jet of X-rays sweeps out an arc as the electron bunch passes through a bending magnet (BM). Thus the horizontal divergence is increased by the arc subtended by the magnet at the centre of curvature (Figure 5.4(b)). The critical energy ε_c of bending-magnet radiation from a synchrotron ring is given by

$$\varepsilon_c = \frac{2.218E^3}{\rho} = 0.665BE^2 \qquad (5.1)$$

where ρ is the radius in metres of the electron orbit at the BM and B is the magnetic field in tesla. To reach the ultraviolet/soft X-ray region, the electrons in a circular ring with inexpensive conventional 1 tesla bending magnets need to be accelerated up to about 1 GeV. For hard X-rays (>5 keV), more than 2 GeV is needed. Thus, NSLS with 2.5 GeV and 1.2 T has $\varepsilon_c = 5.0$ keV, and the APS with 7 GeV and 0.6 T has 19.6 keV.

(d) Ring design parameters

Hard X-ray rings must be larger and more expensive than the soft X-ray ones. A synchrotron storage ring is not a simple circle. It consists of a near-circular/polygonal vacuum tube with alternating straight and curved sections. A pair of magnets above and below the vacuum tube, which constitute a bending magnet, forces the electrons to follow a curved horizontal arc. Each arc from a bending magnet is joined by a straight section between adjacent bending magnets. A user sees none of this technology because the ring and magnets are sealed within a radiation shield.

Electrons accelerated in a booster ring are injected in bunches spaced around the vacuum ring so that they can be controlled. The straight sections contain focusing magnets which keep the bunches tight. Skilled machine physicists have devised mathematically beautiful ways of squeezing the bunches as tightly as possible to give a brighter source of lower emittance without loss of stability during the repetitive circuits. Their success in designing new rings and upgrading old rings is the second reason why synchrotron X-ray beams are becoming so brilliant. The electron bunches repeat the circuit about once per microsecond, and each flash of X-rays from a tight bunch of electrons may last only a fraction of a nanosecond. The circuit time is a fraction of a microsecond for 2 GeV rings and 3.7 μs for the 7 GeV APS.

The electrons are slowly lost by collision with atoms and molecules of the residual vacuum and it is customary to dump the beam when about half the electrons are lost. New bunches are then accelerated in the booster ring and injected into the synchrotron storage ring. The electrons in a bunch repel each other and this, together with other factors, limits the electron current attainable in a ring. The lifetime before disposal and re-injection is strongly governed by the quality of the vacuum. Switching from electrons to positrons reduces collisions with positively charged species in the residual vacuum.

The early synchrotron pioneers suffered extreme frustration from the short lifetime (a maximum of few hours) and the spatial instability (up to 1 mm) of the beam. Current rings are delivering beams with a reliability well over 90% for a lifetime (before disposal) close to 1 day, and with lateral and vertical stability, controlled by sensors and fast feedback circuits, approaching the micrometre range. Some days are reserved for special experiments and instrumental upgrades, leaving between 200 and 250 days per year for standard experiments. The positron beam current of the Advanced Photon Source may be replenished continuously. This is extremely important because the optical elements of the beamlines will become hot, and constancy of temperature will reduce the temporal variation caused by thermal expansion and mechanical strain.

(e) First-generation rings

First-generation hard X-ray sources were 'parasitic' on accelerators used for high-energy physics (Jones, 1992: Table 5.1 lists 11 rings), and provision of time for materials scientists was erratic. As an accelerator became obsolete for high-energy physics, more time was released for X-ray research experiments as long as the accelerator was not closed down. Several are still functioning well and expanding operations after important upgrades (e.g. CHESS, the Cornell High Energy Synchrotron Source on the 5.4 GeV CESR ring at Cornell University, New York, and SSRL, the Stanford Synchrotron Research Laboratory on the 3 GeV SPEAR ring at Stanford University, California, now dedicated to X-ray generation). The bending magnets for X-ray production were fitted into parts of an existing ring and the electron bunches tended to be wide and laterally unstable. Before upgrades to tighten the beam, these sources were best suited for experiments on large samples.

(f) Brightness of bending-magnet X-rays

For bending magnets, the key parameter for experiments using a collimator is the brightness (photon s^{-1} mrad^{-2} per 0.1% bandwidth)

$$B_{BM} = \text{flux/vertical angle} = 1.33 \times 10^{13} E^2 I H_2(\varepsilon/\varepsilon_c) \qquad (5.2)$$

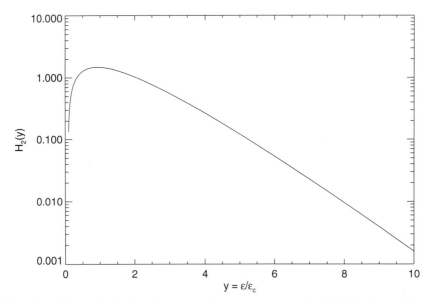

Figure 5.5 Photon distribution from a bending magnet expressed as $H_2(\varepsilon/\varepsilon_c)$. The brightness function, $H_2(\varepsilon/\varepsilon_c)$, initially increases with the energy of the X-ray photon ε and then falls as ε becomes larger than the critical energy ε_c. (Rivers, 1988).

where the storage ring energy E is in GeV, the current I in the ring is in ampere, and $H_2(\varepsilon/\varepsilon_c)$ is a function plotted in Figure 5.5. Note that B_{BM} increases linearly with the ring current and with the square of the ring energy.

(g) Second-generation rings
Second-generation synchrotrons were designed for fully dedicated operation using an array of bending magnets spaced around the entire ring. Typical examples are SRS, the 2 GeV Synchrotron Radiation Source at Daresbury, England; the 0.7 and 2.5 GeV rings at the NSLS and the 2.5 GeV Photon Factory at Tsukuba, Japan. The 2–2.5 GeV energy was chosen because of the efficient generation of 1–25 keV X-rays from the ~1 tesla bending magnets. The NSLS rings were designed to achieve the small beam diameter needed for high brilliance, and the SRS ring was retrofitted to improve the brilliance.

(h) Brightness of bending magnet X-rays from all types of rings
Figure 5.6 compares the brightness of bending-magnet sources from selected first-, second- and third-generation sources, whose features are summarized in Table 5.1. Note how the high-energy rings provide similar brightness

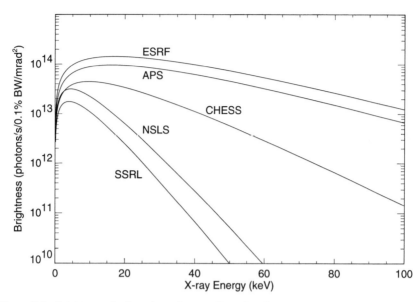

Figure 5.6 Brightness of selected synchrotron X-ray bending-magnet sources. (Rivers, 1988).

Table 5.1 Parameters for bending magnet synchrotron X-ray sources

Source	Energy (GeV)	Current (mA)	Peak field (T)	Radius (m)	ε_c (keV)
APS	7.0	100	0.6	39.0	19.6
CHESS	5.4	75	0.6	32.0	11.5
ESRF	6.0	100	–	–	19.2
NSLS-X	2.5	250	1.2	6.9	5.1
SRS	2.0	300	1.2	5	3.2
SSRL	3.0	100	0.8	12.6	4.7

below 5 keV to the low-energy rings. Above 10 keV, the brightness of the high-energy rings drops off much less than for the low-energy rings. For the NSLS bending-magnet station X26-A, the K spectrum is accessible for XRF up to middle-order atomic number elements in the periodic table (e.g. 37 keV for Ba K lines with $Z = 56$). An APS bending magnet will permit efficient excitation of K electrons for all elements (U at $Z = 92$ requires 116 keV).

5.2.3 Multi-magnet insertion devices

(a) Insertion devices and their influence on the design of third-generation sources

The next factor important for increasing the brightness and brilliance of synchrotron sources was the invention of insertion devices (ID). Each ID consists of an array of N pairs of magnets inserted into a straight section between two bending magnets (Figure 5.7). The magnet pairs alternate in polarity causing the electrons to follow a symmetric snake-like track which restores the emerging electrons back to the original linear track.

The design of the third-generation sources has been driven by emphasis on positioning IDs in the straight sections of the storage ring to advance the state of the art. However, the BMs on third-generation rings are so powerful that they will enable analyses that currently challenge the capability of the second-generation sources.

(b) Wiggler source of white X-rays

The wiggler type of ID generates independent bremsstrahlung beams which merely add up to give a combined flux with an intensity N times higher than for a BM with the same magnetic field. In some old synchrotrons, retrofitted wigglers had only a few magnets because available straight sections were short. Many new wigglers have $N = 20-50$, with a magnetic period of 5–15 cm.

The relativistic physics at each wiggle is the same as for the instantaneous

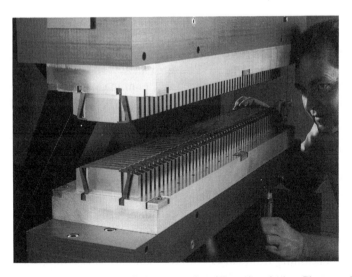

Figure 5.7 Alternating pairs of bending magnets in an insertion device. Photograph of prototype provided by Susan Barr of the Advanced Photon Source, Argonne National Laboratory.

Table 5.2 Parameters for wiggler synchrotron X-ray sources

Source	Energy (GeV)	Current (mA)	Peak field (T)	Radius (m)	ε_c (keV)	Poles number
APS-A	7.0	100	1.0	23.3	32.6	20
CHESS	5.4	75	1.5	12.1	30.2	6
NSLS-X17	2.5	250	5.0	1.6	20.8	5
NSLS-X25	2.5	250	1.2	6.9	5.0	30
SSRL	3.0	100	1.0	9.9	6.0	54

deflection of a bending magnet. The vertical divergence of a wiggler is the same as for a BM with the same field strength ($1/\gamma$). Because the beams from the left and right wiggles are not collinear, the horizontal divergence of a wiggler is greater than the vertical divergence by a factor K_{id}. The deflection parameter, K_{id}, is related to the peak magnetic field B_0 (tesla) by

$$K_{id} = 0.934 \lambda_{id} B_0 \tag{5.3}$$

where λ_{id} is the period of the individual magnets of the insertion device in centimetres. For a wiggler, K_{id} is generally between 10 and 60.

Note that the horizontal divergence of an X-ray from a wiggler is controlled by the size of the wiggle, whereas that from a BM is controlled initially by the size of the bending arc and secondly by any slit system inserted to restrict the arc.

Conventional magnets used as BMs operate at between 0.6 and 1.2 T. Superconducting magnets operating at 5–6 T can shift the spectrum to higher energy, but are expensive to build and maintain. They need liquid helium coolant and are tricky to run because of heavy strain on the fragile coils. The energy-shifting wiggler on the X17 beamline of the second-generation NSLS has a critical energy of 20.8 keV, slightly greater than the 19.6 keV for the APS bending magnet.

Figure 5.8 shows the brightness of selected wiggler sources whose parameters are listed in Table 5.2. The APS 'A' wiggler is about 10 times brighter than the CHESS wiggler and is still yielding a brightness of 10^{15} photons/s/0.1% band width/mrad^2 units at 100 keV. Indeed, some experiments will be made at 200 keV or higher. The brightness curve in Figure 5.8 is for the magnets moved as close as possible, just outside the vacuum tube carrying the positron beam. Moving the racks of magnets symmetrically apart allows the critical energy of the APS wiggler to be reduced from 38 to 10 keV. Such tuning from closed gap to open gap will be important for many experiments. The amount of the detuning is limited.

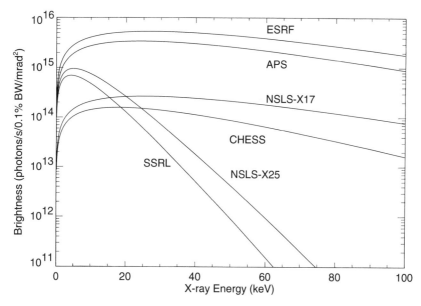

Figure 5.8 Brightness of selected synchrotron wiggler sources. ESRF, 22 pole, 1.2 tesla; APS, 56, 1.0 T; NSLS-X17, 5, 5.0 T; CHESS, 6, 1.5 T; NSLS-X25, 22, 1.2 T; SSRL, 54, 1.0 T. (Modified from Rivers, 1988).

A simple wiggler produces strongly polarized radiation because the electron/positron beam is essentially travelling in coplanar arcs, just as for a bending magnet. The polarization from a wiggler can be modified deliberately by geometrical changes to the magnet arrays. For example, two arrays at 90° can produce circularly polarized X-rays.

(c) Undulator sources of tuneable quasi-monochromatic X-rays

The second type of ID (undulator) uses weaker magnetic fields. The bremsstrahlung beams overlap in the same phase space, and interfere with each other. The spectrum has a lower mean energy than that of a BM or a wiggler, but with harmonic peaks of greatly enhanced brightness and brilliance. Each undulation generates a forward jet out of phase with the next jet by the time taken to reach the next magnet (Figure 5.7). Only those photons with wavelength equal to this difference, or a subharmonic (n) thereof, are reinforced, and the white spectrum from a bending magnet and a wiggler is replaced by peaks at the harmonic wavelengths (Figure 5.9). The X-ray beam is highly collimated in both the horizontal and vertical directions because the undulations are 'locked together', unlike the independent wiggles of the wiggler.

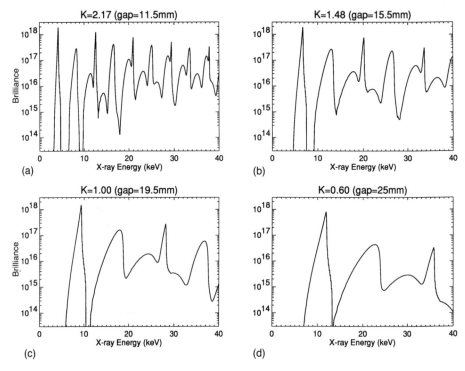

Figure 5.9 Peaked spectrum from APS Undulator 'A' at four different gaps corresponding to different deflection parameters K_{id}. The ordinate axis is in units of brilliance (Figure 5.2). Calculated by MLR.

The energy of the nth harmonic in keV is

$$\varepsilon_n = \frac{0.95 n E^2}{\lambda_{id}(1 + K_{id}^2/2)} \qquad (5.4)$$

where λ_{id} and K_{id} are the same parameters as for a wiggler.

Note that reduction of the vertical spacing between each magnet pair causes B_0 to increase, and the consequent increase in K_{id} drives ε_n to lower energy. It is important to emphasize this inverse relation between increasing magnetic field and decreasing photon energy of the harmonics.

As K_{id} increases, the higher harmonics become more important and more closely spaced in energy until the undulator effectively becomes a wiggler for $K_{id} > 10$. Note that the total energy of all harmonics does increase as K_{id} increases.

The brightness of the nth harmonic is

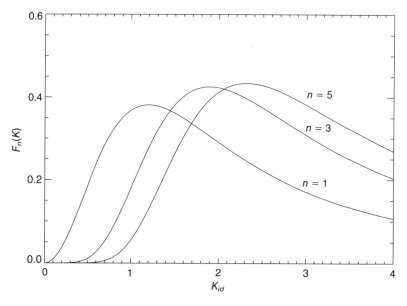

Figure 5.10 Brightness function [$F(K)$] of the odd nth harmonics versus the deflection parameter K_{id}. See text for definitions. (Rivers, 1988).

$$B_n = 1.74 \times 10^{14} N^2 E^2 I F_n(K_{id}) \quad (5.5)$$

where $F_n(K_{id})$ is plotted in Figure 5.10. Observe how the first harmonic loses its dominance over the third harmonic as K_{id} reaches 1.5, and how the fifth and higher harmonics take over for $K_{id} > 2$.

The characteristic opening angle in both the horizontal and vertical directions of the nth harmonic for a perfect ring with infinitely narrow bunches of electrons is

$$\sigma_{r'} = \left(\frac{\lambda_n}{L}\right)^{1/2} \quad (5.6)$$

where λ_n is the photon wavelength and L is the length of the undulator. For APS Undulator 'A', with a harmonic at, say, 12.4 keV ($= 1\,\text{Å} = 10^{-8}\,\text{cm}$) and $L = 158$ periods $\times$ 3.3 cm spacing, $\sigma_{r'}$ becomes 4.5 μrad. This is 15 times less than the intrinsic opening angle for the instantaneous acceleration in a bending magnet at the APS. In an imperfect ring with the electrons/positrons rolling slightly around the perfect circle like a drunken sailor, $\sigma_{r'}$ must be increased considerably. For the APS, $\sigma_{r'}$ is increased 5-fold to $\sigma_{x'} = 25\,\mu\text{rad}$ in the horizontal plane and doubled to $\sigma_{z'} = 10\,\mu\text{rad}$ in the vertical

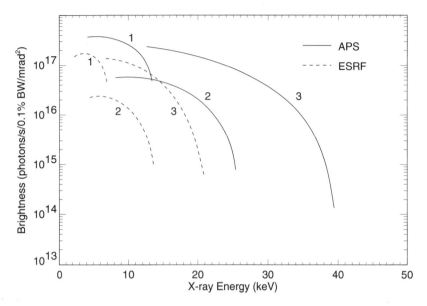

Figure 5.11 Peak brightness of first, second and third harmonics from selected synchrotron X-ray undulator sources. (Modified from Rivers, 1988).

plane. These enlarged values are still some 3 and 7 times smaller than for BM and wiggler sources respectively. The broadening of the opening angle reduces the peak brightness by $\sigma_{x'}\sigma_{z'}/\sigma_{r'}^2$. For the above parameters, the reduction is about 14-fold.

The number of poles for an undulator can be about 2–3 times higher than for a wiggler in a given length of straight section. Thus, the Wiggler A at APS is limited to 54 poles because a higher number would enforce undulator behaviour under certain operating conditions.

Odd harmonics ($n = 1, 3$) are ideally on-axis and even ones are off-axis. For a real ring with non-zero emittance, however, the phase envelopes of the harmonics overlap, and all are accessible on-axis. The formal bandwidth of each harmonic wavelength is $1/nN$, but deviation of the magnetic field from that of a mathematically perfect set of point deflectors causes a complex structure analogous to ghosts from an imperfect optical grating. The polarization is complex and is coupled with the detailed shape of the harmonic peaks in the X-ray spectrum. In general, the odd harmonics have similar polarization but non-perfection causes subtle features. The even harmonics have a polarization different to that of the odd harmonics.

Figure 5.9(d) shows the calculated spectrum for the APS Undulator 'A' with the magnet arrays as far apart as possible to generate the highest energy of the first harmonic. An observed spectrum taken with the undulator at CHESS was similar to the calculated spectrum. The first harmonic is represented by a slightly ragged peak near 13 keV. The peak splays out asymmetrically to 1% of peak brightness at about 10 and 13.5 keV. The second harmonic is nearly flat-topped from 20 to 26 keV and the third harmonic is rather irregular.

The wavelength (and energy) of the set of harmonics can be changed merely by varying the magnetic field. This action corresponds to rotating a monochromating crystal in visible-light optics! The simplest way is to mount the upper and lower magnets on racks which are moved symmetrically about the vacuum tube containing the electron beam.

Figure 5.11 shows the parameters of selected undulators. The APS will operate at 7 GeV to allow Undulator 'A' with 158 periods to be tuned in the first harmonic from about 3.5 to 13 keV, and in the third harmonic from 10.5 to 39 keV. The range of the second harmonic from 7 to 26 keV overlaps those of the first and third harmonics. Please consult the technical literature for the ESRF (6 GeV) and the Spring-8 (8 GeV) rings for the reasons for the choice of acceleration energy.

The harmonics can be broadened by either tilting the magnet racks symmetrically (giving a tapered undulator) or sliding them sideways in opposite directions. This broadening is important for XAS which requires monochromator tuning through a band width of about 10%.

Extremely important for successful operation of an undulator is positional stability of the electron/positron beam. First reports from the ESRF indicate superb stability of the electron beam, and one can hope for good relative stability of the narrow jets of the undulator harmonics. The BM and wiggler beams are less sensitive to positional instability of the electron beam.

The success of brilliance-driven experiments is dependent on the development of (i) optical elements that do not warp under a high heat load (enough to melt any uncooled material); (ii) fast-cycling, multi-element detectors, preferably with energy discrimination; (iii) rapid on-line data processing and (iv) user-friendly, two-way visual communication and data transfer. For example, terabyte storage is needed for certain experiments including tomography.

Very generally, but not exactly, brilliance-driven experiments on small samples must be undertaken at the third-generation sources while flux-driven experiments on large samples can remain cost-effective at the earlier rings.

5.2.2 Figures of merit

Before describing the design of beamlines and experimental stations, it is important to further describe the parameters for evaluating the suitability of a synchrotron source for a particular experiment.

Three figures of merit are used (APS convention – beware of other definitions).

- The flux is the number of photons/second/horizontal angle (θ)/bandwidth, integrated over the entire vertical angle (ψ). It is relevant for a large sample intercepting the entire beam in the vertical direction. For a bending-magnet source, only a fraction of the horizontal divergence is used. Typically for a third-generation source, the outer 1–2 mrad portions of the total horizontal fan of 6–10 mrad supply two independent beamlines. The inner portion is blocked.
- The brightness, which is the flux/vertical angle (ψ), is the number of photons per solid angle. It is appropriate for an experiment which uses a collimator or pinhole to select a small beam. The beam spot may be smaller than the sample area, thus allowing line profiling or area scanning.
- The brilliance is the brightness/source area. It is relevant for a beamline with focusing optics which effectively images the entire source of X-rays onto the sample. Thus, the smaller the spatial cross-section of the bunches of electron/positrons (i.e. from a storage ring of lower emittance), the better the brilliance. In order to be able to image the entire source with low aberration, the design of the optical imaging system is much easier if the beam divergence is small. Hence, the third-generation sources with low emittance and low divergence yield much higher brilliance than the earlier sources.

These three figures of merit should be compared directly only when account is taken of the type of experiment. Thus, an undulator is very powerful at its harmonic energies, especially if a harmonic is used just above a particular absorption edge to excite fluorescence from a chosen element. However, it is not so outstanding if the entire white beam from a bending magnet can be used in a flux-driven experiment. For micro-spot analysis, the brightness is particularly relevant for an experiment with a pinhole, and the brilliance for one with a focusing mirror effectively matched to the sample.

5.3 Beamline and experimental stations

5.3.1 Synchrotron beamlines: general aspects

This section concentrates on general design features. It ignores many technical details, which are critically important to the success of a beamline. These details are generally unnoticed, and indeed should be, by the analyst visiting a user-friendly, general service facility.

Nearly all beamlines, including those for XRF, XAS and XRD, contain a similar set of components. The X-rays from a BM or an ID are transported to the experimental station by a sophisticated set of components designed both to protect the ring from catastrophic vacuum failure, and the scientists from X-rays.

The front end, which is typically behind the main concrete shielding wall, contains vacuum isolation valves and/or Be windows to protect the storage ring from vacuum mishaps on the beamline, and water-cooled apertures to limit the horizontal extent of the beam. Most importantly, it contains a thick metal safety shutter which, when closed, prevents X-rays from passing further down the beamline. The safety shutter cannot be opened unless the beamline and the experimental stations are interlocked.

The beam transport typically consists of one or more Be windows, to isolate the beamline vacuum from that of the storage ring, and evacuated beam pipes ~10–20 cm in diameter to conduct the beam into the experimental stations. Along the beam transport, between the storage ring and the experimental stations, there may be one or more enclosures, which contain X-ray optical devices such as a monochromator and a focusing mirror.

Most synchrotron beamlines are equipped with monochromators which consist of two Si or Ge crystals which diffract a single wavelength. The output beam is parallel to, but slightly offset vertically from, the incoming white beam. All beamlines for XAS require a monochromator. XRF experiments, on the other hand, can use either the original white synchrotron beam or a monochromatized derivative.

Many synchrotron beamlines are equipped with a focusing mirror to increase the photon flux on the sample. Such a mirror works by the principle of total external reflection at the grazing incidence angle (3–8 mrad). At such a shallow angle, a mirror 500–1000 mm long is required to collect the full vertical fan of radiation from the synchrotron, which is typically 3–5 mm tall at the position of the mirror. A mirror placed halfway between the source point in the storage ring and the experimental station is called a 1 : 1 mirror. In principle, it produces a focused beam of X-rays of the same size as that of the electron or positron beam in the storage ring. At the NSLS, for example, the size would be 0.12 mm (vertical) by 0.4 mm (horizontal). To produce the desired spot size, the mirror must be fabricated with a slope error of less than a few microradians.

The experimental station is typically a steel or steel/lead enclosure. The experimenter has access to the station to change samples, and align equipment so long as the shutter is closed. When ready for data collection, the station is interlocked to prevent access and the safety shutter can then be opened.

The XRF station at beamline X26A of NSLS has been configured in different ways, but the overall setup in Figure 5.12 illustrates the essential arrangement. The white X-ray beam from a BM may or may not be monochromatized by a channel-cut monochromator, and may or may not be focused by a mirror before passing through beam-defining apertures. The sample might be a petrographic thin section mounted on a digital x–y–z stage at 45° to the beam. It can be examined and positioned by use of the video signal from a long-working-distance optical microscope. The fluorescent X-rays are detected by either a wavelength- or an energy-dispersive

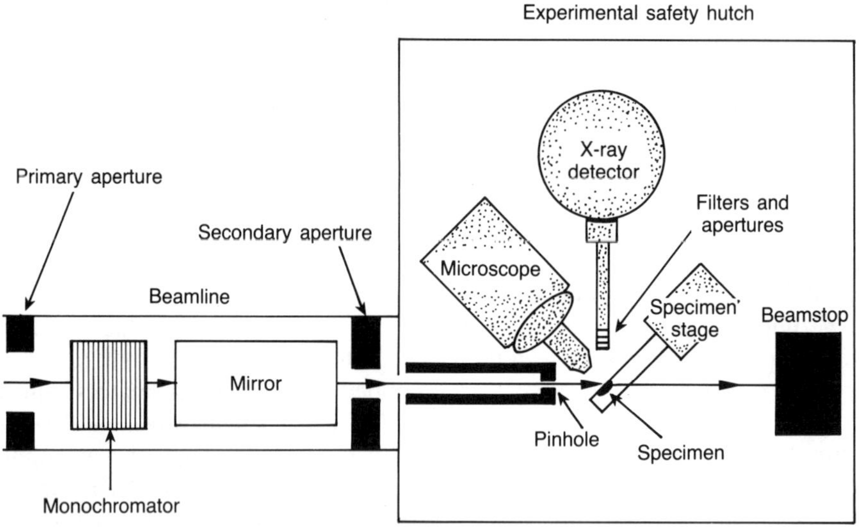

Figure 5.12 Schematic of the X-ray microprobe at the NSLS-X26A beamline. The modular design allows increasing sophistication from incident white X-rays defined by apertures and fluorescent X-rays collected by an energy-dispersive Si(Li) detector to monochromatized and focused incident X-rays and a wavelength detector for the fluorescence. (Based on Sutton et al., 1993(b)).

detector. An XAS or XRD beamline would be similar, with only significant differences in the sample mounting, manipulating equipment and detectors. The XRF microprobe (Figure 5.13) sits on a table in a lead-lined safety hutch, and all positioning motors are operated by remote control.

The design of the Geo/Soil/EnviroCARS sector at the APS (Figure 5.14) is given because it demonstrates what can be expected within a few years at the third-generation sources. This sector is being designed by a collaborative access team to serve the combined Geo/Soil/Enviro community. It will contain two beamlines fed by the outer parts of the fan from a BM and a single beamline from a straight section containing both an undulator and a wiggler. Five safety hutches contain a range of experimental stations optimized to take advantage of the high brightness and brilliance of the APS. The Center for Advanced Radiation Sources (CARS) at the University of Chicago is managing the construction on behalf of Geo/Soil/Enviro scientists who attend workshops and management meetings. Access by peer review will be strongly determined by the need for high brilliance, and the impossibility to do the experiment at an earlier (and cheaper!) synchrotron source. One-quarter of the time is open to any user. At ESRF, the ID-beamlines are being designed and constructed by the ESRF staff, and access will be available to all users, irrespective of discipline, by peer review. Some BM beamlines at the ESRF are assigned to outside teams.

Figure 5.13 Photographs of the collimated X-ray microprobe in the NSLS-X26A lead-lined safety hutch. (a) Close-up of X-ray beam transport (WSW from upper right), sample mounted in orthogonal frames translatable by stepping motors (one shown at bottom left), long-working-distance objective (lower centre), and EDS detector (just showing at bottom centre). The walls of the lead-lined safety hutch are faintly discernible at the back. (b) Wider view almost towards the incoming X-ray beam. The orthogonal frames of the sample holder are visible at the upper left of centre. The EDS detector is at upper right accepting fluorescent X-rays in the horizontal plane at 90° to the incoming beam. The enclosure for the WDS detector occupies the foreground, and is tilted 20° below the EDS detector. (Commissioned by S.R. Sutton).

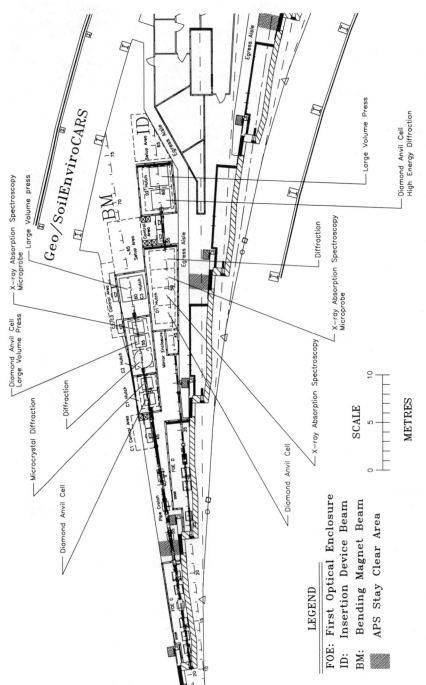

Figure 5.14 Plan of the layout of the Geo/SoilEnviroCARS sector at the APS. (Current drawing in CARS files (Feb. 11, 1994)).

The visible part of the beamline consists of various vacuum chambers connected by metal tubes. Figure 5.15 shows a plan and elevation of the major components of the first optical enclosure and beam transport of the Geo/Soil/EnviroCARS ID beamline at the APS. The first optics enclosure contains ring vacuum isolation equipment, beam-defining slits and heat-reducing filters, Laue/Bragg and Bragg–Bragg monochromators and safety shutters.

The mirror enclosure contains a flat Rh-coated Si mirror for removal of higher-energy harmonics from the monochromator. It will be 60 cm long, and polished to a slope error of less than 5 μrad.

The BM beamline will have a Bragg–Bragg monochromator and a water-cooled toroidal mirror with slope errors of less than 3 μrad longitudinally and 80 μrad transversely.

All the monochromators and mirrors can be moved into and out of the X-ray beam with high positional reproducibility. This, together with the choice of three synchrotron sources, allows flexibility in selecting analytical conditions.

5.3.2 XRF experimental stations

(a) Microbeam optics

For XRF microprobe experiments, the primary requirement is a small X-ray beam. There are several means of producing such a small beam, and hard X-ray microfocusing optics remains a very active area of research and development.

The simplest method of obtaining a small beam is not to focus the beam at all, but just to use a small pinhole. This is in fact the system generally used at beamline X-26A at the NSLS. A pinhole is relatively easy to fabricate, it covers a large energy range and the divergence from the pinhole is governed only by the size of and distance to the X-ray source. For synchrotron beamlines the divergence is typically less than 0.1 mrad. Hence, for a pinhole placed 10 mm from the sample, the beam will only increase in size by 1 μm after it exits the pinhole.

Several more sophisticated schemes have been developed for producing small X-ray beams. They include ellipsoidal mirrors, double elliptical mirror (Kirkpatrick–Baez) systems, Fresnel zone plates and tapered glass capillaries. All except the ellipsoidal mirrors have yielded beams narrower than a few micrometres, with 10–100 times gain in photons/second over a pinhole. Unfortunately, none is yet entirely satisfactory, due to some combination of limitations in energy bandwidth and working distance. For XRF experiments in particular, it has proven difficult to improve on the pinhole microprobe at X-26A. Thus, the pinhole at X-26A is only 9 m from the source, while microfocusing optics must typically be at least 20 m from the source in order to achieve demagnification to a spot diameter of a few micrometres. The

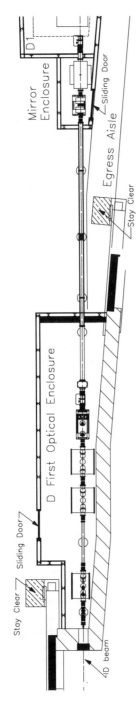
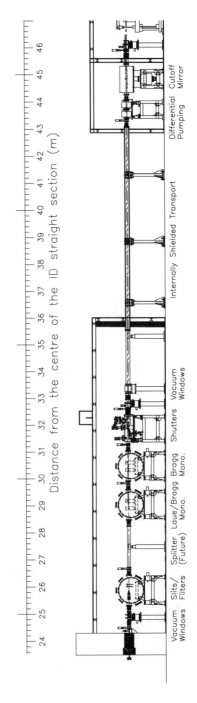

D Line Layout

Figure 5.15 Plan and elevation of the first optical enclosure and beam transport of the GeoCARS ID beamline at APS. (Current drawing in CARS files (Feb. 11, 1994)).

flux gain must be fivefold just to compensate for the greater distance from the source. Furthermore, many microfocusing schemes (zone plate and multilayer coated mirrors for example) can pass only a relatively narrow (0.1–1%) energy bandwidth. For XRF, which can benefit from the 100% energy bandwidth provided by the bending magnet and the pinhole, the signal loss due to reduction in bandwidth may be greater than the signal gain from the focusing optics. This is not so for XRD and XAS experiments, which require a monochromatic beam and for which narrow bandwidth focusing optics are perfectly suitable.

Focusing optics which use total reflection at grazing incidence have a high energy cut-off beyond which the optic does not reflect X-rays. The exact value of the cut-off depends upon the grazing incidence angle and the atomic number and density of the mirror coating. Cut-off at 15 keV is typical for Pt-coated mirrors at 4 mrad glancing angle. The high energy cut-off limits the K-edge range of elements which can be excited by the X-ray beam. A 15 keV cut-off, for example, precludes measurement of the K lines of Sr, Zr and Nb, which are geochemically important elements. Using such a mirror, these elements could only be analysed using the L lines, which are much weaker and overlap the intense K lines of more abundant low-Z elements.

Microfocusing optics must be placed inside the experimental station, rather than upstream along the beamline. The reason is that the electron or positron source, which is typically 500 μm in diameter, must be demagnified by factors of 100–500 to produce the desired spot size on the sample. Since the sample is typically situated 10–50 m from the source, the distance from the focusing optics to the sample must be only 2–50 cm in order to achieve the required geometric demagnification. The focusing optics must be well shielded so that stray scattered radiation does not reach the sample or detector.

Figure 5.16 summarizes the various optical systems used at both the NSLS-X26A beamline and other sources to achieve a small, high-intensity beam on the sample.

The other essential components of an XRF microprobe include a sample stage, visible light microscope, X-ray detectors and computer control system.

(b) Sample stage

The sample stage must have at least $x-y-z$ translations, all with mechanical precision better than the X-ray beam diameter. Stepper motor or DC motor actuators are needed for remote positioning of the sample while the X-ray beam shutter is open and the experimental station is interlocked. Typically, the sample stage is oriented at 45° to the incident beam and the fluorescence detector is at 90° to the beam, in the horizontal plane of the ring. The x and y stage movements position the desired spot on the sample into the beam, while the z stage moves the sample into the focus of the optical microscope.

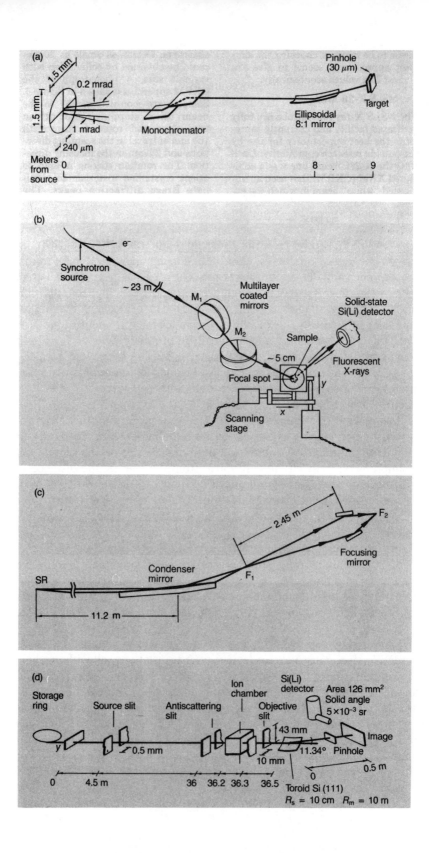

An inner stage rotating about the vertical axis is very useful for dealing with the problem of diffraction peaks. If a set of Bragg planes in a crystalline sample is parallel to the surface of the sample, and the sample is being excited with a white X-ray beam, some set of wavelengths in the beam will satisfy Bragg's law and diffract into the X-ray detector. By rotating the sample about the vertical axis by a few degrees, the position (and energy) of the diffracted beam can be changed, so that it no longer strikes the detector.

For existing X-ray microprobes, the resolution of motor driven stages (0.1–1 µm) is sufficient. For third-generation sources, which offer the prospect of even smaller X-ray beams, piezoelectric stages with a precision of 10 nm or better may be required. Such stages are currently in use in soft X-ray microscopes, as at beamline NSLS-X1-A.

(c) Optical microscope

A visible light microscope equipped with a TV camera is essential to monitor the specimen position. The microscope must have good resolution and contrast to facilitate finding one's way around the specimen. A poor-quality optical system would be very frustrating, since precious beam time would be wasted as experimenters search for the correct position on their samples. The microscope objective needs a sufficiently large working distance to avoid obstructing the incident or fluorescent beams. At the X-26 microprobe, Nikon 5× M-plan (16 mm working distance) and 20× and 40× ELWD (10 mm working distance) objectives are used with excellent results.

The alignment of the optical microscope with respect to the X-ray beam is accomplished with a specimen which fluoresces brightly in the visible spectrum. A polished sample of CsI, benitoite or scheelite is adjusted so that its surface is in optimum optical focus. The microscope position is adjusted to bring the fluorescent spot near to the centre of the field of view. The precise position of the X-ray beam on the specimen is marked on the TV monitor. When a non-fluorescent sample is inserted, the analytical position is inferred from the monitor. Any imprecision in focusing will result in a left–right error because the X-ray beam strikes the sample at 45°, not at normal incidence. Hence, a high-magnification objective with shallow depth of focus is needed for precise sample positioning.

Figure 5.16 Schematic of different approaches to the concentration of a synchrotron X-ray beam into a small spot on the sample. (a) A white beam from a BM is focused in two dimensions by an ellipsoidal mirror with 8:1 asymmetry. A 30 µm pinhole discriminates against scattered radiation. A double-crystal monochromator can be inserted, and rotated without beam displacement. Used at NSLS-X26A (Figure 5.12). (b) Two crossed spherical multilayer mirrors in the Kirkpatrick–Baez configuration. System designed and built at Lawrence Berkeley Laboratory, and tested at NSLS-X26A. (c) Wolter focusing mirror system with combined ellipsoidal and hyperboloidal surfaces of revolution used at the Photon Factory. (d) Bent crystal system used at SRS, Daresbury. (Jones and Gordon, 1989).

(d) Lift table

Because experimenters cannot adjust the position of the X-ray beam from the synchrotron, the entire experimental apparatus is adjusted to the optimum position. The entire experimental setup is mounted on a heavy-duty, motorized lift table. As the lift table is scanned vertically without changing the internal alignment of the microprobe components, the ratio of fluorescence peak to the scattered background is monitored for a maximum at the vertical centre of the beam. This alignment step is normally performed each time the electron bunches are re-injected. At X-26A, the peak-to-background ratio is measurably worse 0.1 mm from the optimum position.

(e) Detectors

The available technology is basically the same as that for the electron and proton microprobes. Solid-state detectors (Si(Li) or intrinsic Ge), also called energy-dispersive spectrometers (EDS), and bent crystal spectrometers, also called wavelength-dispersive spectrometers (WDS), have advantages and limitations.

An EDS detector records the entire spectrum, and hence can quickly measure many elements in the specimen. It can be positioned near the specimen to collect a relatively large solid angle, perhaps capturing as much as 10% of the total fluorescence. The cost of EDS detectors is only one-third that of WDS detectors.

A serious disadvantage of EDS detectors is their relatively poor energy resolution, typically 150 eV at Mn Kα, compared to the natural line width of only 2 eV. The poor resolution leads to problems with overlapping peaks, including Kα/Kβ pairs from adjacent elements, and L lines and K lines from widely separated elements. The inferior resolution also results in a relatively poor peak-to-background ratio when using incident white X-rays, since there are scattered photons from the incident beam with the same energy as that of the fluorescence line of interest. An EDS detector will have a peak-to-background ratio typically 30–50 times lower than that of an ideal detector whose resolution is equal to the natural line width of the fluorescence emission.

Another problem with EDS detectors is the limited count-rate capability. To achieve 150 eV resolution, EDS detectors typically have to be run with an amplifier time constant of 4–8 μs. The consequent limitation on the maximum count rate to less than 10 000 counts s^{-1} would not be serious if it only applied to the trace elements of interest. However, since an EDS detector measures the fluorescence of all elements in the specimen at once, strong major element fluorescence can make it difficult to measure trace elements. Usually, this problem can be reduced by placing a filter (typically a thin plastic or Al foil) in front of the EDS detector to reduce the intense fluorescence of the abundant low-Z elements, for example Ca in calcite or Na-Ca in feldspar. The higher energy K line fluorescence from trace elements with higher Z, for example Cu or Sr, will be hardly affected by the filter.

This strategy can be used only if the major element fluorescence is significantly lower in energy than the trace element fluorescence of interest. Measurement of Cr in Fe-rich olivine, for example, is very difficult or impossible with an EDS detector if white radiation is used as the excitation source. However, an EDS detector would suffice if a monochromatic excitation beam were used. By setting the energy of the incident beam between the absorption edges of Cr (6.0 keV) and Fe (7.4 keV), Cr would be excited but not the Fe.

One method of overcoming the count rate limitation of EDS detectors is to combine the output of a multiple array of individual detectors, each with its own set of electronics. For example, the X-26A beamline is now equipped with a 13-element Ge detector which can collect data 13 times faster than a single-element detector. Because the electronics and detector elements in such systems are not yet monolithic, they tend to be quite expensive. In the future, the cost of such systems can be reduced by fabricating both the detectors and electronics as in large-scale integrated circuits; an experimental 100-element detector is under research and development at the Instrumentation Department of BNL.

Wavelength-dispersive spectrometers typically consist of one or more curved crystals which diffract a single wavelength into a proportional counter. The advantages of WDS detectors include much better energy resolution than EDS detectors, typically 10–25 eV at Mn Kα. Because WDS detectors measure only one wavelength at a time, and can be operated at short time constants, they are not subject to the same count-rate limitations as EDS detectors. One disadvantage of WDS detectors is that mechanically scanning for the determination of multiple elements is slow. Another disadvantage is that WDS detectors collect a smaller solid angle than EDS detectors. Typical WDS detectors might collect 0.1–1% of the total fluorescence of a given wavelength.

For ideal performance, an XRF microprobe should be equipped with both EDS and WDS detectors, capable of simultaneous use, as at beamline X-26 (Figure 5.13(b)). The Si(Li) detector is normally mounted at 90° to the incident beam in the horizontal plane, to take maximum advantage of the polarization of the source. A WDS spectrometer is mounted below the Si(Li) detector, collecting radiation 20° below the horizontal plane. The WDS is much less sensitive to the polarization of the source because its energy resolution is so much better.

Figure 5.17 compares the X-ray emission from Zn metal collected with WDS detectors on an electron microprobe and the X-26 XRF microprobe. The background is much higher on the electron microprobe due to the bremsstrahlung radiation from the decelerated electrons generated within the sample (see section 2.7.4).

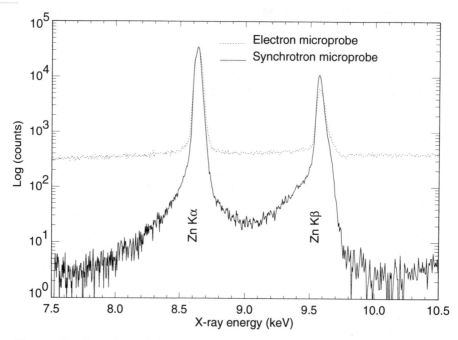

Figure 5.17 Comparison of electron- and synchrotron-microprobe energy scans for Zn K fluorescence radiation from zinc metal measured with a wavelength-dispersive spectrometer.

Figure 5.18 compares the WDS and EDS spectra for apatite and meteoritic iron on the XRF microprobe at X-26A. Both the energy resolution and peak-to-background ratios are better with the WDS. In particular, the WDS enables the partial resolution of L spectra of rare-earth elements – K spectra can of course be resolved with both detectors.

(f) Computer control system

The data acquisition and computer control system must be easy to use, and flexible enough to permit incorporation of new detector systems, scan algorithms and computer-controlled instrumentation.

The basic tasks required of the beamline-control computer include:

- Positioning the beamline components, including slits, mirrors and monochromators.
- Moving the components inside the experimental station, including pinholes, micro-focusing optics, lift table and sample stage. The latter, in particular, is moved very frequently and the efficient operation is much improved if the operator can easily perform tasks such as saving the coordinates of the current stage position for a repeat analysis.

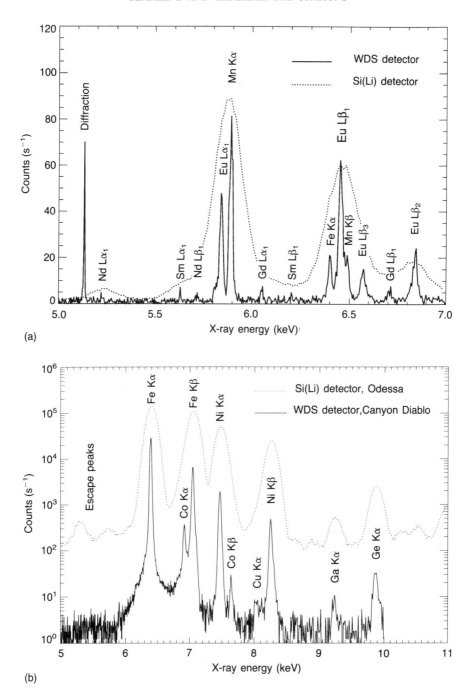

Figure 5.18 Comparison of EDS and WDS spectra for (a) Bolivian apatite and (b) iron meteorites taken at NSLS-X26A.

- Data acquisition. This includes collection of single spectra at data points and scanning of any of the motorized components on the beamline (one-dimensional scans of the lift table for alignment, two-dimensional scans of the sample to produce elemental maps, one-dimensional scan of the monochromator to measure XANES spectra, etc.).
- Data analysis. On-line analysis and data reduction is very important to ensure efficient use of beam time. The required steps are covered below.
- Data display of spectra, one-dimensional scans and two-dimensional scans displayed as images. The software should be very flexible to include the ratio of two images and other types of image processing and enhancement.

5.3.3 XRD experimental stations

Let it suffice that current technology allows the determination of crystal structures for single crystals as small as $(20\,\mu m)^3$, and that anomalous scattering and high-energy X-rays allow considerable advances over the data obtainable with conventional generators. Crystals that are perfect can be studied very rapidly by Laue diffraction and area detectors. Imperfect ones must be studied by Bragg techniques, and single-axis rotation appears to offer the best mechanical stability. In principle, the size of the single crystal can be reduced below $(1\,\mu m)^3$ on third-generation sources, but a host of technical problems arises. To obtain good peak-to-background ratios, a narrow collimator must be used, thus forcing high accuracy in positioning the crystal during rotation. The chemical zoning from the inside to outer surface of many crystals will provide a particularly challenging problem. It may be necessary to carry out diffraction tomography.

Powder diffraction appears to be less challenging technically, but its application to microanalysis is definitely not straightforward. To obtain powder 'lines' of uniform intensity, it is necessary to have many randomly oriented small crystals in the sample. Hence, the sample size cannot be reduced to the micrometre range permitted in principle by the high power of a third-generation source. To take advantage of the high parallelism of a synchrotron beam, the sample must be near-perfect. Nevertheless powder diffraction will prove very useful for structural identification of samples whose chemical composition and bonding state have been determined by XRF or XAS (or a non-X-ray technique). Perhaps the most useful role will be the identification of the components of well-crystalline multiphase samples for which the $\sim 0.01°$ half-width of the diffraction peaks from a perfect crystal structure presents an excellent basis for resolving adjacent peaks.

Also important with respect to conventional sources is the much higher peak-to-background ratio in diffraction patterns obtained at 90° to the synchrotron plane of polarization. Careful design allows this ratio to be at least 4000.

5.3.4 XAS experimental stations

For major elements, an XAS station contains beam monitors upstream and downstream of the sample. The beamline monochromator is scanned through the absorption edge of the element of interest and a plot of absorption versus energy yields the XAS data. For minor and trace elements, the absorption cannot be measured accurately in transmission because the modulation of the transmitted signal is too small and is swamped by noise. Hence, the change of absorption is determined from the variation in the intensity of the fluorescent X-rays or photoelectrons.

This may be accomplished by using an energy-sensitive detector to pull out the fluorescent X-rays efficiently from the background. However, a single energy-sensitive detector is very slow because of the limitation on count rate caused by the long time constant. This limitation is not such a disadvantage for XANES, which requires only a narrow energy range and for which 1% counting statistics (10^4 counts per point) is sufficient, but is very serious for EXAFS which requires many more data points and 0.1% precision (10^6 counts per point) deep into the absorption tail. Hence micro-XAS has been greatly advanced by the development of multiunit, energy-dispersive detectors, each unit of which processes the photons independently to give a signal which is summed with the other signals. A 13-element detector is currently available off-the-shelf, giving better statistical precision by a factor of 3.6. Future prospects include a 100-element detector on a segmented Si chip, with miniaturized electronics to reduce the cost per element. Auger electrons can also be measured, and indeed are especially useful for the analysis of light elements in a surface layer.

At the NSLS-X26A beamline, micro-XANES can currently be carried out with a single-element detector for first-series transition elements at the concentration level of a few hundred parts-per-million for a beam spot 0.1 mm across. The combined sensitivity increases about threefold for the 13-element detector.

5.4 Specimen preparation

The relatively simple techniques for preparation of specimens used at NSLS and other synchrotron sources must become more complex to accommodate the greater sensitivity and spectral resolution expected at the third-generation sources.

5.4.1 X-ray diffraction

For micro-diffraction, it is difficult to mount crystals some $20-50\,\mu m$ across but is achievable by several methods. A crystal can be dropped into a tapered silica glass capillary and wedged by a tuft of fibres. Alternatively, it can be mounted at the end of a tapered fibre of low scattering power. Such a small crystal can usually be picked up with a fine needle, to which it sticks temporarily by surface attraction. Selection of a crystal from a fine powder is an art. One way is to disperse the powder very thinly and check the morphology using a high-quality microscope with long working distance. Accidental intergrowths and contact twins may be detectable. The powder can be dispersed in a liquid and studied by polarized transmitted light. Crystals with poor extinction can be rejected. Ethanol is a suitable liquid which evaporates cleanly. The mounting of crystals smaller than $10\,\mu m$ will be very challenging. One possibility is to sprinkle a little powder on a sticky fibre so that a dozen or so crystals become attached. The assemblage is then coated with a carbon film in an evaporator to achieve electrical conductivity. A secondary electron micrograph should provide information on the morphology to permit selection of the most promising crystals. If the crystal is part of a rock, it may be possible to prepare a polished thin section. With high-energy X-rays, only the forward diffraction cone is needed, and precession or oscillation of (say) 10–30° may yield enough diffraction data without the primary beam hitting adjacent grains. A fine powder may be mixed with epoxy or another low-Z binder, and a thin section made to isolate crystals sufficiently from their neighbours. Since small crystals are commonly zoned, especially at the outer rim, it may be desirable to scan a crystal with a micrometre or even sub-micrometre beam.

5.4.2 X-ray fluorescence

For micro-XRF, small fragments may be treated as in section 5.4.1, but without the concern of searching for crystalline perfection.

With second-generation sources, most rocks are studied from doubly polished thin sections. Grinding and polishing is a black art, and care is needed to reduce cross-contamination. Diamond powder is suitable for fine grinding and polishing. It is generally assumed that each abrasive particle chops an irregular groove with a depth of perhaps about one-tenth of the radius of the particle. A brittle material should not 'smear' onto its neighbours as much as a soft one. A hard component will tend to stand higher than a soft one, and a curved surface will develop at the boundary. The problem of uneven absorption of the fluorescent X-ray beam, as a boundary is scanned, becomes less important and perhaps even trivial for hard X-rays. Double polishing is achieved by mounting a rock slice onto a glass slide with a soluble adhesive such as Canada balsam; levelling, fine grinding and polishing the free surface; cleaning the free surface with solvents and ultrasonic vibration; fixing the free surface to a second glass slide with a heat-resistant epoxy cement, and releasing the unpolished surface by heating the balsam; levelling, fine grinding and polishing the newly freed surface;

and finally cleaning. For analysis at the part-per-million level, extreme attention to cleanliness and minimization of cross-contamination is essential. Final polishing might be done with $1\,\mu m$ diamond powder mixed in pure petroleum jelly. To reduce curvature, soft laps should be avoided. The diamond–jelly mixture might be smeared on a flat plate made from an 'innocuous' material such as pure silica.

The initial rock slice can be obtained by cutting into the rock specimen with two paper-thin, diamond-impregnated copper discs separated by a shim. Careful feeding of the saw may yield a slice whose surface requires only final grinding and polishing. A multiple gang of saws is useful for salami slicing of specimens containing small inclusions. A friable rock may be impossible to saw into thin coherent slices until it is vacuum impregnated with a resin.

A hard X-ray beam will penetrate through the rock slice into the glass slide. Normal commercial glass slides contain high levels of many elements, and commercially available pure silica slides should be used for the glass backing. To obtain an even lower background, the doubly polished thin section can be prepared using only an adhesive which loosens upon heating, and the specimen can be removed from the glass slide after polishing. It can then be attached to a Kapton film stretched tightly across a frame. A record should be kept of the batches of materials used for specimen preparation. The polished thin section should be stored in a jewel box in a desiccator. Most specimens will be studied with an electron probe before synchrotron XRF analysis. The carbon film will help to reduce atmospheric corrosion of sulphide and other unstable minerals and does not interfere with the SXRF analysis, provided that the carbon evaporator is used only for carbon (not also for Au or other metals). The carbon coat may make optical examination more difficult.

Small particles may be mounted on Kapton film. Atmospheric aerosols may be collected on filters for which a blank analysis should be done on an adjacent clean area.

A specimen may be enclosed in a cell with walls that are almost transparent to X-rays. Chemical reactions can be followed in real time. For soils, and roots growing in soils, cells with Kapton or Mylar walls mounted in a Lexan plastic frame have been devised (e.g. Tokunaga, Sutton and Bajt, 1994, Figure 3). Only those elements which are accessible by analysis of hard X-rays (e.g. Se) can be studied easily in these cells.

Most analyses of surface composition and structure have been made on highly polished metals or simple inorganic compounds in a hard vacuum. Experience needs to be gained with mineral surfaces, and the study of the thickness of the adsorbed layer on the cleavage surface of calcite (Chiarello, Wogelius and Sturchio, 1993) provides useful pointers.

As the sensitivity of third-generation sources increases, current methods for sample preparation and mounting will become inadequate. Microtoming with a diamond knife may prove valuable. Tomography may be necessary to obtain a 'clean' analysis inside some samples.

5.4.3 XAS

Broad-beam XAS is generally undertaken on a powder thinly dispersed on backing film to achieve an overall absorption that gives the optimum statistics (commonly for an effective thickness of about a micrometre). For transmission XAS, a concentration of the chosen element of about 1% is often optimum. For narrow-beam XAS of minor elements, whose fluorescence signal must be used, one is generally forced to use whatever signal can be obtained. The essential factors are implied in sections 5.4.1 and 5.4.2.

5.5 Analysis

5.5.1 XRF

Because synchrotron XRF is identical in many respects to conventional XRF, the reader is referred to the many excellent texts for the basic theory and data analysis techniques.

The basic steps for synchrotron XRF data are:

- Analyse spectra to remove background and fit peak profiles to yield a net area (intensity) for each element of interest.
- Use the net peak areas to estimate concentration. Either compare with standards of known composition, or refer to an internal standard, typically a major element whose composition is known from mineral stoichiometry or following analysis by another technique, such as EPMA.

(a) Background removal

The background in SXRF spectra arises from several sources, including:

- Rayleigh and Compton scattered photons from the incident beam. This effect is significant only for white beam excitation. The background from Compton and Rayleigh scattering is minimized in synchrotron radiation by placing the detector at 90° to the incident beam in the polarization plane. For a low-emittance ring such as the NSLS, the reduction can be greater than 50 times relative to an unpolarized excitation source. The magnitude of the scattered background measured at the detector depends upon the spectrum of the incident beam, the atomic number of elements in the sample (which controls the relative magnitudes of the scattering and self-absorption cross-sections) and the mass thickness.
- Broad scattering from short-range order and any deviations from a regular lattice. This effect occurs for amorphous materials and disordered crystals.

- Diffraction from a regular crystalline lattice. A second spectrum with the sample in a different orientation will not have the same diffraction peaks as the first spectrum.
- Incomplete charge collection in the detector, which causes some fraction of the counts from high-energy peaks to be distributed throughout the lower-energy range of the detected spectrum. This is the primary source of background for monochromatic excitation.
- Compton scattering in the detector. The magnitude and position of this background depends upon the details of the spectrum striking the detector and the size and composition of the detector.

The actual background in XRF is sufficiently complex that it is best fitted and removed by empirical techniques, rather than by calculation of all the above contributions.

At the X-26 beamline, a background-fitting algorithm, based upon extensions of the polynomial tangent technique described by Kajfosz and Kwiatek (1987), is normally used. Figure 5.19 illustrates the background fitted using this technique to a spectrum of a fluid inclusion in quartz from Bingham, Utah.

(b) Peak fitting

After removal of the background, the peaks are fitted by a non-linear, least-squares procedure. At X-26, a commercial mathematical subroutine is used. The spectrum is modelled as a sum of Gaussian peaks. The fit parameters can include the energy, amplitude and full-width at half-maximum (FWHM) of each peak, plus two energy calibration coefficients which relate energy to channel number:

$$\text{energy} = A_1 * \text{channel} + B_1 \qquad (5.7)$$

and two FWHM calibration coefficients which relate the variation in FWHM to energy:

$$\text{FWHM} = A_2 * (\text{energy})^{1/2} + B_2 \qquad (5.8)$$

Typically, the number of fitted parameters is reduced by taking advantage of the known physics of the experiment. The energies of all the fluorescent peaks to be fitted are known, and the FWHM of each peak is assumed to fall on the curve described by equation 5.8, which is an excellent approximation for Si(Li) and Ge detectors. Thus, the number of fitted parameters is reduced to two energy calibration coefficients, A_1 and B_1, two resolution calibration coefficients, A_2 and B_2, and the amplitude of each peak. Some of the peak amplitudes can then be removed from the list of fitted parameters. This is normally done for $K\beta$ and $L\beta$ peaks, whose amplitude is constrained

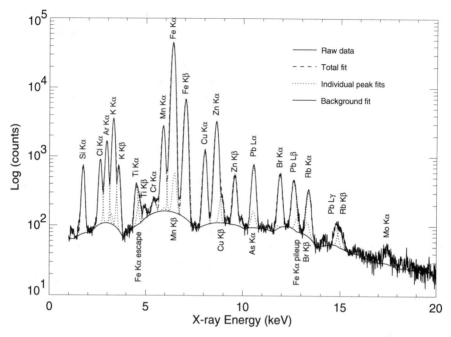

Figure 5.19 Spectrum of fluid inclusion in quartz from Bingham, Utah, showing fitted background and peaks.

to be a fixed fraction of the amplitude of the corresponding Kα or Lα peak. The only peaks for which the energy and FWHM are typically left as free parameters are the diffraction peaks, which can arise when a single crystal sample is illuminated with white beam. These peaks need only be fitted if they overlap a fluorescent peak of interest.

Figure 5.19 shows a spectrum of a fluid inclusion with the fitted profiles of each peak and the overall fitted spectrum (sum of the peaks plus fitted background). The agreement with the measured spectrum is excellent.

(c) Quantitative analysis

Once the net peak areas have been determined, the elemental concentrations in the sample can be calculated. There are several approaches, depending upon the type of sample and availability of standards.

For thin films, i.e. those for which the absorption of the fluorescent X-rays within the sample is insignificant, one can simply use standard thin films to calibrate the sensitivity of the instrument (counts/(μg cm^{-2})). Good thin-film standards are available in the USA from NIST (SRM 1832 and 1833). Although these films are not certified to be homogeneous on the micrometre scale, we find that they are very uniform on at least the 5 μm scale. It is, however, advisable to move a standard during analysis to collect a representative spectrum. Note that this technique only produces a composition in $g\,cm^{-2}$. To determine a concentration in weight percent or ppm it is necessary to know the total mass per unit area of the film in $g\,cm^{-2}$.

For thicker samples, it may still be possible to find standards which closely match the sample so that a simple ratio of counts can be used. Remember that not only the composition but also the thickness of the sample and the standard must be matched. The only exception is when both the sample and the standard are 'infinitely thick', which means that fluorescent X-rays from the back side of the sample make a negligible contribution to the observed count rate because of self-absorption in the sample.

Most samples are neither 'thin film' nor 'infinitely thick' and it is necessary to compute the concentration of the elements according to the following factors:

- The spectrum of the excitation source. The spectrum from the synchrotron is known accurately. The effect of Be windows and air absorption before the X-rays strike the sample must also be computed.
- Absorption of the excitation X-rays as they penetrate the sample. For white beam excitation, the lower energy X-rays are absorbed closer to the surface, while the higher energy X-rays penetrate further.
- Excitation of the target atoms by the X-rays at various depths within the specimen. With white beam, the relative yield will decrease with depth due to the beam hardening effect.
- Absorption of the fluorescent X-rays on their way out of the sample towards the detector.
- Secondary fluorescence, which is fluorescence of atoms, not by the incident radiation, but by the X-rays produced by fluorescence of other elements in the sample.
- Absorption in air and any filters placed between the sample and the detector.
- Variation of detector efficiency with energy.

The physics of these processes has been well understood for some time, and some excellent computer codes exist. At X26 we use the Naval Research Laboratory X-ray fluorescence program (NRLXRF) which can predict from first principles the absolute fluorescent intensities of all elements in a specimen. The inputs are only the spectrum of the exciting radiation and the composition and thickness of the specimen. In practice, neither the absolute intensity of the incident beam nor the detector solid angle are usually sufficiently well known to use the absolute intensities predicted by NRLXRF to measure accurately elemental concentrations. One can use, however, the relative intensities predicted by NRLXRF to determine the composition providing that the actual concentration of at least one element in the sample is known by some independent technique. In practice this is often easy: the Ca content of calcite, for example, is known quite accurately simply from stoichiometry. The Fe content of olivine can be measured on the electron microprobe. By using the measured ratios of the trace element fluorescence intensity to the major element fluorescence, and the predicted trace element to major element ratios for a sample of an identical major element composition and thickness and arbitrary assumed trace element concentrations (say, 100 ppm), it is possible to obtain quantitative trace element analysis with no external standards at all. Normally one also uses whatever standards are available to check and perhaps slightly correct the results of NRLXRF. Figure 5.20 shows the measured and predicted sensitivity for thin film standards at X-26. The fit is excellent except for Zn, which appears to be inhomogeneous in the standard.

Figure 5.21 shows the predicted and measured sensitivities for thick samples of Corning glasses prepared as standards for EPMA. The fit for K lines is superb, except for Rb and Y, for which checking is needed. The predicted values for the $L_{\alpha 1}$ lines are consistently higher than the measured values by a factor of about 2, and an appropriate adjustment of the NRLXRF

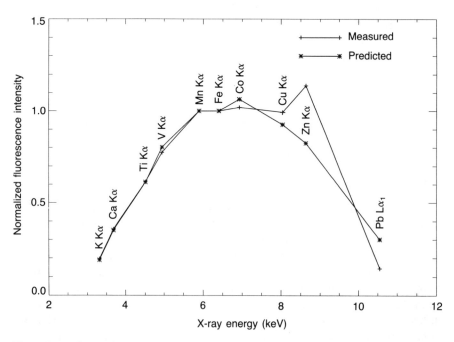

Figure 5.20 Comparison of measured and predicted sensitivity for thin films. The measured intensities are for the NIST thin film ($<1\,\mu m$) glass standards SRM 1832 and SRM 1833. Measured and predicted intensities are counts/concentration normalized to unity for Mn (SRM 1832) and Fe (SRM 1833). Measurements made for white beam at NSLS-X26A with no filters in front of the Si(Li) detector. Zn appears inhomogeneous in SRM 1833.

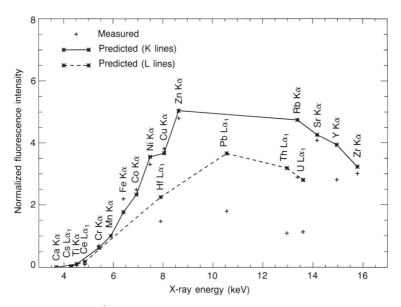

Figure 5.21 Comparison of measured and predicted sensitivity for thick samples. Each of the four 300 μm thick Ca-aluminosilicate glasses contains ~0.6 wt% of several minor elements. A 490 μm thick Kapton film in front of the Si(Li) detector reduced the Ca count rate. Measured and predicted (NRLXRF program) intensities are counts/concentration normalized to unity at Mn. Note the good agreement for K lines (except for Rb and Y) but the approximately twofold higher predicted values for the L lines.

program is needed. This problem with L lines was also found by Lu et al. (1989) for feldspars calibrated by other techniques.

The accuracy of the results for intermediate thickness depends directly upon the accuracy with which the thickness and density of the sample can be determined. For samples 30–50 μm thick the thickness can be measured with an optical microscope to perhaps 1 μm, or 2–3%. For thinner samples the relative error grows larger, and it will be necessary to develop more accurate techniques to measure the thickness. This is particularly true when beam sizes are reduced to 1 μm or less, since to obtain this spatial resolution samples must be only a few micrometres thick.

The NRLXRF program works well for an idealized geometry of a thin homogenous slab of material. This model is an excellent approximation for many specimens, for example polished thin sections with mineral grains which go all the way through from the front to back. However, there are some specimens which are very interesting, for example those with irregular subsurface fluid inclusions, for which it is a poor approximation. Further programming is needed to obtain convenient calculations for such samples. The physics is not difficult, particularly if Monte-Carlo calculations are used.

However, being able to input conveniently the three-dimensional shape and composition information into a program is not trivial.

5.5.2 XRD

The reader is recommended to consult a standard text such as Coppens (1992).

5.5.3 XAS

Please refer back to Figure 5.3 for the distinction between XANES and EXAFS. First consider XANES. Examples of interpretations are given in Brown and Parks (1989). It is important to emphasize that most published XAS spectra were measured on a fine powder sprinkled on an adhesive tape. This form of sample presentation tends to average out any structural anisotropy in the polarized signal if the sample is crystalline. For micro-XAS of a single crystal in (say) a thin section, the signal will depend on any structural anisotropy. Thus, a signal from a Cu atom in square-planar coordination in a high-temperature superconductor is highly anisotropic (Alp, Mini and Ramanathan, 1990). Rotation of a micro-specimen in the holder may reveal anisotropy as the diffraction peaks flash in and out. For brevity, the remaining discussion ignores any crystalline anisotropy.

An expert familiar with XANES spectra from crystalline powders and glasses of many reference materials can often 'read' the peaks for a known element immediately. Figure 5.22 shows how the variation of XANES spectra was used to calibrate a procedure for measurement of the ratio of hexa- to trivalent Cr in glass used to encapsulate nuclear waste (Bajt et al., 1993). Trivalent Cr displays only minor perturbations from a monotonic absorption whereas hexavalent Cr shows a pronounced pre-edge peak very suitable for analysis. Calibration with a suite of standards of different atomic structure indicated an accuracy $\sim 5\%$ in measurement of the percentage of hexavalent Cr. The detection level at NSLS-X26A is 10 ppm for a $150\,\mu m$ synchrotron beam.

Another example is the assignment of Ti in staurolite to a distorted octahedral environment on the basis of comparison of the XANES spectrum with those of other compounds (Figure 5.23). The detailed text in Henderson et al. (1993) references the extensive earlier literature, and explains the details of the spectra. Particularly important for the specific assignment for staurolite is the good match of its XANES spectrum with those for rutile and amphiboles (not shown).

Turning now to the EXAFS part of the absorption specturm, again consult Brown and Parks (1989) for examples and review papers. In principle, all that is necessary is to subtract the sloping background to reveal the 'ripples' which, when inverted, yield the radial distribution function of the nearer

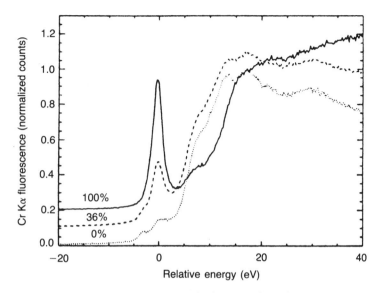

Figure 5.22 Cr XANES spectra for Na_2CrO_4 (0%), Cr_2O_3 (100%) and a 36% mixture. Plot of Cr Kα normalized counts vs. energy relative to the pre-edge energy peak defined as zero energy. The spectra are labelled with the percentage of total Cr that is hexavalent. Counts are normalized to the total intensity in a chosen energy range above the edge, and are offset for clarity. (Bajt et al., 1993).

neighbours. Figure 5.24 shows the background-subtracted K edge XAS spectra for Fe, Mn and Zn in staurolite. A strong qualitative similarity is an excellent guide to the conclusion from a detailed mathematical analysis that all three elements occupy the same type(s) of structural site(s). Since X-ray and neutron diffraction have shown without doubt that Fe is divalent in a tetrahedrally coordinated site, and because the Fe and Mn spectra are so similar, at least most of the Mn must follow the Fe. The Zn spectrum is generally similar, but there are detailed differences which would permit some of the Zn to occupy a different site or sites.

In contrast to this easy qualitative comparison, rigorous mathematical analysis is very difficult, and is just slightly a 'black art'. The problem is that the interfering waves have subtle differences of phase angle. Although they can be calculated to a considerable degree of accuracy, they depend on assumptions about the electronic wave functions. Simple electronic theory of electrons in atoms deals with individual orbitals, but accurate theory must consider the linkage between the individual probabilities. Hence, phases deduced from analysis of spectra from compounds of known structure are useful in improving the accuracy of a quantitative prediction of a radial distribution function. The details are very complex but, generally speaking, the number of atoms in the first coordination can be determined with an

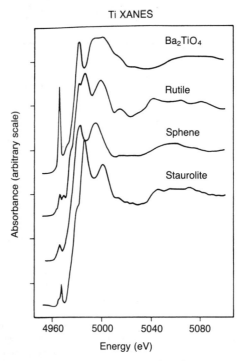

Figure 5.23 Ti XAS spectra used by Henderson et al. (1993) to show that Ti in staurolite is in a distorted octahedron of oxygen atoms.

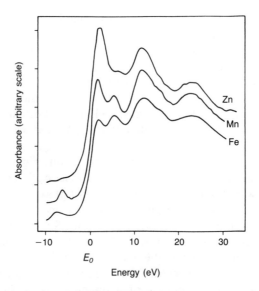

Figure 5.24 Background-subtracted K edge XAS spectra of Fe, Mn, and Zn in a powder of staurolite. The energy scale is referenced to the formal absorption edge for an isolated atom. (From Henderson et al., 1993).

accuracy of ~0.5–1 and the bond length to ~0.01–0.03 Å. Thus for Fe and Mn respectively in staurolite, the values of 3.6, 1.99 Å and 4.0, 2.01 Å quantified the qualitative conclusion already given of divalent atoms occupying a 4-coordinate site. The values of 5.1, 1.95 Å for Zn are somewhat ambiguous, but consistent with predominant divalency and 4-coordinate occupancy.

To conclude, here is another reminder of the likelihood of anisotropy for micro-XAS analysis of a crystal. Simultaneous collection of diffraction data could of course lead to a detailed structural conclusion.

5.6 Illustrative applications and future advances

This section is deliberately brief. The main aim is to show how important geochemical (*sensu lato*) problems are being tackled with synchrotron-based techniques. Many are at the current limits of technology at the existing synchrotron sources.

5.6.1 Coal

Trial XRF measurements with a millimetre beam at CHESS showed that the vitrinite from a drill core in the Emery coalfield, Utah, has a systematic depth trend involving S, Ca, Ti, Fe, Zn, Br and Sr (Chen *et al.*, 1984). However, detailed microanalysis of the individual components has not been undertaken.

Systematic study of petrographically identified plant remains might prove useful to palaeobotanists. To obtain micrometre resolution, the advanced capability of a third-generation source would be needed.

The mineral components of coal are a useful indicator of the chemical deposition from fluids, which have entered the coal bed and its adjacent sedimentary rocks, including shales. In a joint study between the Universities of Sheffield and Chicago, White *et al.* (1989) obtained detection levels of 5–50 ppm for elements with $Z > 25$ (Mn) for a $20 \times 20 \mu$m spot on sulphide minerals (pyrite and marcasite) in East Midlands (UK) coals. A simple energy-dispersive system was used with the NSLS-X26A beam limited by crossed tantalum jaws. Ni, Cu, Zn, Ga, Ge, As (up to 3.4 wt%), Se (up to 1250 ppm), Rb, Sr, Mo (up to 1430 ppm), Tl (up to 370 ppm) and Pb (up to 4100 ppm) were detected. As and Se were closely related. Cu, Ni, As, Se and Pb occured in almost all the sulphides, Zn, Mo and Tl in many, while Ge, Sr and Rb were accidental contaminants. These data showed that environmentally hazardous elements are significant substituents. Determination of their spatial distribution across the highly faulted East Midlands coal basin would require analysis of many more specimens.

The Illinois coal basin is much less disturbed. Small millimetre-sized fractures contain a sequence of minerals which were deposited as early saline fluids were followed by near-surface ground water. Kolker and Chou (1994) characterized the complex trend from measurements of Mn (780–9500 ppm), Sr (<5–460 ppm) and Fe (52–16 700 ppm) in compositionally zoned calcite of the cleats. Detection levels of 5 ppm were obtained for a <10 μm beam at the NSLS-X26A energy-dispersive XRF microprobe.

5.6.2 Cosmic dust

The Earth is continually being bombarded by small particles, many of which have bulk compositions generally similar to those of carbonaceous chondrites. However, the concentrations of certain 'volatile' elements in the chondritic particles are quite different, and may indicate the existence of a more primitive class of chondritic material (Sutton, 1994) closer to the supposed primary composition of stars. Particles may be collected on special 'shock-absorbing' panels carried by high-flying aircraft and satellites, or their survivors after atmospheric heating may be collected from lake beds fed by melting Greenland ice. Interpretation is complicated by the atmospheric heating and perhaps by scavenging of stratospheric aerosols carrying residues from volcanic eruptions and other materials.

The synchrotron energy-dispersive XRF technique is very suitable for determination of the bulk composition of the irregular particles, as long as hard X-rays with low absorption are measured. The following papers report data for S, Cl, K, Ca, Ti, Cr, Mn, Fe, Ni, Cu, Zn, Ga, Ge, Se, Br, Rb, Sr: Chevallier et al. (1987); Flynn and Sutton (1990, 1991, 1992); Flynn, Sutton and Bajt (1993); Klöck et al. (1994); Sutton and Flynn (1988, 1990). The bulk content of Zn generally decreases in particles that have thicker magnetite rims consequent on atmospheric heating. The implication that Zn is lost during the heating is currently being tested. Whether loss of halogens, especially Br, by heating contributes significantly to the photochemical catalytic processes responsible for ozone depletion in the stratosphere is purely speculative at this time. To characterize further the composition of individual components, including heating rims containing magnetite and possible interstellar components, the present femtogram sensitivity level needs to be improved to the attogram level.

5.6.3 Element partitioning and diffusion couples

The partitioning of elements among coexisting phases is a key parameter for deciphering many geological processes, including crystal–liquid differentiation, metamorphic re-equilibration, ore deposition, etc. The SXRF microprobe is particularly useful for trace elements inaccessible to the other microanalytical methods. Furthermore, micro-XAS yields information on the electronic state and stereochemical bonding. Because element partitioning is controlled by diffusion, the SXRF microprobe is used to measure the diffusion profile across an interface. The interface of the diffusion couple should be oriented at 45° to the surface of a polished thin section.

Some exemplary studies are: Baker (1989), tracer versus trace element diffusion applied to Sr isotopes; Baker (1990), chemical interdiffusion between dacite and rhyolite; Baker and Watson (1988), diffusion in complex CI- and F-bearing silicate melts; Carroll *et al.* (1993), Kr diffusion and solubility in silicic glasses; Dalpe, Baker and Sutton (1992), partition of Cr, Rb, Ga, Sr, Y, Nb and Zr between pargasitic amphibole and nephelinic melt at 1.5 GPa; Mavrogenes *et al.* (1993), metal distribution between immiscible fluids.

Sector zoning in minerals is the product of crystallographic control of elemental partitioning across surfaces during anisotropic growth. Differences in trace element composition between portions of symmetrically non-equivalent sectors that were contemporaneous during growth (sector zoning) and between regions contained within a given sector (intrasectoral zoning) have been found in geologically important minerals (apatite, calcite, dolomite, quartz, topaz, zoisite, etc.). Intrasectoral zoning corresponds directly to symmetrically non-equivalent steps that comprise growth hillocks on certain crystal faces. The XRF microprobe allows the zonation of trace elements to be studied with high spatial resolution, as exemplified by a study of topaz (Northrup and Reeder, 1994).

5.6.4 Environmental

The synchrotron microprobe is useful for a wide range of analytical challenges needed for the quantification of environmental problems.

The migration of an element in nuclear waste depends strongly on how it is chemically bonded. Hence coupled XRF and XAS analyses are valuable, especially if high spatial resolution allows pin-pointing of a particular host mineral in a complex matrix. The electronic state of U and other high-Z elements in lake sediments may change during the year in response to chemical events triggered by overturning of the water layers. The bonding of an element in a glass and the resistance to leaching, depend on the electronic state. These and other phenomena can be quantified by synchrotron microprobe studies of carefully chosen samples. The following pioneering studies involve fascinating science, and have important societal implications.

- Chromium and technetium are important in low-level waste disposal. The oxidized species diffuse faster than the reduced ones. Bajt *et al.* (1993) used XANES to determine the ratio of hexavalent to total chromium, and obtained an accuracy of 5% with a minimum detection level of 10 ppm for a 150 μm X-ray beam on the NSLS-X26A microprobe. Much higher combined sensitivity will be obtainable on a third-generation source.
- Sediments around uranium processing facilities generally contain

hexavalent U, but some tetravalent U is also present. Using combined micro-XRF–XANES at 50–300 μm spatial resolution, Bertsch et al. (1994) found that hexavalent U is associated mainly with Fe and Mn, perhaps on the surface of oxyhydroxide minerals. The tetravalent U, and some of the hexavalent U, was correlated with Cu, perhaps indicating the presence of a discrete Cu, U-containing phase. The analyses were made on soils and sediments sprinkled lightly on Kapton adhesive tapes.

A second area involves aerosols and fly ash in the atmosphere. Careful choice of samples, both in space and time, is important to maximize the value of the synchrotron analyses. The following studies point the way to comprehensive programmes.

- Tuniz et al. (1994) determined the concentrations of Ti, V, Cr, Mn, Fe, Ni, Cu, Zn, As, Pb and Sr in particles (1–200 μm) of fly ash from power stations in Italy and Hungary. Maps were made with a 4 × 7 μm beam on 15 μm thin sections at the NSLS-X26A microprobe. Indications that some elements are concentrated at the surface require further evaluation (see also Giauque, Garrett and Goda 1977(b)).
- The chemical composition of atmospheric aerosols is important for human health and the vigour of plants and trees. The days of 'pea-souper' fogs have gone in the UK and the USA (although not in all countries), as the burning of soft coal in open fires has been replaced by natural gas and as some coal power stations have been retrofitted with scrubbers. Smog from photochemical oxidation of hydrocarbons from automobiles and other sources remains a major problem in the industrialized world. The residue from volcanic eruptions descends from the stratosphere into the troposphere to increase the haze from industrial and other sources. As frontal systems move around the world the chemistry of the aerosols changes. Atmospheric chemists need to monitor the changes in the aerosols, and synchrotron-XRF provides a useful tool. At the simplest level, a filter loaded with captured aerosol can be analysed chemically with a broad beam to the ppm level. At a more advanced level, one can envisage chemical identification of individual droplets from an aerosol. Grant, Schulze and Sutton (1994) measured K, Ca, Ti, Cr, Mn, Fe, Ni, Cu and Zn in aerosols from rural Indiana and found that the concentrations were higher when air moved in from the Gulf of Mexico. Anthropogenic elements appear to be higher when snow covers the incoming path, while elements associated with soil minerals are lower. Török et al. (1994) compared urban and background aerosols from Hungary. In addition to XRF analysis, one can envisage XAS study of the electronic state and chemical bonding.

5.6.5 Inclusions in minerals

Analysis of trapped melts and vapours in minerals provides major analytical challenges. Ingenious methods include mechanical release into a mass spectrometer or electron probe microanalysis of a frozen sample. Synchrotron XRF and XAS provide interesting possibilities that are particularly useful for medium- to high-Z elements accessible to hard X-rays. Ideally, the host mineral should have a simple composition with a uniform content of light elements. A polished thin section or cleavage fragment should be prepared with some inclusions intact and just under the surface. Electron probe analysis of a frozen inclusion less than $2\,\mu$m or so from the surface would provide useful complementary information to the SXRF data. The following are representative of SXRF studies.

- Frantz et al. (1988) tested the technique on synthetic inclusions $\sim 10\,\mu$m across of known fluid composition, containing Ca-, Mn- and Zn-chlorides, enclosed by quartz.
- Lowenstern et al. (1991) analysed melt inclusions in late-stage quartz phenocrysts from the Pantelleria rhyolite. Devitrified inclusions of sanidine, cristobalite and alkali amphibole were remelted at 1098 K and quenched to form a glass with small vapour bubbles. SXRF analysis with a $10\,\mu$m scanning beam at NSLS-X26A revealed 3 ppm Cu in the glasses, and local concentrations some hundred times greater in the vapour bubbles. The strong partitioning of copper sulphide into the CO_2-rich vapour is important for modelling the formation of porphyry copper deposits from vapours rising from silicic magmas.
- Hypersaline fluids containing Fe, S, Pb, P, Li, Sn, Cu, Be, Sr and Be were studied by laser-ICP and SXRF analysis of inclusions in topaz of the Sn-W-Cu-Pb-Zn-mineralized Mole granite (Rankin et al., 1992).
- SXRF analyses of brine inclusions from rocks dredged at mid-oceanic ridges (Vanko et al., 1993, 1994) are important because the major transfer of heat from the Earth's interior is associated with hydrothermal fluids. Black smokers are the most dramatic consequence of the escaping fluids, and reconnaissance SXRF analyses indicate the potential for comprehensive studies (Campbell, Rivers and Sutton, 1989; Campbell et al., 1990). The inclusions in metagabbros from the dormant Mathematician Ridge in the eastern Pacific are $NaCl > FeCl_2 > CaCl_2$-brines rich in Fe (10^4–10^5 ppm) and with significant Zn (200–1500 ppm) and Cu (100–1400 ppm). Those from the Oceanographer Transform in the Mid-Atlantic Ridge have variable Fe (0–90 000 ppm), generally higher Zn (650–6600 ppm) and lower Cu (0–500 ppm). Both sets of inclusions show positive correlation between chlorine and Br, Fe, Mn, Cu, K, Ca and Na. The general implication of these exploratory data is that the overall geochemical signature arising from interaction of hot basalt with sea

water is coupled with local signatures related to mantle source regions of basalts and associated fluids.
- The first XAS study of synthetic fluid inclusions by Anderson *et al.* (1993) augurs well for application to selected natural inclusions.

5.6.6 Earth's mantle

Some rock and mineral fragments in kimberlites and alkali-basalts contain inclusions which have been preserved from alteration during transport. Particularly interesting are inclusions in diamonds. Sutton *et al.* (1992, 1993(a)) measured the Cr^{2+}/Cr^{3+} ratio in several olivine inclusions by synchrotron micro-XANES. These reconnaissance data indicate that the redox state varies among the olivine inclusions and that a systematic study of carefully selected diamonds appears worthwhile.

SXRF data were obtained at NSLS-X26A for many sulphide inclusions carefully selected from peridotite and other xenoliths characterized by J.B. Dawson and J.V. Smith. Unfortunately, the presence of numerous micrometre spots (possibly altered oxides derived from O dissolved in the primary sulphide liquid) containing Rb and other elements indicative of infiltration by kimberlitic fluid casts doubt on the value of the SXRF data taken with a spatial resolution of $\geq 10\,\mu$m. When better spatial resolution is obtainable at the APS, these sulphide inclusions should be re-examined to determine whether 'primary' compositions can be extracted from modal summation of the individual sulphides.

5.6.7 Meteorites and the Moon

SXRF and SXAS are very useful for the analysis of meteorites and lunar rocks and minerals because of the complex micro-textures and the rarity of samples. The following are exemplary studies.

- Nickel and copper correlate positively with associated kamacite and taenite lamellae in iron meteorites (Sutton, Rivers and Smith, 1987). The Cu distribution between troilite and metal changes greatly for the band width of kamacite, ranging from 10-fold higher in troilite at 0.05 mm to threefold higher in kamacite at 1–10 mm. Nickel shows a similar variation, but shifted overall about sixfold in favour of kamacite. Since magmatic evolution and partial melting should produce Cu-enriched troilite only, based on experimental distribution coefficients, these SXRF measurements indicate migration of Cu to exsolving metal during subsolidus equilibration. Saito *et al.* (1993) determined ruthenium by SXRF using a monochromatized beam 0.25 mm across at 22.5 and 23 keV. The concentration variation from 3 to 33 ppm with respect to Ni allows assignment of iron meteorites to groups IA or IIA.

- Diverse groups of coloured spheres of overall Mg-rich basaltic composition at the lunar surface have been ascribed to volcanic fire fountains driven by deep-seated eruptions. SXRF analyses (Delano, Sutton and Smith, 1991) reveaded a higher Zn content than for the crystalline mare basalts that flooded the impact basins. Hence, it is not unreasonable to speculate that the volcanic source regions in the Moon are richer in Zn and other volatile elements at greater depth. Further SXRF and SXAS studies are needed to characterize the inferred multiple reservoirs. Perhaps the currently fashionable hypothesis that the Moon is essentially the product of a single glancing impact that tore off part of the Earth may need to be extended to cover multiple reservoirs related to events that include capture of planetesimals and proto-moons.
- The presence of Fe, Ni inclusions in plagioclase of coarse-grained lunar rocks is just one of the many pieces of evidence for a strongly reduced state of the accessible exterior. A wide range of redox conditions is implied by the coexisting mineral assemblages in the many types of meteorites (e.g. enstatite to ordinary chondrites), including those thought to have been blasted off Mars and the Moon. Synchrotron-XAS allows quantification of the electronic state and chemical bonding of transition elements in carefully selected small regions. Because each transition metal of the first series has its own characteristic redox property, an oxygen fugacity grid was proposed by Delaney, Sutton and Bajt (1994) for use in nebular and planetary geochemistry. The demonstration that the Cr in olivine from an Apollo 15 basalt is entirely divalent (Sutton *et al.*, 1993(c)) was particularly accessible to current S-XANES instrumentation.
- Other examples of straightforward SXRF analyses are trace elements in plagioclase from Apollo 16 breccias (Delaney, Sutton and Smith, 1989); abundances of Ni, Cu, Zn and Ga in magmatically zoned pyroxene from the Zagami shergottite basalt, believed to have crystallized on Mars prior to being shock-metamorphosed during ejection (Treiman and Sutton, 1992). The analysis of Ni, Cu, Zn, Ga, Ge and Se in the rims of chondrules in an unequilibrated CO3 carbonaceous chondrite (Brearley, Bajt and Sutton, 1994) is illustrative of the increasing technical achievement. The rims are compositionally homogeneous, and their richness in volatile elements implies a source distinct from that of the chondrule cores.

5.6.8 Ore mineralogy

The SXRF microprobe is particularly valuable for the study of sulphide ore minerals because of its good sensitivity for medium- to high-Z elements, including Fe-Se, Mo-Te and Re-Bi. This sensitivity will improve as the intense, high-energy beams from the third-generation synchrotron sources become available. Most sulphide minerals in igneous and metamorphic rocks occur as tiny grains in the interstices of the coarser silicate grains, and are often aggregated or intergrown with other minerals. Thus, the high-temperature M-type Fe-Ni-Cu solid solution is typically represented at low temperature by a micrometre-scale intergrowth of pyrrhotite and pentlandite rimmed by atolls of chalcopyrite. Metasomatic ore deposits commonly have very complex textures and intergrowths ranging down to the sub-micrometre level. SXRF analysis of minor and trace elements in associated gangue minerals such as calcite is generally easier than for the sulphide minerals because of the lightness of the major elements.

The following give a useful guide to the growing potential of SXRF analysis.

- Analysis of pyrite and marcasite in coal (section 5.6.1).
- Analysis of fluid inclusions related to porphyry copper deposits (section 5.6.5; Bodnar *et al.*, 1993).
- Location of gold in Carlin-type ore deposits. Chen *et al.* (1987) showed that the gold is not spatially correlated with the pyrite. (Other techniques with sub-micrometre sensitivity have identified tiny Au grains, mostly dispersed through the matrix).
- Zoned carbonate gangue in Mississippi Valley-type deposits. Kopp *et al.* (1990) correlated Fe and Mn with the cathodoluminescence. A combination of cathodoluminescence microscopy, which is very rapid, with selective SXRF analysis, which is slower and more expensive, would allow rigorous testing of the idea that the gangue-forming fluids have a systematic spatial and temporal trend across the sedimentary basin.
- Ore metals in titanate minerals from the Tuolomne Intrusive Suite, California (Candela, Piccoli and Rivers, 1991).
- Rare earth elements, with reconnaissance analysis of oxides in the Bayan Obo deposits, China (Chen *et al.*, 1993).

5.6.9 Soil mineralogy and agriculture

A soil is a complex assemblage of inorganic and organic constituents whose properties depend on the source rocks, climate and meteorology and the time of year. Soils used for agriculture undergo modification depending on the type of crop and intensity of cultivation. Trace elements in soils are affected by microbial activity and by external factors, including the chemistry of rain and aerosols (section 5.6.4). Pioneering applications include the following.

- Take-all disease of wheat (rotting of root and foot) is caused by the fungus *Gaeumannomyces graminis* var. *tritici*. The hypothesis that the fungus oxidized Mn in the rhizosphere prior to invasion by the pathogen was tested by micro-SXAS (Schulze *et al.*, 1993). Synchrotron-XANES was also used to measure the redox state of Mn in soils (Schulze, Sutton and Bajt, 1994). Roots of wheat seedlings grown in potato dextrose agar with up to 100 ppm $MnSO_4$ were studied by micro-XANES with a resolution of 200 μm. Mn was found to be mainly divalent in clear agar and uninfected roots and mainly tetravalent and 10 times more concentrated in the infected roots. This spectacular demonstration is at the limit of current technology and the need for a third-generation synchrotron source is absolutely clear. Combined synchrotron XRF and XAS techniques should yield interesting chemical information on the Mn-rich nodules that occur in certain soils. Comparison with data for nodules in oceans and lakes should also be fascinating.

- The Se content in soils has become of environmental concern. Prolonged flow of water into an evaporation basin without an outflow can cause build-up of elements such as Se to toxic levels, as at the Kesterson Reservoir, California. Tokunaga, Sutton and Bajt (1994) demonstrated the potential of synchrotron XRF microprobe mapping for characterization of heterogeneity of Se in soil aggregates. In an experimental cell, containing an artificial soil aggregate soaked in saline solution with 240 mg kg^{-1} Se(VI), decaying roots of *Scirpus robustus* and *S. californicus* were found to concentrate the Se up to 20-fold. The Se becomes reduced to the metallic state.

Acknowledgements

It has been our privilege to collaborate with both physicists at the Department of Applied Science of Brookhaven National Laboratory (particularly Keith Jones), geoscientists of GeoSync, the national user's organization in the United States of America, and soil and environmental scientists (particularly Darrell Schulze) belonging to Soil/EnviroCARS. This chapter benefits from the advances in both technology and research that resulted from the combined efforts that are now moving to an even higher level. Special thanks go to Alan Gaines and Daniel Weill (NSF-EAR-Instrumentation). William Luth (DOE–Basic Energy Sciences–Geochemistry) and Donald Bogard (NASA–Planetary Science) for their keen insight into the scientific basis for funding the instrumental development and accompanying pioneering research. Stephen Sutton has worked wonders for the users of the NSLS-X26A beamline and has provided detailed help with the manuscript: he should really have been a co-author. Instead, he has been preparing a review article 'Earth science research using synchrotron radiation' for *Reviews in Geophysics*. Finally, the development programme at NSLS-X26A under the CARS umbrella could not have been successful without the firm support of Dean Stuart Rice and Fred Stafford at the University of Chicago. We also thank our many colleagues in CARS, particularly Nancy Weber, who provided the key administrative support in the pioneering days.

Invitation

Contact JVS if you would like to join one of the GeoSync organizing groups that represent geoscientists using synchrotron facilities around the world. We need to get in touch with research/graduate students and advanced undergraduates so that they can be referred to appropriate mentors, invited to hands-on workshops and helped to learn which analytical techniques are appropriate to their research interests.

Appendix 5.A Notation and abbreviations

Abbreviations

AEM	analytical electron microscopy
APS	Advanced Photon Source at Argonne National Laboratory
BM	bending magnet
CHESS	Cornell High Energy Synchrotron Source
DAFS	diffraction absorption fine structure
EPMA	electron probe microanalysis
EXAFS	extended X-ray absorption fine structure
ESRF	European Synchrotron Radiation Facility at Grenoble
ID	insertion device (linear array of magnets)
NRA	nuclear reaction analysis
NSLS	National Synchrotron Light Source at Brookhaven National Laboratory
PIGE	particle (proton) induced gamma-ray emission
PIXE	particle (proton) induced X-ray emission
PRT	Participating Research Team of scientists at an NSLS beamline
RBA	Rutherford back-scatter analysis
SIMS	secondary ion mass spectrometry
SRIXE	synchrotron radiation induced X-ray emission
SRS	Synchrotron Radiation Source at Daresbury Laboratory
SSRL	Stanford Synchrotron Radiation Laboratory
SXFMA	synchrotron X-ray fluorescence microanalysis
SXAS	synchrotron X-ray absorption spectroscopy
SXRD	synchrotron X-ray diffraction
SXRF	synchrotron X-ray fluorescence
XANES	X-ray absorption near edge spectroscopy
XRF(A)	X-ray fluorescence (analysis)

REFERENCES

Notation

c	velocity of electromagnetic radiation in vacuum
B	magnetic field in tesla (T)
B_0	peak magnetic field of an undulator
B_n	brightness of nth harmonic of an undulator
E	kinetic energy, expressed in electron volt
$F_n(K_{id})$	function relating brightness of nth undulator harmonic to K_{id}
GeV	giga electron volt
$H_2(\varepsilon/\varepsilon_c)$	function expressing photon distribution from bending magnet
h	Planck's constant
I	current in electron ring, expressed in ampere (A)
K	used for series of X-ray spectra related to K electron shell
K_{id}	deflection parameter of insertion device
keV	kilo electron volt
L	series of X-ray spectra related to L electron shell
L	length of undulator = $N \times$ spacing of magnets (λ_{id})
M	series of X-ray spectra related to M electron shell
N	number of magnet pairs in an insertion device
n	order of harmonic in an undulator X-ray beam
m_0	rest mass of particle (e.g. electron)
Z	atomic number
γ	normalized energy parameter
ε	energy of photon
ε_c	critical energy of photons from synchrotron bending magnet
ε_n	photon energy of nth harmonic from undulator
θ	angle of divergence in horizontal direction
λ_n	photon wavelength of nth harmonic of undulator
λ_{id}	period of insertion device = spacing of magnets in cm
p	radius of electron orbit
$\sigma_{\gamma'}$	*characteristic opening angle of conical X-ray beam from undulator*
$\sigma_{x'}$	actual opening angle for undulator beam along horizontal x
$\sigma_{z'}$	actual opening angle for undulator beam along vertical z
ν	frequency of photon
Ψ	angle of divergence in vertical direction

References

Alp, E.E., Mini, S.M. and Ramanathan, M. (1990) X-ray absorption spectroscopy: EXAFS and XANES – a versatile tool to study the atomic and electronic structure of materials, in *Synchchrotron X-ray Sources and New Opportunities in the Soil and Environmental Sciences: Workshop Report* (eds D.G. Schulze and J.V. Smith), Argonne National Laboratory/Advanced Photon Source/TM-7, pp. 25–36.

Anderson, A.J., Mayanovic, R.A., Bodnar, R.J. *et al.* (1993) The first study of aqueous species in inclusion fluids using X-ray absorption spectroscopy. *Geol. Soc. Am. Abstr. Progr.*, **25**, A316.

Bajt, S., Clark, S.B., Sutton, S.R. *et al.* (1993) Synchrotron X-ray microprobe determination of chromate content using X-ray absorption near edge structure (XANES). *Anal. Chem.*, **65**, 1800–4.

Baker, D.R. (1989) Tracer *versus* trace element diffusion: diffusional decoupling of Sr concentrations from Sr isotope composition. *Geochim. Cosmochim. Acta*, **53**, 3015–23.

Baker, D.R. (1990) Chemical interdiffusion of dacite and rhyolite: anhydrous measurements at 1 atm and 10 kbar, application of transition state theory and diffusion in zoned magma chambers. *Contrib. Mineral. Petrol.*, **104**, 407–23.

Baker, D.R. and Watson, E.B. (1988) Diffusion of major and trace elements in compositionally complex Cl- and F-bearing silicate melts. *J. Non-Crystalline Solids*, **102**, 62–70.

Bertsch, P.M., Hunter, D.B., Sutton, S.R. *et al.* (1994) *In situ* chemical speciation of uranium in soils and sediments by micro X-ray absorption spectroscopy. *Environ. Sci. Technol.*, in press.

Bodnar, R., Mavrogenes, A., Anderson, A. *et al.* (1993) Synchrotron XRF evidence for the sources and distributions of metals in porphyry copper deposits. *Eos*, **74**, 669.

Brearley, A.J., Bajt, S. and Sutton, S.R. (1994) Metamorphism in the CO3 chondrites: trace element behaviour in matrices and rims. *Lunar Planet. Sci.*, **XXV**, 165–6.

Brown, G.E. Jr and Parks, G.A. (1989) Synchrotron-based X ray absorption studies of cation environments in Earth materials. *Rev. Geophys.*, **27**, 519–33.

Campbell, A.C., Rivers, M.L. and Sutton, S.R. (1989) Trace element composition of hydrothermal plume particles by synchrotron X-ray fluorescence spectroscopy. *Eos*, **70**, 494.

Campbell, A.C., Cramer, S., Rivers, M.L. and Sutton, S.R. (1990) Applications of synchrotron X-ray spectroscopy to the geochemistry of hydrothermal plume particles. *Eos*, **71**, 1621.

Candela, P.A., Piccoli, P.M. and Rivers, M.L. (1991) Synchrotron X-ray fluorescence study of ore metals in titanates from the Tuolomne Intrusive Suite, California. *Geol. Soc. Am. Abstr. Progr.*, **23**, A465.

Carroll, M.R., Sutton, S.R., Rivers, M.L. and Woolum, D.S. (1993) An experimental study of krypton diffusion and solubility in silicic glasses. *Chem. Geol.*, **109**, 9–28.

Chen, J.R., Chao, E.C.T., Minkin, J.A. *et al.* (1987) Determination of the occurrence of gold in an unoxidized Carlin-type ore sample using synchrotron radiation. *Nucl. Instrum. Methods. Phys. Res.*, **B22**, 394–400.

Chen, J.R., Chao, E.C.T., Back, J.M. *et al.* (1993) Rare earth element concentrations in geological and synthetic samples using X-ray fluorescence analysis. *Nucl. Instrum. Methods Phys. Res.*, **B75**, 576–81.

Chen, J.R., Martys, N., Chao, E.C.T. *et al.* (1984) Synchrotron radiation determination of elemental concentrations in coal. *Nucl. Instrum. Methods. Phys. Res.*, **B231**, 241–5.

Chevallier, P., Jehanno, C., Maurette, M. *et al.* (1987) Trace element analyses of spheres from the melt zone of the Greenland Ice Cap using synchrotron x-ray fluorescence. *J. Geophys. Res.*, **92**, E649–56.

Chiarello, R.O., Wogelius, R.A. and Sturchio, N.C. (1993) *In-situ* synchrotron X-ray reflectivity measurements at the calcite-water interface. *Geochim. Cosmochim. Acta*, **57**, 4103–10.

Coppens, P. (1992) *Synchrotron radiation crystallography*, Academic Press, London, 316 pp.

Cousins, C.S.G., Gerward, L., Laundy, D. *et al.* (1993) Pile-up phenomena in solid-state detection. Effects due to synchrotron sources. *Nucl. Instrum. Methods Phys. Res.*, **A324**, 598–608.

Dalpe, C., Baker, D.R. and Sutton, S.R. (1992) Partition coefficient of Cr, Rb, Ga, Sr, Y, Nb, and Zr between pargasite and nephelinic melt at 1.5 GPa. *Eos*, **73**, 607.

REFERENCES

Delaney, J.S., Sutton, S.R. and Bajt, S. (1994) An oxygen fugacity grid for nebular and planetary geochemistry. *Lunar Planet. Sci.*, **XXV**, 323–4.

Delaney, J.S., Sutton, S.R. and Smith, J.V. (1989) Trace elements in plagioclase from three Apollo 16 breccias. *Lunar Planet. Sci.*, **XX**, 238–9.

Delano, J.W., Sutton, S.R. and Smith, J.V. (1991) Lunar volcanic glasses: trace element abundances in individual spheres using synchrotron X-ray fluorescence. *Lunar Planet. Sci.*, **XXII**, 311–2.

Flynn, G.J. and Sutton, S.R. (1990) Synchrotron X-ray fluorescence analyses of stratospheric cosmic dust: new results for chondritic and low-nickel particles. *Proc. 20th Lunar Planet. Sci. Conf.*, Lunar and Planetary Institute, Houston, pp. 335–42.

Flynn, G.J. and Sutton, S.R. (1991) Chemical characterization of seven large area collector particles by SXRF. *Proc. 21st Lunar Planet. Sci. Conf.*, Lunar and Planetary Institute, Houston, pp. 549–56.

Flynn, G.J. and Sutton, S.R. (1992) Trace elements in chondritic stratospheric particles: zinc depletion as a possible indicator of atmospheric entry heating. *Proc. 22nd Lunar Planet. Sci. Conf.*, Lunar and Planetary Institute, Houston, pp. 171–84.

Flynn, G.J., Sutton, S.R. and Bajt, S. (1993) Trace element content of chondritic cosmic dust: volatile enrichments, thermal alterations, and the possibility of contamination. *Lunar Planet. Sci.*, **XXIV**, 495–6.

Frantz, J.D., Mao, H.K., Zhang, Y.-G. *et al.* (1988) Analysis of fluid inclusions by X-ray fluorescence using synchrotron radiation. *Chem. Geol.*, **69**, 235–44.

Giauque, R.D., Garrett, R.B. and Goda, L.Y. (1977(b)) Determination of forty elements in geochemical samples and coal fly ash by X-ray fluorescence spectrometry. *Anal. Chem.*, **49**, 1012–17.

Grant, R.H., Schulze, D.G. and Sutton, S.R. (1994) Influence of snow cover on the background trace element aerosol of rural Indiana. *Atmospheric Environment*, submitted.

Henderson, C.M.B., Charnock, J.M., Smith, J.V. and Greaves, G.N. (1993) X-ray absorption spectroscopy of Fe, Mn, Zn, and Ti structural environments in staurolite. *Am. Mineral.*, **78**, 477–85.

Jones, K.W. (1992) Synchrotron radiation-induced X-ray emission, in *Handbook of X-Ray Spectrometry* (eds R.E. Van Grieken and A.A. Markowitz), Marcel Dekker, New York, pp. 411–52.

Jones, K.W. and Gordon, B.M. (1989) Trace element determinations with synchrotron-induced X-ray emission. *Anal. Chem.*, **61**, A341–58.

Kajfosz, J. and Kwiatek, W.M. (1987) Non-polynomial approximation of background in X-ray spectra. *Nucl. Instrum. Methods*, **B22**, 78–81.

Klöck, W., Flynn, G.J., Sutton, S.R. *et al.* (1994) Heating experiments simulating atmospheric entry of micrometeorites. *Lunar Planet. Sci.*, **XXV**, 713–4.

Kolker, A. and Chou, C.-L. (1994) Cleat-filling calcite in Illinois Basin coals: trace-element evidence for meteoric fluid migration in a coal basin. *J. Geol.*, **102**, 111–6.

Kopp, O.C., Reeves, D.K., Rivers, M.L. and Smith, J.V. (1990) Synchrotron X-ray fluorescence analysis of zoned carbonate gangue in Mississippi Valley-type deposits (USA). *Chem. Geol.*, **81**, 337–47.

Lowenstern, J.B., Mahood, G.A., Rivers, M.L. and Sutton, S.R. (1991) Evidence for extreme partitioning of copper into a magmatic vapor phase. *Science*, **252**, 1405–9.

Lu, F.-Q., Smith, J.V., Sutton, S.R. et al. (1989) Synchrotron X-ray fluorescence analysis of rock-forming minerals; 1. Comparison with other techniques, 2. White-beam energy-dispersive procedure for feldspars. Chem. Geol., 75, 123–43.

Mavrogenes, J., Bodnar, R., Anderson, A. et al. (1993) Metal distributions in immiscible fluids: evidence from synchrotron XRF analyses of synthetic fluid inclusions. Eos, 74, 669.

Northrup, P. and Reeder, R. (1994) Evidence for the importance of growth-surface structure to trace element incorporation in topaz. Am. Mineral., submitted.

Potts, P.J. (1987) A Handbook of Silicate Rock Analysis, Blackie, Glasgow and London.

Rankin, A.H., Ramsey, M.H., Coles, B. et al. (1992) The composition of hypersaline, iron-rich granitic fluids based on laser-ICP and synchrotron-XRF microprobe analysis of individual fluid inclusions in topaz, Mole granite, eastern Australia. Geochim. Cosmochim. Acta, 56, 67–79.

Rivers, M.L. (1988) Characteristics of the Advanced Photon Source and comparison with existing synchrotron facilities, in Synchrotron X-ray Sources and New Opportunities in the Earth Sciences: Workshop Report (eds J.V. Smith and M.H. Manghnani), Argonne National Laboratory/Advanced Photon Source/TM-3.

Saito, J., Nakamura, T., Yamaguchi, A. et al. (1993) Analysis of trace ruthenium in meteoritic FeNi minerals by microbeam XRF using synchrotron radiation. Mineral. J., 16, 258–267.

Schulze, D.G., Sutton, S.R. and Bajt, S. (1994) Measurement of Mn oxidation state in soils using X-ray absorption near-edge structure (XANES) spectroscopy. Soil Sci. Soc. Am. J., submitted.

Schulze, D.G., McCay-Buis, T., Sutton, S.R. and Huber, D.M. (1993) Manganese oxidation states in the rhizosphere of wheat roots infected with the take-all fungus, Gaeuannomyces graminis var. tritici. Agronomy Abstracts, S9–18.

Sutton, S.R. (1994) Chemical compositions of primitive solar system particles. Proc. Interplanet. Dust Particles Workshop, preprint.

Sutton, S.R. and Flynn, G.J. (1988) Stratospheric particles: synchrotron X-ray fluorescence determination of trace element contents. Proc. 18th Lunar Planet. Sci. Conf., Cambridge University Press, pp. 607–14.

Sutton, S.R. and Flynn, G.J. (1990) Extraterrestrial halogen and sulfur content of the stratosphere. Proc. 20th Lunar Planet. Sci. Conf., pp. 357–61.

Sutton, S.R., Rivers, M.L. and Smith, J.V. (1987) Applications of sychrotron X-ray fluorescence to extraterrestrial materials. Nucl. Instrum. Methods, B24/25, 405–9.

Sutton, S.R., Spanne, P., Rivers, M.L. and Jones, K.W. (1992) Computed microtomography (CMT) of extraterrestrial objects using a linear photodiode detector. Lunar Planet. Sci., XXIII, 1393–4.

Sutton, S.R., Bajt, S., Rivers, M.L. and Smith, J.V. (1993(a)) X-ray microprobe determination of chromium oxidation state in olivine from lunar basalt and kimberlitic diamond. Lunar Planet. Sci., XXIV, 1383–4.

Sutton, S.R., Delaney, J., Bajt, S. et al. (1993(b)) Microanalysis of iron oxidation state in iron oxides using X-ray absorption near edge structure. Lunar Planet. Sci., XXIV, 1385–6.

Sutton, S.R., Jones, K.W., Gordon, B. et al. (1993(c)) Reduced chromium in olivine grains from lunar basalt 15555: X-ray absorption near edge structure (XANES). Geochim. Cosmochim. Acta, 57, 461–8.

Tokunaga, T.K., Sutton, S.R. and Bajt, S. (1994) Heterogeneous selenium distributions within

individual soil aggregates: evidence for anaerobic microsites obtained with synchrotron X-ray fluorescence microprobe mapping. *Soil Sci.*, in press.
Török, Sz., Sindor, C., Xhoffer, C. *et al.* (1994) X-ray microprobe studies of Hungarian background and urban aerosols. *Adv. X-ray. Anal.*, **35**, in press.
Treiman, A.H. and Sutton, S.R. (1992) Petrogenesis of the Zagami meteorite: inferences from synchrotron X-ray (SXRF) microprobe and electron microprobe analyses of pyroxenes. *Geochim. Cosmochim. Acta*, **56**, 4059–74.
Tuniz, C., Jones, K.W., Rivers, M.L. *et al.* (1994) Elemental characterization of individual fly ash particles from power plants using the NSLS synchrotron radiation X-ray microprobe. *Environ. Sci. Technol.*, submitted.

Vanko, D.A., Sutton, S.R., Rivers, M.L. and Bodnar, R.J. (1993) Major-element ratios in synthetic fluid inclusions by synchrotron X-ray fluorescence microprobe. *Chem. Geol.*, **109**, 125–34.
Vanko, D.A., Sutton, S.R., Ghazi, A.M. and Cline, J.S. (1994) Synchrotron X-ray fluorescence microprobe analyses of individual fluid inclusions from two oceanic gabbros and the Questa molybdenum deposit. *Eur. J. Mineral.*, submitted.

White, R.M., Smith, J.V., Spears, D.A. *et al.* (1989) Analysis of iron sulphides from UK coal by synchrotron radiation X-ray fluorescence. *Fuel*, **68**, 1480–6.

Further reading

Note: only a few papers are listed in the text. A full complement of peer-reviewed papers is referenced so that individual details can be located. The literature review is essentially comprehensive to March 1994. Later references will be listed in a near-weekly newsletter of news items and scientific papers of interest to members of CARS, the Consortium for Advanced Radiation Sources. This is sent out only by e-mail from SMITH@GEO1. UCHICAGO.EDU. Send your address to receive the newsletter automatically. Please do not request a written copy. JVS welcomes preprints and reprints.

Ahmedali, S.T. (ed.) (1989) *X-ray Fluorescence Analysis in the Geological Sciences: Advances in Methodology*. Geol. Assoc. Can. Short Course, **7**, 297 pp.
Arnold, J.R., Testa, J.P. Jr, Friedman, P.J. and Kambic, G.X. (1983) Computed tomographic analysis of meteorite inclusions. *Science*, **219**, 383–4.
Artioli, G., Sacchi, M., Balerna, A. *et al.* (1991) XANES studies of Fe in pumpellyite-group minerals. *Neues. Jahrb. Mineral. Monatsh.*, 413–21.
Attwood (1992) New opportunities at soft-x-ray wavelengths. *Physics Today*, August, 24–31.

Bajt, S. (organizer) (1993) *Workshop on X-ray Microfocusing Techniques and Applications.* NSLS, May 19. (Contact SB at Bldg 815, Brookhaven National Laboratory, Upton NY 11973 for copy of graphics.)
Bajt, S., Sutton, S.R. and Delaney, J.S. (1994) Microanalysis of iron oxidation states in silicates and oxides using X-ray absorption near edge structure (XANES). *Geochim. Cosmochim. Acta*, in press.
Baruchel, J., Hodeau, J.-L., Lehmann, M.S. *et al.* (1993) *Neutron and Synchrotron Radiation for Condensed Matter Studies. Vol. 1. Theory, Instruments and Methods*, Springer, Berlin.
Bassett, W.A. and Brown, G.E. Jr (1990) Synchrotron radiation in the Earth Sciences. *Ann. Rev. Earth Planet. Sci.*, **18**, 387–447.
Beno, M. and Rice, S. (eds) (1989) *Chemical Applications of Synchrotron Radiation: Workshop Report*. Argonne National Laboratory/Advanced Photon Source/TM-4.

Bidoglio, G., Gibson, P.N., O'Gorman, M. and Roberts, K.J. (1993) X-ray absorption spectroscopy investigation of surface redox transformations of thallium and chromium on colloidal mineral oxides. *Geochim. Cosmochim. Acta*, **57**, 2389–94.

Bockman, R.S., Repo, M.A. Warrell, R.P. Jr *et al.* (1990) Distribution of trace levels of therapeutic gallium in bone as mapped by synchrotron X-ray microscopy. *Proc. Natl Acad. Sci. USA*, **87**, 4149–53.

Bowen, D.K., Davies, S.T. and Ambridge, T. (1985) Quantitative analysis of arsenic-implanted layers in silicon by synchrotron-radiation-excited X-ray fluorescence. *J. Appl. Phys.*, **58**, 260–3.

Brissaud, I., Wang, J.X. and Chevallier, P. (1989) Synchrotron radiation induced X-ray fluorescence at LURE. *J. Radioanalyt. Nucl. Chem.*, **131**, 399–413.

Brown, G. and Moncton, D.E. (eds) (1991) *Handbook of Synchrotron Radiation*, **3**, North Holland, Amsterdam.

Brown, G.E. Jr (1990) Spectroscopic studies of chemisorption reaction mechanisms at oxide/water interfaces. *Reviews in Mineralogy*, **23**, 309–63.

Brown, G.E. Jr, Calas, G., Waychunas, G.A. and Petiau, J. (1988) X-ray absorption spectroscopy and its applications to mineralogy and geochemistry. *Rev. Mineral.*, **18**, 431–512.

Carlson, W.D. and Denison, C. (1992) Mechanisms of porphyroblast crystallization: results from high-resolution computed X-ray tomography. *Science*, **257**, 1236–9.

Charnock, J.M., Garner, C.D., Pattrick, R.A.D. and Vaughan, D.J. (1989) EXAFS and Mössbauer spectroscopic study of Fe-bearing tetrahedrites. *Mineral. Mag.*, **53**, 193–9.

Charnock, J.M., Garner, C.D., Pattrick, R.A.D. and Vaughan, D.J. (1989) Coordination sites of metals in tetrahedrite minerals determined by EXAFS. *J. Solid State Chem.*, **82**, 279–89.

Cheetham, A.K. and Wilkinson, A.P. (1992) Synchrotron X-ray and neutron diffraction studies in solid-state chemistry. *Angew. Chem. Int. Ed. Engl.*, **31**, 1557–70.

Chen, J.R., Chao, E.C.T., Minkin, J.A. *et al.* (1990) The uses of synchrotron radiation for elemental and chemical microanalysis. *Nucl. Instrum. Methods*, **B49**, 533–43.

Chong, N.-S., Norton, M.L. and Anderson, J.L. (1990) Multielement trace metal determination by electrodeposition, scanning, electron microscopic X-ray fluorescence and inductively coupled plasma mass spectrometry. *Anal. Chem.*, **62**, 1043–50.

Cohen, L.H. and Smith, D.K. (1989) Thin specimen X-ray fluorescence analysis of major elements in silicate rocks. *Anal. Chem.*, **61**, 1837–40.

Cooke, P.M. (1990) Chemical microscopy. *Anal. Chem.*, **62**, R423–41.

Cressey, G., Henderson, C.M.B. and van der Laan, G. (1993) Use of L-edge X-ray absorption spectroscopy to characterize multiple valence states of 3d transition metals: a new probe for mineralogical and geochemical research. *Phys. Chem. Mineral.*, **20**, 111–19.

Dilmanian, F.A., Garrett, R.F., Thomlinson, W.C. *et al.* (1991) Computed tomography with monochromatic X rays from the National Synchrotron Light Source. *Nucl. Instrum. Methods*, **B56/57**, 1208–13.

Emura, S. and Maeda, H. (1994) Simple removal method of diffraction peaks in x-ray absorption fine structure spectra from a single crystal in the fluorescence mode. *Rev. Sci. Instrum.*, **65**, 25–7.

Farges, F., Sharps, J.A. and Brown, G.E. Jr (1993) Local environment around gold (III) in aqueous chloride solutions: an EXAFS spectroscopic study. *Geochim. Cosmochim. Acta*, **57**, 1243–52.

Finger, L.W. (1989) Powder diffraction. *Reviews in Mineralogy*, **20**, 309–31.

Flannery, B.P., Deckman, H.W., Roberge, W.G. and d'Amico, K.L. (1987) Three-dimensional x-ray microtomography. *Science*, **237**, 1439–44.

Giauque, R.D., Garrett, R.B. and Goda, L.Y. (1977(a)) Energy-dispersive X-ray fluorescence spectrometry for determination of twenty-six trace and two major elements in geochemical samples. *Anal. Chem.*, **49**, 62-7.

Giauque, R.D., Garrett, R.B. and Goda, L.Y. (1979) Determination of trace elements in light element matrices by X-ray fluorescence spectrometry with incoherent scattered radiation as an internal standard. *Anal. Chem.*, **51**, 511-16.

Giauque, R.D., Jaklevic, J.M. and Thompson, A.C. (1986) Trace element determination using synchrotron radiation. *Anal. Chem.*, **58**, 940-4.

Giauque, R.D., Thompson, A.C., Underwood, J.H. *et al.* (1988) Measurement of femtogram quantities of trace elements using an X-ray microprobe. *Anal. Chem.*, **60**, 855-8.

Gilfrich, J.V. and Birks, L.S. (1984) Estimation of detection limits in X-ray fluorescence spectrometry. *Anal. Chem.*, **56**, 77-9.

Gilfrich, J.V., Skelton, E.F., Qadri, S.B. *et al.* (1983) Synchrotron radiation X-ray fluorescence analysis. *Anal. Chem.*, **55**, 187-90.

Gordon, B.M. (1982) Sensitivity calculations for multielemental trace analysis by synchrotron radiation induced X-ray fluorescence. *Nucl. Instrum. Methods, Phys. Res.*, **B204**, 223-9.

Gordon, B.M. and Jones, K.W. (1985) Design criteria and sensitivity calculations for multielemental trace analysis at the NSLS X-ray microprobe. *Nucl. Instrum. Methods Phys. Res.*, **B10/11**, 293-8.

Goshi, Y., Aoki, S., Iida, A. *et al.* (1987) A scanning X-ray fluorescence microprobe with synchrotron radiation. *Jap. J. Appl. Phys.*, **26**, L1260-2.

Hanson, A.L., Jones, K.W., Gordon, B.M. *et al.* (1987) Trace element measurements using white synchrotron radiation. *Nucl. Instrum. Methods Phys. Res.*, **B24/25**, 400-4.

Hayakawa, S., Iida, A., Aoki, S. and Gohshi, Y. (1989) Development of a scanning X-ray microprobe with synchrotron radiation. *Rev. Sci. Instrum.*, **60**, 2452-5.

Hayakawa, S., Gohshi, Y., Iida. A. *et al.* (1990) X-ray microanalysis with energy tunable synchrotron X-rays. *Nucl. Instrum. Methods Phys. Res.*, **B49**, 555-60.

Hayakawa, S., Gohshi, Y., Iida, A. *et al.* (1991) Fluorescence X-ray absorption fine structure measurements using a synchrotron radiation x-ray microprobe. *Rev. Sci. Instrum.*, **62**, 2545-9.

Hayes, K.F., Roe. A.L., Brown, G.E. Jr *et al.* (1987) *In situ* absorption study of surface complexes: selenium oxyanions on goethite (α-FeOOH) *Science*, **238**, 783-6.

Hazemann, J.L., Manceau, A., Sainctavit, Ph. and Malgrange, C. (1992) Structure of the $\alpha Fe_xAl_{1-x}OOH$ solid solution. I. Evidence by polarized EXAFS for an epitaxial growth of hematite-like clusters of Fe-diaspore. *Phys. Chem. Minerals*, **19**, 25-38.

Hegedüs, F., Winkler, P., Wobrauschek, P. and Streli, C. (1990) TXRF spectrometer for trace element detection. *Advances in X-ray Analysis*, **33**, 581-3.

Howells, M.R. and Hastings, J.B. (1983) Design considerations for an X-ray microprobe. *Nucl. Instrum. Methods*, **208**, 379-86.

Howells, M.R., Kirz, J. and Sayre, D. (1991) X-ray microscopes. *Sci. Am.* February, 88-94.

Iida, A. and Noma, T. (1993) Synchrotron X-ray microprobe and its application to human hair analysis. *Nucl. Instrum. Methods Phys. Res.*, **B82**, 129-38.

Iida, A., Matsushita, T. and Gohshi, Y. (1985) Synchrotron radiation induced X-ray fluorescence analysis using wide band pass monochromators. *Nucl. Instrum. Methods Phys. Res.*, **A235**, 597-602.

Iida, A., Yoshinaga, A., Sakurai, K. and Gohshi, Y. (1986) Synchrotron radiation excited fluorescence analysis using total reflection of X-rays. *Anal. Chem.*, **58**, 394-97.

Jach, T. and Bedzyk, M.J. (1993) X-ray standing waves at grazing angles. *Acta Crystallogr.*, **A49**, 346-50.

Jackson, A. (1990) The challenges of third-generation synchrotron light sources. *Synchrotron Radiation News*, **3-3**, 13–20.

Jaklevic, J.M., Giauque, R.D. and Thompson, A.C. (1985) Quantitative X-ray fluorescence analysis using monochromatic synchrotron radiation. *Nucl. Instrum. Methods Phys. Res.*, **B10/11**, 303–8.

Jaklevic, J.M., Giauque, R.D. and Thompson, A.C. (1988) Resonant Raman scattering as a source of increased background in synchrotron excited radiation X-ray fluorescence. *Anal. Chem.*, **60**, 482–4.

Janssens, K.H., van Langevelde, F., Adams, F.C. *et al.* (1994) Comparison of synchrotron X-ray microanalysis with electron and proton micro microscopy for individual particle analysis. *Adv. X-ray Anal.*, **35**, 1265–73.

Johns, R.A., Steude, J.S., Castanier, L.M. and Roberts, P.V. (1993) Nondestructive measurements of fracture aperture in crystalline rock cores using X-ray computed tomography. *J. Geophys. Res.*, **98**, 1889–1900.

Jones, K.W., Gordon, B.M., Hanson, A.L. *et al.* (1989) X-ray microscopy with the use of synchrotron radiation. *Microbeam Analysis–1989*, San Francisco Press, pp. 191–5.

Jones, K.W., Gordon, B.M., Schidlovsky, G. *et al.* (1990) Biomedical elemental analysis and imaging using synchrotron X-ray microscopy. *Microbeam Analysis–1990*, San Francisco Press, pp. 401–4.

Jones, K.W., Schidlovsky, G., Burger, D.E. *et al.* (1990) Distribution of lead in human bone: III. Synchrotron X-ray microscope measurements, in *Advances in In Vivo Body Composition Studies* (eds S. Yasamura *et al.*) Plenum, New York, pp. 281–6.

Jones, K.W., Kwiatek, W.M. *et al.* (1988) X-ray microscopy using collimated and focused synchrotron radiation. *Advances in X-ray Analysis*, **31**, 59–68.

Jones, R.C. and Schulze, D.G. (1994) A Hawaiian bauxite: a comparison between standard laboratory XRD and synchrotron powder diffraction. Clays and Clay Minerals, submitted.

Kaplan, D., Hunter, D., Bertsch, P. *et al.* (1994) Integrated application of synchrotron X-ray fluorescence (SXRF) spectroscopy and energy-dispersive X-ray (EDX) analysis in the research of colloid-facilitated contaminated transport. *Environ. Sci. Tech.*, in press.

Klockenkämper R., Knoth, J., Prange, A. and Schwenke, H. (1992) Total-reflection X-ray fluorescence spectroscopy. *Anal. Chem.*, **64**, A1115–23.

Knöckel, A., Petersen, W. and Tolkiehn, G. (1983) X-ray fluorescence analysis with synchrotron radiation. *Nucl. Instrum. Methods*, **208**, 659–63.

Koch, E.E. (ed.) (1983) *Handbook of Synchrotron Radiation*, **1**. North Holland, Amsterdam.

Kunz, C. (1979) *Synchrotron Radiation: Techniques and Applications*. Springer, Berlin.

Kwiatek, W.M., Hanson, A.L. and Jones, K.W. (1990) Selection of the experimental conditions for white-light SRIXE measurements. *Nucl. Instrum. Methods*, **B50**, 347–52.

Kwiatek, W.M., Cichocki, T., Galka, M. and Paluszkiewicz, C. (1992) Microanalysis using synchrotron radiation. *Nucl. Instrum. Methods Phys. Res.*, **B68**, 122–4.

Lee, P.L., Beno, M.A. and Jennings, G. (1994) An energy dispersive X-ray absorption spectroscopy beamline, X6A, at NSLS. *Rev. Sci. Instrum.*, **65**, 1–6.

Lindquist, W.B., Kaufmann, A.E., Spanne, P. (1992) Conceptual design for high speed computed tomography at the Advanced Photon Source. *Eos*, **73**, 208.

Manceau, A., Bonnin, D., Stone, W.E.E. and Sanz, J. (1990) Distribution of Fe in the octahedral sheet of trioctahedral micas by polarized EXAFS. *Phys. Chem. Minerals*, **17**, 363–70.

Margaritondo, G. (1988) *Introduction to Synchrotron Radiation*, Oxford University Press, New York.

Markowicz, A.A. and Van Grieken, R.E. (1988) X-ray spectrometry. *Anal. Chem.*, **60**, R28–42.

Muramatsu, Y., Oshima, M., Shoji, T. and Kato, H. (1992) Undulator-radiation-excited X-ray fluorescence analysis system for light elements. *Rev. Sci. Instrum.*, **63**, 5597–600.

Nomura, M. (1992) Trace element analysis using X-ray absorption edge spectrometry. *Anal. Chem.*, **64**, 2711–14.

Nusshardt, R., Bonse, U., Busch, F. *et al.* (1991) Microtomography: a tool for nondestructive study of materials. *Synchrotron Radiation News*, **4-3**, 21–3.

Pantenburg, F.J., Beier, T., Hennrich, F. and Mommsen, H. (1992) The fundamental parameter method applied to X-ray fluorescence analysis with synchrotron radiation. *Nucl. Instrum. Methods Phys. Res.*, **B68**, 125–32.

Paris, E., Mottana, A., Della Ventura, G. and Robert, J.-L. (1993) Titanium valence and coordination in synthetic richterite – Ti-rich amphiboles. A synchrotron-radiation XAS study. *Eur. J. Mineral.* **5**, 455–64.

Pattrick, R.A.D., van der Laan, G., Vaughan, D.J. and Henderson, C.M.B. (1993) Oxidation state and electronic configuration determination of copper in tetrahedrite group minerals by L-edge X-ray absorption spectroscopy. *Phys. Chem. Minerals*, **20**, 395–401.

Pella, P.A. and Dobbyn, R.C. (1988) Total reflection energy-dispersive X-ray fluorescence spectrometry using monochromatic synchrotron radiation: application to selenium in blood serum. *Anal. Chem.*, **60**, 634–7.

Pella, P.A., Newbury, D.E., Steel, E.B. and Blackburn, D.H. (1986) Development of the National Bureau of Standards thin glass films for X-ray fluorescence spectrometry. *Anal. Chem.*, **58**, 1133–7.

Premuzic, E.T., Kwiatek, W.M., Lin, M. and Jones, K.W. (1989) Regional variation in the metal composition of residual brine sludges derived from geothermal brines. *Geothermal Sci. Techn.*, **2**, 125–37.

Prewitt, C.T., Coppens, P., Philips, J.C. and Finger, L.W. (1987) New opportunities in synchrotron X-ray crystallography. *Science*, **238**, 312–9.

Prins, M., Dries, W., Lenglet, W. *et al.* (1985) Trace element analysis with synchrotron radiation at SRS Daresbury. *Nucl. Instrum. Methods Phys. Res.*, **B10/11**, 299–302.

Przbylowicz, W., van Langevelde, F., Kucha, H. *et al.* (1992) Trace element determinations in selected geological samples using a 15 keV synchrotron microprobe at the SRS, Daresbury, UK. *Nucl. Instrum. Methods Phys. Res.*, **B68**, 115–21.

Rarback, H., Shu, D., Feng, S.C. *et al.* (1988) Scanning X-ray microscope with 75-nm resolution. *Rev. Sci. Instrum.*, **59**, 52–9.

Rivers, M.L., Sutton, S.R. and Gordon, B.M. (1989) X-ray fluorescence microprobe imaging with undulator radiation. *Mater. Res. Soc. Symp. Proc.*, **143**, 285–90.

Rivers, M.L., Sutton, S.R. and Jones, K.W. (1991) Synchrotron X-ray fluorescence microscopy. *Synchrotron Radiation News*, **4-2**, 23–6.

Rivers, M.L., Sutton, S.R. and Jones, K.W. (1992) X-ray fluorescence microscopy. *X-ray Microscopy III; Springer Series in Optical Sciences*, **67**, 212–6.

Rivers, M.L., Thorn, K.S., Sutton, S.R. and Jones, K.W. (1992) Wavelength dispersive analysis with the synchrotron X-ray fluorescence microprobe. *Lunar Planet. Sci.*, **XXIV**, 1203–4.

Schulze, D.G. and Smith, J.V. (eds) (1990) *Synchrotron X-ray Sources and New Opportunities in the Soil and Environmental Sciences: Workshop Report.* Argonne National Laboratory/Advanced Photon Source/TM-7.

Smith, J.V. (1992) Present status and future prospects for synchrotron-based microtechniques that utilize X-ray fluorescence, absorption spectroscopy, diffraction and tomography. *Inst. Phys. Conf. Ser.*, **130**, 605–12.

Smith, J.V. and Manghnani, M.H. (eds) (1990) *Synchrotron X-ray Sources and New Opportunities in the Earth Sciences: Workshop Report*. Argonne National Laboratory/Advanced Photon Source/TM-3.
Spanne, P. and Rivers, M.L. (1987) Computerized microtomography using synchrotron X-rays. *Nucl. Instrum. Methods Phys. Res.*, **B24/25**, 1063-7.
Spanne, P., Jones, K.W., Lindquist, W.B. et al. (1993) Determination of the microstructure of Berea sandstone using synchrotron computed microtomography. *Eos*, **74**, 626.
Sparks, C.J. Jr (1980) X-ray fluorescence microprobe for chemical analysis, in *Synchrotron Radiation Research* (ed. H. Winich and S. Doniach), Plenum, New York, pp. 459-512.
Stöhr, J. (1992) *NEXAFS Spectroscopy*, Springer, Heidelberg.
Stragier, H., Cross, J.O., Rehr, J.J. and Sorensen, L.B. (1992) Diffraction anomalous fine structure: a new X-ray structural technique. *Phys. Rev. Lett.*, **69**, 3064-7.
Sutton, S.R. (organizer) (1992) *Workshop on Applications of Synchrotron Radiation to Earth Materials*. (Contact SRS at 5734 S.Ellis Avenue Chicago IL 60637 for copy of hands-on handbook.)
Sutton, S.R., Bajt, S. and Delaney, J. (1994) Microanalysis of iron oxidation states using X-ray absorption spectroscopy. *Lunar. Planet. Sci.*, **XXV**, in press.
Sutton, S.R., Bajt, S., Rivers, M.L. et al. (1992) Oxidation state of chromium in an olivine inclusion within a Kalimantan diamond. *Eos*, **73**, 651.
Sutton, S.R., Delaney, J., Smith, J.V. and Prinz, M. (1987) Copper and nickel partitioning in iron meteorites. *Geochim. Cosmochim. Acta*, **51**, 2653-62.
Sutton, S.R., Rivers, M.L. and Smith, J.V. (1986) Synchrotron X-ray fluorescence: the diffraction interference. *Anal. Chem.*, **58**, 2167-71.
Sutton, S.R., Rivers, M.L., Bajt, S. and Jones, K.W. (1993(e)) Synchrotron X-ray fluorescence microprobe analysis with bending magnets and insertion devices. *Nucl. Instrum. Methods Phys. Res.*, **B75**, 553-8.
Sutton, S.R., Rivers, M.L. Smith, J.V. et al. (1988) Synchrotron sources in the Earth sciences. *Eos*, **69**, 1666-75.
Sutton, S.R., Rivers, M.L., Bajt, S. et al. (1993(d)) Synchrotron X-ray fluorescence microprobe: a microanalytical instrument for trace element studies in geochemistry, cosmochemistry and the soil and environmental sciences. *Proc. 8th Natl Conf. Synchrotron Radiation Instrum.*, in press.

Themner, K., Spanne, P. and Jones, K.W. (1990) Mass loss during X-ray microanalysis. *Nucl. Instrum. Methods Phys. Res.*, **B49**, 52-9.
Thiel, D.J., Bilderback, D.H. and Lewis, A. (1993) Production of intense micrometer-size X-ray beams with tapered glass monocapillaries. *Rev. Sci. Instrum.*, **64**, 2872-5.
Thompson, A.C., Underwood, J.H., Wu, Y. et al. (1988) Elemental measurements with an X-ray microprobe of biological and geological samples with femtogram sensitivity. *Nucl. Instrum. Methods Phys. Res.*, **A266**, 318-23.
Thompson, A.C., Underwood, J.H., Wu, Y. et al. (1990) An X-ray microprobe using focusing optics with a synchrotron radiation source. *X-ray Microscopy in Biology and Medicine* (eds K. Shinohara et al.), Springer, Berlin, pp. 119-27.

Underwood, J.H., Thompson, A.C., Wu, Y. and Giauque, R.D. (1988) X-ray microprobe using multilayer mirrors. *Nucl. Instrum. Methods Phys. Res.*, **A266**, 296-302.

van Langevelde, F. and Vis, R.D. (1991) Trace element determinations using a 15 keV synchrotron X-ray microprobe. *Anal. Chem.*, **63**, 2253-9.
van Langevelde, F., Tros, G.H.J., Bowen, D.K. and Vis, R.D. (1990) The synchrotron radiation microprobe at the SRS, Daresbury (UK) and its applications. *Nucl. Instrum. Methods Phys. Res.*, **B49**, 544-50.

van Langevelde, F., Bowen, D.K., Tros, G.H.J. *et al.* (1990) Ellipsoid X-ray focusing for synchrotron-radiation microprobe analysis at the SRS, Daresbury, UK. *Nucl. Instrum. Methods Phys. Res.*, **A292**, 719–27.

Wakatsuki, M., Hayakawa, S., Aoki, S. *et al.* (1991) Detection of impurities in diamond grown in the metallic solvent by X-ray fluorescence using synchrotron radiation. *New Diamond Sci. Tech., MRS Int. Conf. Proc.*, pp. 143–7.

Waldo, G.S., Carlson, R.M.K., Moldowan, J.M. *et al.* (1991) Sulfur speciation in heavy petroleums: information from X-ray absorption near-edge structure. *Geochim. Cosmochim. Acta*, **55**, 801–14.

Waychunas, G.A. and Brown, G.E. Jr (1990) Polarized X-ray absorption spectroscopy of metal ions in minerals. Applications to site geometry and electronic structure determination. *Phys. Chem. Minerals*, **17**, 420–30.

Waychunas, G.A., Rea, C.C., Fuller, J.A. and Davis, J.A. (1993) Surface chemistry of ferrihydrite: Part 1. EXAFS studies of the geometry of coprecipitated and adsorbed arsenate. *Geochim. Cosmochim. Acta*, **57**, 2251–69.

Williams, K.L. (1987) Introduction to X-ray Spectrometry: X-ray and electron microprobe analysis, Allen & Unwin, London.

Winick, H. and Doniach, S. (eds) (1980). *Synchrotron Radiation Research*, Plenum Press, New York.

Winick, H. and Williams, G.P. (1991) Overview of synchrotron radiation sources world-wide. *Synchrotron Radiation News*, **4**, 23–6.

Wu, X., Zeng, X. and Yao, H. (1993) Analysis of a single strand of hair by PIXE, IXX, and synchrotron radiation. *Nucl. Instrum. Methods Phys. Res.*, **B75**, 566–70.

Wu, Y., Thompson, A.C., Underwood, J.H. *et al.* (1990) A tunable X-ray microprobe using synchrotron radiation. *Nucl. Instrum. Methods*, **A291**, 146–51.

CHAPTER SIX
Ion microprobe analysis in geology
Richard W. Hinton

6.1 Introduction

For many years it has been possible to measure accurately the small-scale variations in the major and minor element chemistry of minerals by the electron microprobe technique, in which the sample surface is bombarded with an electron beam and the X-ray signal is measured (Chapter 2). Abundances of isotopes, trace and ultra-light elements (e.g. H and Li) which cannot be determined by this method can, however, be measured by secondary ion mass spectrometry (SIMS; Shimizu, Semet and Allègre, 1978; Shimizu and Hart, 1982; Reed, 1989; Zinner, 1989). The ion microprobe is a SIMS instrument with a focused primary ion beam, which permits *in situ* microanalysis of minerals in samples prepared as standard polished sections. Nearly all elements from H to U can be detected and many can be analysed quantitatively down to part-per-million levels, or below.

Given these potential advantages, why has this method not replaced all others for the isotopic, major and trace element analysis of minerals? The reasons for this lie with the uncertainties in analytical results caused by the dependence of secondary ion formation on mineral chemistry (i.e. matrix effects), as well as interferences resulting from the complexity of the secondary ion mass spectra created by the sputtering of multi-element geological materials. One should also mention the rather high cost of the equipment.

6.2 Instrumentation

During ion microprobe analysis, the sample surface is bombarded with a beam of focused high-energy ions. The impact of each ion displaces atoms in the sample, creating a 'collision cascade' which causes surface atoms that receive more than their binding energy to be ejected, or sputtered (Figure 6.1). While many of the matrix atoms are sputtered as neutral particles, some are ejected as ions which can be accelerated by an electrostatic field for mass spectrometric analysis. In its simplest form, the ion microprobe

Microprobe Techniques in the Earth Sciences. Edited by P.J. Potts, J.F.W. Bowles, S.J.B. Reed and M.R. Cave. Published in 1995 by Chapman & Hall, London. ISBN 0 412 55100 4

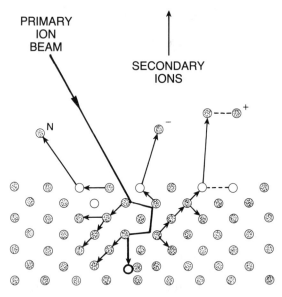

Figure 6.1 Atoms and molecules are sputtered from the surface by a collision cascade. Ions form during bond breaking or while traversing the surface layer.

(Figure 6.2) consists of a source of primary (bombarding) ions, an ion optical column that generates a finely focused primary ion beam, an extraction system which transfers the secondary ions from the surface of the sample to the entrance slit of the mass spectrometer and the mass spectrometer itself. Depending on the polarity of the secondary ion accelerating voltage, either positive or negative ions can be extracted. Not only are ions of individual elements produced, but also ionized molecular combinations of elements present in the matrix (including the implanted bombarding species). These 'molecular' ions (Colton, Ross and Kidwell, 1986) can consist of two, three or more atoms, forming clusters that survive long enough to travel from source to detector. For example, in the positive ion mass spectrum of plagioclase (Figure 6.3), the most intense peaks are those of the major elements, but molecular peaks are also present at virtually every mass. Conversely, the negative ion mass spectrum (Figure 6.4) is dominated by molecular ions, especially combinations with oxygen. There is, therefore, a high probability that in chemically complex minerals, the intensity measured at any given mass will include a contribution from molecular, as well as elemental, ions. However, there are often small differences in mass, owing to the 'mass defect' of the atoms concerned, which can be used to separate atomic from molecular ions. The frequent necessity of separating elemental from molecular species on this basis dictates that secondary ion mass spectrometers used for geological research should have high mass resolution and

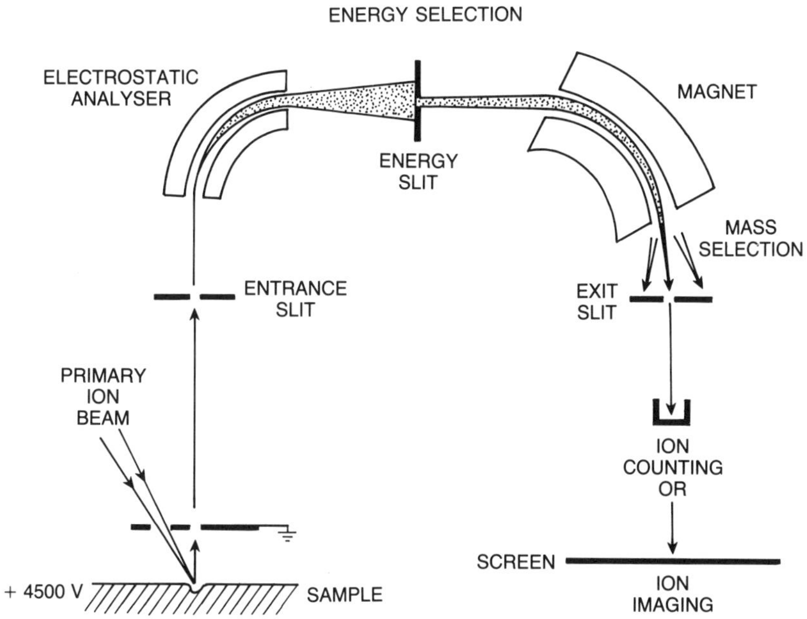

INSTRUMENTATION

Figure 6.2 Diagrammatic representation of ion microprobe: a SIMS instrument with a finely focused point source. Note the angled primary beam and extraction normal to the surface.

they are almost exclusively of the magnetic sector type. As no prior chemical separation of elements (as in conventional mass spectrometry) is carried out, unresolvable isobaric overlaps between certain radioactive isotopes and their daughters (e.g. ^{87}Rb and ^{87}Sr, ^{176}Lu and ^{176}Hf) occur, which rule out the direct application of the ion microprobe to dating by these methods.

The number of ions formed per atom sputtered is highly variable between different elements and matrices. The ionization efficiency is lower than 1% for many elements (for either positive or negative secondary ions). Physical models of the ion-forming process are inadequate to predict ion yields, so that conversion of measured secondary ion count rates to either absolute concentrations or relative isotopic abundances must be made by comparison with well-characterized standards.

The above summary highlights some of the problems which affect SIMS analysis. Ideally, the analytical method used should permit the measurement of each isotope, from a small area and under controlled conditions, free from molecular ion interference, but without a prohibitive loss in ion transmission. The fundamental limit of detection depends on the number of atoms in the sputtered volume, combined with the degree of ionization of each element (the fraction of the sputtered atoms that are ionized) and the

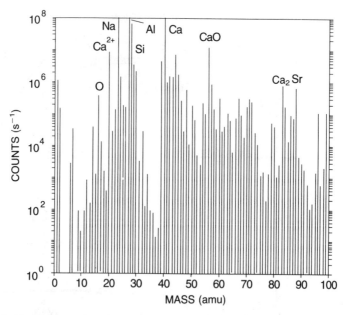

Figure 6.3 Positive secondary ion mass spectrum of plagioclase sputtered by a 15 keV O⁻ primary ion beam. The energy window was set to accept all ions up to 20 eV energy. Al^+ counts are approximately 2×10^8 and Na, Si and Ca, 1×10^8. 5 nA primary beam current.

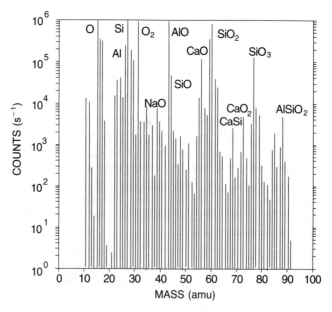

Figure 6.4 Negative secondary ion mass spectrum of plagioclase sputtered by a 15 keV Cs^+ primary ion beam. The energy window was set to accept all ions up to 20 eV energy. Si^+ and O^- counts were approximately 1×10^6 and 1×10^8 respectively.

transmission of the mass spectrometer. However, despite the potential problems outlined above, the importance of the ion microprobe as an analytical tool in geology cannot be overstated. While the accuracy obtained rarely matches that of bulk techniques, the ability to measure a particular grain, free of inclusions and alteration, or a particular zone within a grain, is often critically important in geological applications.

6.2.1 Primary ion source

The primary beam used for the analysis of most geological materials nearly always consists of either O^- or Cs^+ ions, with energies that are generally between 4 keV and 20 keV. The minimum Cs^+ spot size (liquid metal Cs source) is about 0.1–0.2 μm, compared to about 0.5 μm for O_2^+ and 5 μm for O^- (duoplasmatron source). Gallium has been used for imaging of geological materials and can give beams as small as 0.02 μm (Levi-Setti, Crow and Wang, 1985), but is useful only for major elements, owing to the low beam current under these conditions.

(a) Duoplasmatron (O^-, O_2^+, Ar^+)

The 'duoplasmatron' consists of a hollow Ni cathode, a movable intermediate electrode and an anode plate containing a small hole. An arc discharge is formed by leaking oxygen into the hollow cathode. The intermediate electrode confines, both physically and magnetically, the plasma formed by this arc discharge (hence 'duo'-plasmatron). The plasma bulges through the anode hole and ions extracted from the surface are accelerated into the primary column. Either positive or negative ions can be extracted by appropriate choice of the polarity of the extraction voltage. The most commonly used primary ion is O^-, which is preferable to positive species for insulating samples because charging effects are minimized (section 6.4.2). For some purposes, O_2^+ or Ar^+ are appropriate and higher beam currents can be obtained with these species.

Primary ions of other elements present in the source region may be produced, including vacuum gases (especially water) and metals which form internal parts of the source (principally Ni). If these are not removed from the primary beam they will be implanted into the sample surface and will be sputtered together with the matrix elements. For this reason, most ion microprobes use a mass-filtered primary beam to ensure that such impurity ions are removed before they reach the sample.

(b) Cs source

Cs^+ ions are produced by heating a reservoir of Cs metal and diffusing liquid Cs down a capillary (Storms, Brown and Stein, 1977). Cs is evaporated and ionized on the hot metal surface at the tip of the capillary tube and then accelerated into the primary beam column. To avoid the necessity of handling

Cs metal, a source which uses Cs chromate has been developed (Slodzian et al., 1991). Beam densities produced by the Cs$^+$ source are similar to those of O_2^+ produced by the duoplasmatron, but approximately 10 times higher than those of O^-.

6.2.2 Probe-forming system

The ions extracted from the source are focused by a series of electrostatic lenses. The final size of the primary beam is dependent on the initial source size and the demagnification in the primary column. Increasing the strength of the lenses increases the demagnification but leads to greater loss of current at the apertures (Figure 6.5). Curves of current loss versus beam size are given in Figure 6.6. Typical beam densities are $50\,\text{mA cm}^{-2}$ for Cs$^+$ and O_2^+ but only $5\,\text{mA cm}^{-2}$ for O^-. The maximum total currents obtainable are, however, similar for both O^- and Cs$^+$.

For trace element and isotopic analyses, beam currents of 1 to 10 nA are typically used. At these currents, a beam diameter of $1-10\,\mu\text{m}$ is possible for Cs$^+$ but only $10-25\,\mu\text{m}$ for O^-. (Note, however, that the quoted diameter is commonly that containing 90–95% of the current, whereas in some applications the extent of the peripheral region may be important). Volumes sputtered in measurements lasting around 30 minutes are of the order of $500\,\mu\text{m}^3$.

The apertures and lenses in the primary column can be arranged so that the final lens focuses an image of a uniformly illuminated aperture rather than the source itself. This 'Kohler illumination' gives a lower beam density than that obtained by normal focusing but has the advantage of very even current distribution. The sputter pits produced are circular (essentially an image of the aperture), with flat bottoms and sharp edges. The size of the sputter pit is fixed by the aperture size rather than by aberrations of the primary ion optical system and can be changed by selecting an aperture of different size, the beam current being directly proportional to the area of the aperture.

6.2.3 Secondary ion extraction system

The secondary ions are sputtered from the surface of the sample with a variety of energies, in a range of directions and from different positions within the area sputtered (which in some cases may be up to $500\,\mu\text{m}$ across). Whilst most ions have energies of less than 20 eV, a significant proportion of the atomic ions have much higher energies. Despite the spatial, energy and angular variations, the mass spectrometer must be capable of resolving small differences in mass to overcome isobaric interferences from molecular ions.

Secondary ion extraction systems are therefore designed to transfer the maximum possible number of sputtered secondary ions through the entrance slit of the mass spectrometer (Slodzian, 1980, 1988). The sample is held at

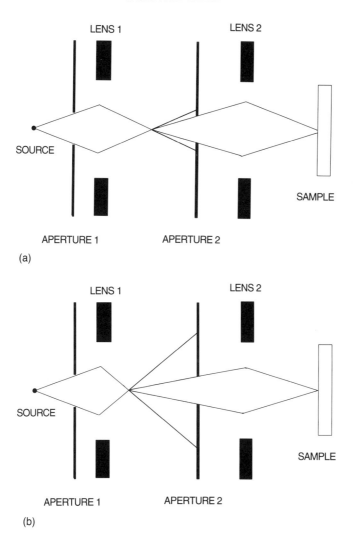

Figure 6.5 Primary beam-forming system with two lenses and apertures. Actual instruments may have three lenses. (a) Lens 1 set to produce crossover close to aperture 2, thereby ensuring high beam current but resulting in large primary beam diameter. (b) Lens 1 set to give crossover close to the lens and further from aperture 2, resulting in large intensity loss on the aperture but a well-focused beam.

high voltage (1–10 kV) and the ions are accelerated towards an extraction electrode held at ground potential. In some instruments the initial extraction voltage may be less than 200 V, but the ions are subsequently accelerated to high energy (several kV). Secondary ions are focused onto the entrance slit of the mass spectrometer by a second 'transfer lens' (or lenses). In the case of ion microprobes, *sensu stricto*, the original (approximately circular) source image is converted into a line focus to match the entrance slit of the

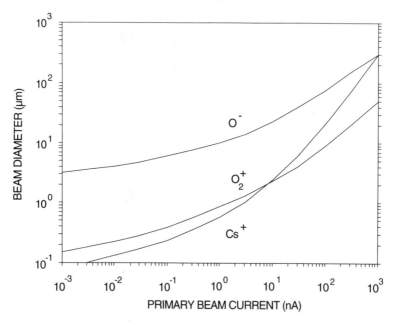

Figure 6.6 Nominal curves of the relationship between primary beam current and beam diameter.

mass spectrometer and thus maximize transmission at high mass resolution. Apertures in the extraction system eliminate ions with extreme trajectories (i.e. those leaving the surface with a high divergence angle and energy), thereby causing some energy filtering prior to the ions entering the mass spectrometer.

The large energy spread of secondary ions is an important factor affecting the ability of the ion-optical system to form a sharp focus on the entrance slit of the mass spectrometer. A sharper focus can be obtained by reducing the energy spread of the secondary ions accepted for transmission. However, owing to differences in the energy distribution between different elements, fractionation of one element relative to another is likely. The focusing of secondary ions is also limited by the size of the secondary ion source, i.e. the primary beam spot size. Decreasing the primary beam diameter reduces the size of focus at the source slit of the mass spectrometer and hence improves the transmission.

6.2.4 Mass spectrometer

The motion of ions in a uniform magnetic field is governed by their charge, energy and mass. The energy ($E = eV$) of a given ion is the sum of that provided by the extraction potential (1–10 kV) and any energy imparted by the sputtering process (0– > 150 eV). The radius, r, (in cm) of the circular path of a singly charged ion of mass m, accelerated by potential difference V, in a magnetic field of strength B (in gauss) is given by:

$$r = \frac{143.95}{B} \times \sqrt{mV} \tag{6.1}$$

In a mass spectrometer with a single collector, r is fixed, therefore the mass of the ion which passes through the exit slit is dependent on B (for a given V).

The difference between the absolute mass of an isotope, given in atomic mass units (1 amu =1/12 of the mass of ^{12}C) and the nominal mass, or 'mass number' A (equal to the total number of nucleons), is known as the mass defect and is usually given in milli-mass-units, or mmu (= 10^{-3} amu). The mass defect is related to the nuclear binding energy of the atom and hence the stability of the nucleus. It decreases slowly up to mass 80, is relatively constant from 80 to 140, then increases up to 238. Therefore, at low masses molecular ions tend to have higher absolute masses than atoms of the same mass number (Figure 6.8 below) and at high masses the situation is reversed. This offers an opportunity to resolve isobaric interferences, providing the mass spectrometer has a sufficiently high mass resolution. Mass resolving power is given by the expression $M/\Delta M$, where ΔM is the difference in mass between two species of mass number M, such that the valley between two peaks of equal intensity is 10% of the peak height. Most secondary ion mass spectrometers used for geological applications are capable of mass resolution of at least 10 000. At this resolution, it is possible to separate molecular species containing two atoms below mass 70 and those containing three or more atoms throughout virtually the whole mass range.

It follows from equation 6.1 that the resolution (peak width) of a simple mass spectrometer is limited by the energy spread of the ions, which is quite large in the case of secondary ionization.

High mass resolution is achieved with a 'double focusing' mass spectrometer, which combines an electrostatic filter with a magnet sector. The secondary ions are dispersed according to their velocity in the electrostatic sector and then focused according to their momentum by the magnetic sector (in some instruments the magnet precedes the electrostatic sector). The mass resolving power is increased by decreasing the width of the entrance and exit slits and by restricting the transmission of the higher energy component of the secondary ion beam. However, there is a simultaneous decrease in the transmission of ions through the mass spectrometer (Figure 6.7).

For ion microprobes with small magnets, secondary ion transmission diminishes rapidly when the mass resolution exceeds 1000, in contrast with the largest instruments, where no intensity loss occurs until the mass reso-

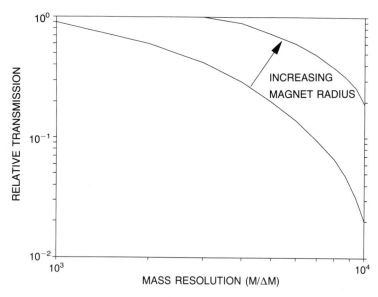

Figure 6.7 Transmission losses with increasing mass resolution. The transmission, for a given mass resolution, increases as the instrument (magnet) size increases.

lution exceeds 5000 (Figure 6.7). There are a number of ways by which the transmission of the mass spectrometer may be increased. If its physical size (magnet radius) is increased, the slit width for a given mass resolution will be greater, allowing more secondary ions to enter. Increasing the extraction voltage reduces the initial angular spread of the secondary ions sputtered from the sample surface; the energy spread is then a smaller proportion of the total ion energy. This leads to greater transmission for the same mass resolution, but there are practical limits on the size of the extraction voltage which can be applied.

Most mass spectrometers used for geological applications have a single exit slit. The secondary ion intensities at each mass under investigation are recorded by stepping the magnetic field under computer control such that each mass in turn is focused on the exit slit. In some instruments, isotope ratios can be measured by simultaneous collection of two or more isotopes using an array of slits and detectors.

An example of a spectrum recorded at mass 56 in garnet is given in Figure 6.8. The two major peaks consist of Fe^+ and CaO^+, which have mass defects of -65.09 mmu (mass 55.93491) and -42.05 mmu (mass 55.95795) respectively. The difference in mass is therefore 23.04 mmu and a mass resolution of 2430 (56/0.02304) is required to resolve these peaks. At the mass resolution used here (approximately 3000), there is a clear valley between the two peaks. The minor peak on the tail of the CaO^+ peak is MgO_2^+, with a mass defect of -25.04 mmu. A minor Si_2^+ peak with a mass defect of -46.14 mmu is unresolved beneath the CaO^+ peak. It must be

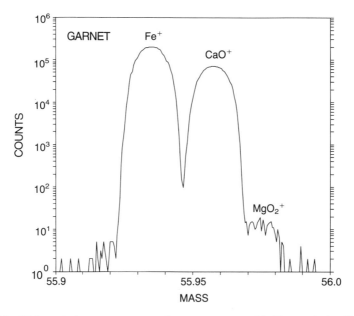

Figure 6.8 High-resolution mass spectrum for garnet at mass 56. The resolution ($M/\Delta M$) is approximately 3000. Note that a minor Si_2^+ peak occurs beneath the CaO^+ peak.

noted that the nominal mass resolution required to separate two peaks is based on the assumption that the peaks are of equal height. If the peaks have very different intensities, the tail of the major peak may significantly interfere with its smaller neighbour. In this case, either measurements must be made at higher mass resolving power, or a correction applied for the overlap, based on the known or assumed peak shape.

6.2.5 Secondary ion detection

Secondary ions which pass through the exit slit of the mass spectrometer can be detected by a number of different devices. If secondary ion intensities are very high ($>10^6$ counts s^{-1}), a Faraday cup collector is appropriate and the current can be measured directly. However, the measured intensity of ions is usually lower than 10^6 counts s^{-1} and single-ion detectors are employed. The detectors used convert the impact of each ion into a charge pulse which can be recorded individually by the data acquisition system. The most commonly used detector for single ion counting is the electron multiplier.

In the electron multiplier, incident ions are converted to electrons on striking a 'dynode' with an activated surface (e.g. Cu/Be or Al). These electrons are accelerated to hit further dynodes, the number multiplying at each stage. The output pulses are amplified and passed to the counting system. The efficiency of ion–to–electron conversion varies from element to element and between isotopes of the same element (Rudat and Morrison, 1978; Zinner, Fahey and McKeegan, 1986; Fahey *et al.*, 1987; Harte and Otter, 1992). In general, as a detector ages, such variations become more extreme. Provided standards and unknowns are recorded under the same detector conditions, the effect will cancel out. However, it should be noted that the absolute ratios of elements or isotopes measured on different instruments (or even the same instrument on different dates) may differ. This is particularly true for hydrogen (Deloule, France-Lanorde and Albarède, 1991; Deloule, Chaussidon and Allé, 1992).

The detection of very low concentrations is ultimately dependent on the detector background. The background count rate for an electron multiplier is normally less than 0.1 counts s^{-1} and is often less than 0.01 counts s^{-1}. If the observed count rate on a peak is close to that of the background, i.e. the analytical signal is approaching the detection-limit level, it is necessary to measure the background at the same time as the peak. This procedure usually involves the measurement of a dummy mass within the measurement cycle, at a position where no secondary ion peaks are expected.

6.2.6 *Ion microscopy*

The discussion so far has been concerned with the ion microprobe. A closely related instrument is the 'ion microscope', in which the secondary ion extraction system permits retention of spatial information (Slodzian, 1980, 1988). In this type of instrument a relatively large area (e.g. several hundred micrometres diameter) is bombarded with primary ions. Having passed through the mass spectrometer exit slit, the secondary ions are focused onto the final image plane; an image of the surface is produced by converting the ions to electrons on a multiple channel plate and accelerating these electrons onto a fluorescent screen. A 'field' aperture in the intermediate image plane permits the selection of ions from only a limited part of the bombarded area. The removal of the conductive coating from large areas of insulating samples (e.g. silicates) by the primary beam leads to charging of the sample and distortion of the secondary ion trajectories, and hence the image. Thus, while this imaging capability is of value in some applications, geological analyses are normally made using a finely focused beam (i.e. in ion microprobe mode). The imaging capability is, however, extensively used in instrument alignment.

6.3 Secondary ion production

6.3.1 *Sputter rates*

The sputter rate using a Cs^+ primary ion beam is substantially greater than that obtained using O^- or O_2^+, which contributes to the higher count rates observed using Cs^+. Measurements made by Wilson et al. (1988) demonstrated that the sputter rate of SiO_2 for Cs^+ (at 14.5 keV) is 2.5 times greater than for O^- (at 4 keV), corresponding to sputter rates of $0.25\,\mu m^3\,nA^{-1}\,s^{-1}$ for Cs^+ and $0.1\,\mu m^3\,nA^{-1}\,s^{-1}$ for O^-. These values are only approximate, since sputter rates vary between matrices and may be affected by crystal orientation (Jull et al., 1980). Where sample volume is not critical, and negative secondary ions are required, the higher sputter rate and ion yield per atom obtained using Cs^+, compared to O^-, might be expected to lead to a preferential use of the Cs^+ source. However, other practical limitations dictate the use of an O^- ion source in some applications (i.e. no need for charge neutralization, control of matrix effects, cost and difficulty in the operation of the Cs gun etc.).

Primary ions implanted into the sample not only change the surface chemistry but also are sputtered to form part of the secondary ion mass spectrum. The purity of the primary ion beam is therefore very important. The observed $^{18}O/^{16}O$ ratios from both positive and negative secondary ion spectra generated by the bombardment of silicates by $^{16}O^-$ ions are usually lower than natural values by approximately a factor of three, indicating that over 60% of the secondary O atoms (either O^- or O^+) originate from the primary beam implant.

6.3.2 Secondary ion yields

Large variations exist in the efficiency of formation of secondary ions. The ion yields for positive and negative secondaries generated by Cs^+ and O_2^+ primary ion beams from silicon are given in Figure 6.9 (Stevie et al., 1988; Wilson and Novak, 1988). Included in this figure are data from silicate matrices (Hinton, 1990) and from borosilicate glass. The curves are similar to those obtained for pure metals (Storms, Brown and Stein, 1977). In general, elements with the lowest ionization potential have the highest positive ion yields and those with the highest ionization potential (highest electron affinity) have the highest negative ion yields. Thus, the alkali metals have very high positive ion yields, possibly exceeding 30% in silicates (section 6.4.8), whereas C, O and F have high negative ion yields. Some elements give reasonably high ion yields for both negative and positive ions (e.g. H, F, S) and can be measured using either species. In contrast, other elements give very low yields for both positive and negative secondary ions (e.g. N, Zn, rare gases).

Depending on the nature of the matrix and of the bombarding beam, the ion yield for a given element may vary by orders of magnitude. Furthermore, the ratio of one ion intensity to another also varies. The overall range in ion yields from K to Zn is less in silicates than in silicon (Figure 6.9). As noted below (section 6.3.6), the presence of electronegative elements in the matrix, or implanted from the primary beam, enhances the positive

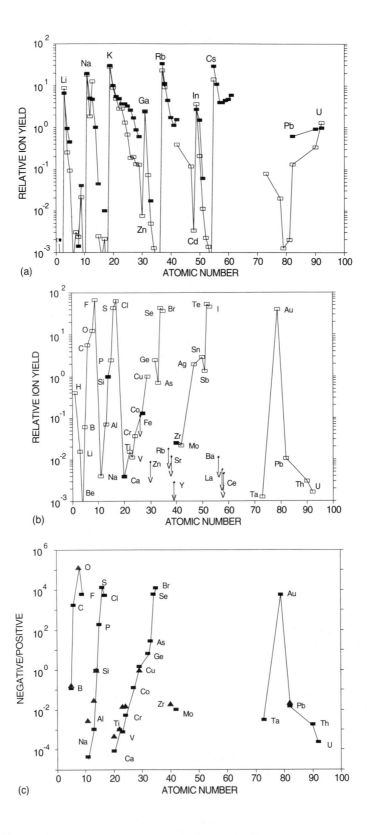

Table 6.1 Ion yield dependence on primary beam species and beam energy

Primary beam species	Energy (keV)	Si$^+$ (counts s^{-1}wt%$^{-1}$nA^{-1})	Pb$^+$ cps/ppm/nA
		NBS 610 Glass	
O$^-$	18.8	7.7×10^5	
O$^-$	15.1	7.8×10^5	6.5
O$^-$	10.0	6.8×10^5	
O$_2^-$	15.1	1.5×10^6	
O$_2^-$	10.0	$1.1 \times 10^{6*}$	
O$_2^-$	15.1	$5.7 \times 10^{4\dagger}$	
		Zircon ANU SL-1	
O$^-$	15.1	8.3×10^5	7.9 (11‡)
O$_2^-$	15.1	1.7×10^6	18
		Diopside	
O$^-$	15.1	6.1×10^5	
O$_2^-$	15.1	1.3×10^6	

*Low beam density, †75 ± 20 eV secondary ion energy, ‡oxygen flooding. Diopside O$^+$ 630 cps/%wt/nA using 10.6 keV O$^-$ 2070 using 10.6 keV O$_2^-$. Primary beam energy includes contribution from 4.5 keV secondary accelerating voltage.

secondary ion yield. Thus, metals have lower ion yields than oxides, even if sputtered with an oxygen primary beam (Storms, Brown and Stein, 1977). The high oxygen content of silicates, coupled with the implantation of oxygen from the primary beam, results in relatively high ionization efficiencies for positive secondary ions. However, even under these conditions, the addition of oxygen into the sample chamber can further enhance positive ion yields in silicates (Table 6.1).

The physical processes which lead to the formation of secondary ions are poorly understood. No model accurately predicts either the observed variations in the degree of ionization for a single element between different matrices, or between different elements in the same matrix. Most models explain only observed variations in simple systems, e.g. the bond-breaking model of Yu (1987). The local thermal equilibrium (LTE) model of Andersen and Hinthorne (1973) treats the ionization process in terms of a plasma. Although this model can be applied to complex systems, the work of Engström *et al.* (1987) demonstrates that it predicts (low-energy) ion yields for glasses only to within a factor of two or three and cannot be used if only high energy ions are recorded.

Figure 6.9 Relative ion yields for ions sputtered from a silicon matrix (open squares: Stevie *et al.*, 1988) and from silicate glass (filled squares: Hinton, 1990, and this work). (a) Positive secondary ions generated by an O$_2^+$ or an O$^-$ primary beam. (b) Negative secondary ions generated by a Cs$^+$ primary beam. Upper limits are derived from silicate glasses. (c) Ratio of negative to positive ions. Note that for a silicate matrix the Si$^-$ to Si$^+$ ratio is approximately 1.

Analyses can be made by establishing working curves based on well-characterized standards (e.g. Ray and Hart, 1982). To eliminate gross matrix effects and variations caused by changes in primary beam current and secondary ion transmission, secondary ion intensities are usually normalized to a major element (frequently Si). Working curves of Si-normalized concentrations based on M^+/Si^+ ratios (or M^-/Si^- for negative secondaries), derived from measurements on trace element standards of similar major element chemistry, give linear arrays passing through the origin. The slope is proportional to the ion yield relative to Si. Non-zero intercepts, e.g. for Sc (Sisson, 1991; Ray and Hart, 1982), indicate the presence of a significant interference. Provided the interference is constant (as is often the case when caused by a major element), the element can still be determined accurately. This approach works well where there is close correspondence between the major element composition of the standard and the 'unknown', but may fail where significant major element zoning (especially in the Fe content) occurs in the unknown sample. However, trace element ion yield variations may be correlated with one of the other major elements and a simple correction factor can then be applied to unknowns, based on measured major element concentrations. For example, Shimizu, Semet and Allègre (1978) demonstrated that Ca and Si ion yields for pyroxene can be related to changes in Fe content. Since natural minerals are frequently inhomogeneous and may not have a suitable range of concentrations for many elements, a number of studies have used glasses as calibration standards. Glasses of composition matched to minerals have been shown to give similar ion yields (Muir, Bancroft and Metson, 1987; Hinton, 1990; MacRae *et al.*, 1993); differences observed by Ray and Hart (1982) were possibly caused by differences in major element chemistry, especially the Fe content.

6.3.3 Molecular ions

Oxides are the dominant molecular ions in both the positive and negative mass spectra of silicate matrices. The MO/M ratio for low energy ions is approximately correlated with the bond strength of the metal–oxide pair. The MO/M ratio increases approximately from 0.002 to 2 as the bond strength increases from 1 to 9 eV. If only high-energy ions of similar oxide bond strength are considered, the correlation between MO/M ratio and oxide bond strength is only observed for elements in the same period. For elements of the same oxide bond strength, the MO/M ratio increases by approximately a factor of three from one period to the next (Hinton, 1990). Thus, at high mass, even oxides of trace elements may produce significant molecular interferences. The correction for oxide overlap of the REE is discussed below (section 6.5.3).

The change in mass defect with increasing mass up to $m = 90$ is sufficiently large that the oxides can be resolved at relatively modest mass resolutions of between 1000 (at mass 26) and 8000 (at mass 88). In the mass range 90–150, the mass defect does not change significantly and resolutions in excess of 10 000 are required to resolve the oxide interferences. For masses greater than 150 the mass defect becomes progressively more positive and resolutions of 7000–9000 are sufficient.

Molecular combinations of all major elements present in the matrix occur in the mass spectrum (Figure 6.3). Thus, the mass range 46 to 60 contains overlaps of two-fold combinations of the major elements Na, Mg, Al and Si. These species, combined with the oxides of Ca and Ti, overlap the first-row transition elements and can only be effectively removed by using high mass resolution.

As noted below (section 6.3.5), hydrides have very similar energy distributions to elemental ions and can only be removed by extreme energy filtering (section 6.3.5) or high mass resolution. High mass resolution is not routinely used to resolve hydrides above mass 60. At high masses the presence of hydrides must be corrected either by direct measurement of the hydride peak where elemental peaks do not interfere or by comparison with hydrides of other elements which can be measured (Long and Hinton, 1984).

6.3.4 Multiply charged species

Doubly charged ions (e.g. M^{2+}) can be present in the mass spectrum below mass 119 ($^{238}U^{2+}$). The M^{2+}/M^+ ratio varies considerably between elements and, for a given element, between different matrices. The 2+ ion yield is inversely related to the second ionization potential, as a rule, but there are some notable exceptions: for example, the alkali elements form 2+ ions despite their very high second ionization potentials (Hinton, 1990). The most abundant 2+ ions are those generated by the alkaline earths and the Ba^{2+} ionization efficiency approaches that of Si^+ (Hinton, 1990). While M^{3+} ions have much lower intensity than M^{2+} ions of the same element, they cannot always be neglected (e.g. $^{27}Al^{3+}$ overlap on $^9Be^+$ can be significant).

The M^{2+}/M^+ ratio is strongly matrix dependent; for example, carbonates and sulphates give very low M^{2+} yields compared to silicates. The absence of a 2+ ion in one matrix therefore cannot be used as a guide to its presence in an unrelated matrix. Multiply charged ions can usually be avoided in trace element analysis by the choice of the isotope used for measurement. Where overlap is unavoidable, corrections can often be applied by using the intensity of doubly charged ions of odd-mass isotopes.

6.3.5 Secondary ion energy distribution

The collision cascade caused by primary ion bombardment creates secondary ions with a wide range of energies. While most ions sputtered from silicate

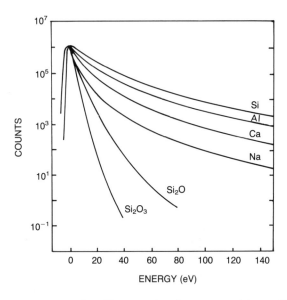

Figure 6.10 Energy spectrum of positive secondary ions sputtered by an O^- primary beam from a silicate matrix. The energy window is narrow (<2 eV). Si_2O and Si_2O_3 are approximate curves based on Crozaz and Zinner (1985).

matrices have energies of only a few eV, a significant number have much higher energies and the tail of the distribution extends to over 1 keV (Metson, Bancroft and Nesbitt, 1985; Thompson, 1987). Most secondary ion mass spectrometers are capable of accepting a wide range of ion energies; however, the energy can be restricted to a small 'window', which may be scanned to enable the secondary ion energy distribution to be studied in detail. The energy spectra of Na, Al, Si and Ca sputtered from a borosilicate glass are given in Figure 6.10. It can be seen from this figure and the tables of Rudat and Morrison (1979) that the width of the energy distribution varies significantly between elements. The energy distribution for a single element also varies between matrices (Slodzian, Lorin and Havette, 1980).

As the energy window is narrowed, the elements with the broadest energy distributions suffer greater losses than those with narrow distributions (Slodzian, 1980). For example, Si has a broad energy distribution, and so suffers greater losses than Na, which has a narrow one. It follows that if the energy window is very narrow, the position of the energy maximum must be located accurately, or not only will the absolute intensity change but also element ratios (Hinton, 1990).

Since molecular ions have a narrow energy distribution, which does not extend to high energies, they can be effectively discriminated against if

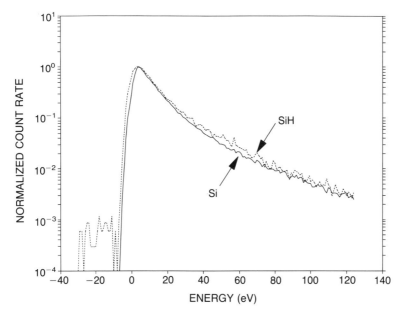

Figure 6.11 Energy spectrum of Si^+ and SiH^+ sputtered by an O^- primary beam from a silicate matrix. The energy window is narrow (<2 eV).

only high energy ions are measured. In general, the greater the number of atoms in the molecular ion species, the narrower the energy distribution and therefore the greater the suppression of intensity achieved by energy filtering. For example, Crozaz and Zinner (1985) demonstrated that $Si_2O_3^+$ has a much narrower energy distribution than Si_2O^+; the intensity of the former is reduced by five orders of magnitude at an energy of 25 eV, while an energy of 75 eV is required to reduce the latter by the same amount (Figure 6.10). In contrast, at 75 eV, SiO^+ is reduced by two orders of magnitude and Si^+ by only about one. The overlap of CaO^+ on Fe (Figure 6.8) is only reduced by a factor of 10 by similar energy filtering and, although corrections can be made (in Ni-free minerals) using measured $^{44}CaO^+$ intensities, detection limits for Fe^+ are poor.

The formation of hydrides can represent a serious problem if vacuum conditions are poor or if hydrated phases are analysed. Hydrides cannot be removed by conventional energy filtering since the MH^+ energy distributions are very similar to those of M^+ (Figure 6.11). Molecular 2+ ions, like singly charged molecular ions, can also be reduced by energy filtering. However, atomic 2+ ions have similar energy distributions to M^+ ions and therefore energy filtering is not an effective method for reducing these interferences (Hinton, 1990).

The virtual elimination by energy filtering of molecular interferences con-

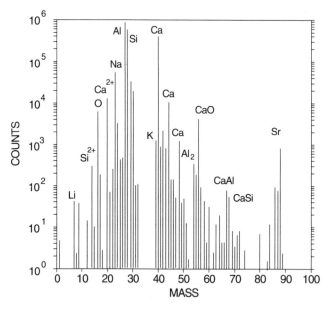

Figure 6.12 'Energy-filtered' positive secondary ion mass spectrum of plagioclase sputtered by 15 keV O^- primary ions, 1 nA beam current. Only secondary ions with energies between 55 and 95 eV were recorded; note suppression of molecular species compared to the low-energy secondary ion mass spectrum (Figure 6.3).

taining three or more species results in a relatively clean mass spectrum at higher masses (>85) for minerals containing the major rock-forming elements, such as Na, Mg, Al, Si, K, Ca and Fe (Figure 6.12). Even when overlaps occur at one of the isotopic masses of an element, it is usually possible to select an alternative isotope which is free of significant molecular interference, if only concentrations rather than isotope ratios are required. Where molecular ion overlaps cannot be avoided, the use of well-characterized standards can still permit corrections to be made (section 6.3.2).

6.3.6 Influence of the primary ion beam

Secondary ions can be generated by bombardment with any energetic ion or atom. The primary beam species used is principally dictated by whether anions or cations are to be used for the analysis. The presence of Cs^+ in the surface layers decreases the surface work function and increases the number of negative secondary ions formed by elements with high electron affinities (Andersen, 1970). For example, the ion yield of O^- is 60 times greater using a Cs^+ primary beam compared to Ar^+ (Slodzian, 1988). In contrast, sputtering using an oxygen primary beam increases the surface work function, thus reducing the number of electrons available for the neutralization and thereby increasing the number of positive secondary ions (Andersen, 1970). The ion yield for Si^+ generated by O^- bombardment is approximately the same as Si^- from Cs^+; the ratios of negative to positive ions in Figure 6.9(c) therefore give a reasonable indication of the influence of the primary beam species. The ion yield is not strongly dependent on the energy of the primary ion (Table 6.1). If a sufficiently intense O_2^- beam could be obtained, the use of this species would have significant advantages for the analysis of cations in insulators, since twice the number of oxygen atoms per unit current would be implanted into the sample (Table 6.1). The amount of oxygen on the surface can also be increased by an oxygen leak near the analysed area (Table 6.1) and increases similar to those obtained by bombarding with O_2^- may be expected.

6.4 Analytical procedures

6.4.1 Specimen preparation

In most geological applications, samples are prepared as thin sections. Sample preparation for ion microprobe analysis is significantly more stringent than for the electron microprobe or SEM. The sample chamber vacuum in ion microprobes is generally several orders of magnitude better than in these instruments, being commonly 10^{-8}–10^{-10} mbar. The mounting medium must therefore be compatible with high vacuum. Prior to coating, samples must be ultrasonically cleaned to remove polishing compounds and chemicals from the surfaces and cracks.

During polishing, some cross-transfer of material between different minerals present is inevitable. In particular, such effects must be taken into consideration in the analysis of the major rock-forming elements (Na, Mg, Al, Si, Ca and Fe) at trace levels. For these elements, the ultimate detection limit may be dependent on surface contamination rather than molecular interferences or detector background. For example, the analysis of Na or Al in quartz requires considerable precleaning/sputtering of the sample surface (Hervig and Peacock, 1989). By this means, the detection limit for Al can be reduced to about 1 ppm, whereas sub-part-per-billion limits could be achieved if no contamination were present.

Industrial cleaning and polishing compounds can contain a variety of elements, including alkali metals and boron (Steele *et al.*, 1981; Shaw *et al.*, 1988(a),(b)). These elements will also be smeared over the surface and may be concentrated in cracks. Polishing contamination is significantly worse if the mineral analysed is soft or very fine-grained and therefore takes a very poor surface polish (e.g. micas, chlorite, clays, etc.). Contamination of the surface can be considerably reduced by using diamond powder in an organic medium such as vaseline, which is free of trace elements (Steele *et al.*, 1981). The interpretation of trace element abundances that appear to increase near crystal boundaries should, therefore, take into account the possibility of beam overlap onto cracks containing contamination introduced during sample preparation. Deliberate analysis of cracked areas can sometimes confirm or eliminate this possibility.

The sample surface must be conducting, as it forms part of the ion optics. An exception to this is analysis using the specimen isolation (SI) technique (section 6.4.4). In ion microscope instruments, the sample is held at a voltage of at least 4.5 kV and must be both coated and flat over an area of a few millimetres. For instruments with low extraction fields, the surface topography is far less critical. In most laboratories the surface is coated with a conducting layer of either gold or carbon. The higher conductivity of gold has the advantage that only a very thin coat (10–30 nm) is required, which can be removed quickly by the primary beam (especially since gold sputters rapidly). Gold is used since it is mono-isotopic, poorly ionizing and of high mass; molecular complexes with gold should not therefore generate significant interferences. Care has to be taken not to introduce contaminants, especially Pb, during coating (Compston, Williams and Meyer, 1984).

6.4.2 Specimen charging

The bombardment of an insulator by energetic particles, including neutral species, induces a surface charge which cannot be dissipated simply by the presence nearby of a conductive coating. Cs^+, Ar^+ or O_2^+ can, however, be used on insulators if the positive charging is compensated by flooding the bombarded area with electrons. If a high-energy electron beam is used, care must be taken that the sample is stable under electron bombardment and that the beam does not impinge on volatile material such as the organic mounting medium. High-energy electron beams can also generate large numbers of secondary electrons, causing even greater charging problems (Hervig, Thomas and Williams, 1989). Sample charging may be avoided if small fragments of material are pressed into gold for analysis. This technique has permitted the trace element and isotopic analysis of small interplanetary dust particles (McKeegan, Walker and Zinner, 1985); however, the fragments must be small (<25 μm) and spatial information within the grains may be lost.

While the O^- beam causes significantly less charging of the sample than O_2^+ (since the incoming and outgoing negative charge is approximately in balance) some still occurs, depending on the size of the area from which the conductive coating is removed. If this area is less than 25 μm^2, charging will generally be less than 5 V. However, potentials as high as 60–100 V may occur in the centre of a 200 μm^2 uncoated area. Ion microscope images of the sputtered area will be severely affected by charging and even a homogeneous mineral may appear zoned. Charging may also distort the focusing of the primary beam and result in poor rastered ion images. High mass resolution operation (section 6.4.4) may require the use of a narrow energy window, in which case charging of only a few volts may lead to intensity losses.

6.4.3 High mass resolution

High mass resolution is frequently required for isotope measurements, to eliminate molecular interferences. The mass spectrometer slits are usually set such that the focus of the entrance slit is narrower than the exit slit, so that if the magnet is scanned, a flat-topped peak is observed (Figure 6.8). Under these conditions, small instabilities in the magnetic field will not lead to changes in the intensity measured for each isotope. Measurements are made by changing the magnetic field such that each isotope in turn passes through the exit slit and onto the ion counter. The positions of the individual peaks are usually checked before or during analysis by recording the whole peak and calculating its centre; slow drift in the calibration of the magnet system can therefore be accommodated.

6.4.4 Energy filtering

Although the energy filtering technique is not highly sensitive to small changes in the secondary ion energy, sample charging will change the average energy of the ions transmitted. It is therefore necessary to position the energy window of the spectrometer accurately. While the window can be set by determining the maximum of the energy distribution (Pan, Holloway and Hervig. 1991), it can be seen from Figure 6.10 that the low-energy tail is very sharp and its position can be defined with great accuracy (<1 eV). The voltage used for energy filtering is often set, not by a fixed voltage above the maximum, but by a fixed voltage above the low-energy tail (Figure 6.13), as described by Zinner and Crozaz (1986(b)). The energy window is usually set relatively wide (approximately 40 eV).

An extreme form of energy filtering has been pioneered by the ion microprobe group at University of Western Ontario (Metson, Bancroft and Nesbitt, 1985; Lau, McIntyre and Metson, 1985). The sample is held in a special holder which masks all but 3 mm of an electrically isolated and uncoated surface. In this SI technique the sample is bombarded with a broad, relatively high current (70–750 nA), O^- primary beam. The negative ion bombardment of the uncoated surface results in charging of the sample to between 450 and 800 V. Only relatively high-energy ions can then escape the kinetic energy barrier created by the surface charging, to be passed into the mass spectrometer and analysed.

Advantages of the SI method include the almost complete suppression of molecular interferences (Lau, McIntyre and Metson, 1985) and the ability to analyse uncoated samples, thereby avoiding a possible source of contamination (Muir *et al.*, 1990). The main disadvantages are the loss of spatial resolution and the large analytical volume associated with the high primary

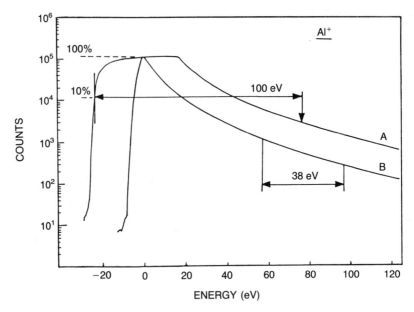

Figure 6.13 Energy spectrum of Al$^+$ sputtered by an O$^-$ primary beam from a silicate matrix. Curve B is the actual energy distribution of the ions, whereas curve A is that recorded with an energy window of 40 eV. The energy window for energy filtering is set with reference to the low-energy tail.

beam current which is required to counteract intensity losses caused by the collection of only very high-energy ions. The SI technique may also give larger differences in Si-normalized ion yields between minerals and glasses of the same composition (MacRae *et al.*, 1993). Although efficient focusing of high-energy ions is claimed for this method, the beam current must be at least 50 times greater than for conventional energy filtering, to give the same secondary ion signal (MacRae *et al.*, 1993).

6.4.5 *Molecular ion corrections*

There are some situations where the loss in intensity caused by either energy filtering or using high mass resolution is unacceptable and the peak must be recorded with an unresolved molecular overlap. Correction for interferences may also be necessary even after the intensities of molecular peaks have been reduced by energy filtering. A simple correction can be applied if the same molecular species is also present uniquely at another mass. If a molecular species overlaps two or more isotopes of an element under investigation it may be possible to use simultaneous equations to correct for the molecular ion overlap. At least one more peak is required than the number of elements to be measured. However, any correction procedure will be valid only if all the overlapping species are known. Interferences from hydrides and, more significantly, hydroxides, which often occur together with other molecular species, can be highly variable and difficult to predict. Correction procedures are normally based on natural isotopic abundances taken from tables. However, it must be recognized that mass fractionation, which occurs in the ionization of elements, also affects molecular ions.

6.4.6 Ion counting

In pulse counting mode, the ultimate precision of a measurement is limited by Poisson counting statistics. To achieve high precision within a reasonable recording time (<1 h), the highest permissible count rate must be used. This consideration is particularly important for single collector pulse counting, where the count rate for the minor isotopes is often limited not by the number of ions, nor transmission losses, but by the maximum permissible count rate for the major isotope. However, at very high count rates some pulses arrive at the detector virtually simultaneously; the counter will not respond to the second pulse if it occurs before the first has been fully processed and so this pulse is lost. The 'dead-time' of pulse counting systems is described by Hayes and Schoeller (1977) and, for electron multiplier counting of secondary ions, by Zinner, Fahey and McKeegan (1986).

Corrections can be made for dead-time losses, according to the approximate formula:

$$C_{meas} = C_{true} \exp(-t \cdot C_{true}) \qquad (6.2)$$

where C is the count rate and t is the dead-time, which is typically in the range 15–40 ns. Ideally, t should be known to ± 1 ns for high precision isotopic measurements, and the count rate should be kept below 10^6 counts s^{-1}. Zinner, Fahey and McKeegan (1986) have shown that different elements, and even isotopes of the same element, have different characteristic dead-times, owing to differences in the average number of electrons generated for each ion and in the detected charge distribution (pulse width). Ideally, the dead-time should thus be measured for each element or isotope. The effect of any error in the dead-time can be minimized by keeping the count rate of the standards and unknowns constant (Valley and Graham, 1992).

6.4.7 Contamination

If the ion microprobe is pumped for extended periods without being vented, the ultimate vacuum will be limited by the outgassing of metal surfaces and O-rings, the efficiency of the vacuum pumps and the presence of any minor leaks etc. Assuming that the base pressure in the sample chamber is low, i.e. $<1 \times 10^{-8}$ mbar, the major source of contamination during analysis is likely to be gases derived from the sample itself. Any gases present in the sample chamber arriving at the sample surface will be sputtered, together with the matrix and implanted primary ions. Obviously, the higher the partial pressure of a particular gas species the greater the likelihood of this occurring. Water is the most likely contaminant to be introduced when the sample is inside the instrument. Even the simple action of exposing the sample to air for a few seconds causes substantial adsorption of water onto the surface. Other contaminants are N_2 and O_2 from the air, and hydrocarbons introduced during sample preparation and cleaning.

Water is particularly difficult to remove from the vacuum system since it remains adsorbed on metal surfaces for an extended period. The energy distribution of secondary H^+ ions is often sharply peaked at low energy (<10 eV), in contrast to a very broad distribution at higher energies. The low-energy peak is probably caused by the desorption and ionization of adsorbed water by secondary electrons generated by ion beam impact (Hervig, Thomas and Williams, 1989; Hervig and Williams, 1986). Interferences with the secondary ion signal from the vacuum component are therefore substantially reduced if only high-energy ions are recorded. Yurimoto, Kurosawa and Sueno (1989) demonstrated that contamination-free measurements of hydrogen as H^+ could be made down to 5 ppm (atomic) by (i) maintaining a base pressure of $<1.5 \times 10^{-9}$ mbar, (ii) using a very intense primary beam, (iii) gating the secondary ion signal to analyse only ions from the centre of the analysed area and (iv) measuring only high energy ions. Under these conditions, good correlations were observed between the H^+/Si^+ ratios determined by ion probe and H concentrations determined by infrared spectroscopy (Yurimoto, Kurosawa and Sueno, 1989). The analysis of CO_2 contents of minerals is also affected by contamination from the vacuum system and contributions from this source can vary from 0.3 wt% CO_2 under poor vacuum conditions (even when using a liquid N_2 cold trap) down to less than 0.02 wt% CO_2 (Pan, Holloway and Hervig, 1991).

Long and Hinton (1984) demonstrated that, although M^+/MH^+ ratios vary from element to element, they tend to increase and decrease together and are predictable. The origin of the hydrogen, whether from the mineral, primary beam or vacuum system, does not appear to change the relationship between M^+/MH^+ ratios of different elements. MOH^+ (and OH^+) peaks are always considerably larger than MH^+ and under poor vacuum conditions can be larger than the corresponding MO^+ molecular peak. MOH^+ species can cause significant interference problems where measurements are made using the oxide, rather than elemental, peaks. For example, in the measurement of Hf isotopes, Kinny, Compston and Williams (1991) used the HfO^+ peaks rather than Hf^+, since they are more intense and the overlap due to Yb^+ is smaller. $HfOH^+$ is a significant problem and corrections must be made for this species; nitrides could also overlap the $HfOH^+$ peaks and N_2 must be kept low in the sample chamber vacuum.

The presence of water (and hydrogen) in the vacuum system has obvious implications for the study of hydrogen isotope ratios or hydrogen abundances. Deloule, France-Lanord and Albarède (1991(b)) and Deloule, Chaussidon and Allé (1992) demonstrated that the isotopic signature and the H_2^+/H^+ ratios differ between hydrogen derived from the mineral matrix and from gas molecules sputtered from the sample surface. The change in the H_2^+/H^+ and H^+/D^+ ratios with time demonstrates that it may take up to 4 days for the water vapour introduced with the sample to be reduced sufficiently to permit the isotopic analysis of hydrogen in amphiboles (Deloule, France-Lanord and Albarède, 1991(a); Deloule, Albarède and Sheppard, 1991(b); Deloule, Chaussidon and Allé, 1992). Baking the sample chamber together with the specimen, or pre-baking the sample in a preparation chamber, should decrease the time required to remove water vapour contamination.

6.4.8 Limitations on quantitative analysis

The precision and detection limit of the ion microprobe are ultimately limited by the number of ions which can be collected from any given sputtered volume. The total number of atoms (N) collected for a given isotope in an analysis is:

$$N = (V\rho/A) \cdot N_A \cdot C_a \cdot I \cdot n \cdot T \cdot D \qquad (6.3)$$

where V is the volume sputtered, ρ is the density and A the average atomic weight, N_A is Avogadro's number, C_a the atomic concentration of the element measured, I the isotopic abundance, n the ionization efficiency, T the proportion of the ions formed which are transmitted and counted, and D the 'duty cycle'. The duty cycle is the time spent measuring each peak as a proportion of the total analysis time (including time required for stabilization of the magnetic field after jumping to each mass in turn). As noted above, in isotopic analysis, the maximum secondary ion count rate may be limited by the use of a single detector. Once this count rate is achieved, improvements in transmission etc. only permit reduction in the volume analysed. Thus in isotope ratio measurements, the count rate of the minor isotope (R) may be restricted to:

$$R = 5 \times 10^5 \cdot (I_{min}/I_{maj}) \qquad (6.4)$$

where I_{maj} and I_{min} are the isotopic abundances of the major and minor isotopes respectively, so placing a limit on the precision with which this ratio can be determined. In trace element analysis there are generally no restrictions on the count rate of a single intense species, since either minor isotopes or multiply charged ions of major elements may be selected for reference. The duty cycle is usually arranged such that the precision of the very low abundance elements is improved. Despite longer counting times on the minor elements, the duty cycle losses may be over 90% for any one element if a large number of elements are analysed.

Measurement of the absolute ion yield is difficult since the transmission of the instrument is not usually known accurately. Measurements can, however, be made using major elements with no significant molecular interferences, under conditions which permit maximum transmission of ions (all apertures fully open, low mass resolution and collection of a large range of ion energies). The effective ion yield, which combines the ionization efficiency and the transmission, can be calculated by comparing measured secondary ion count rates with the volume of sample sputtered (R.W. Hinton, unpublished). Comparison between Na and K count rates from glass used for making thin sections, against known volumes of material sputtered, gave 15.5 Na and 23 K ions collected for every 100 atoms sputtered. Similar effective ionization efficiencies have been calculated for Na and K in SiO_2 by Migeon et al. (1989) using the same instrument. Ionization efficiencies for Na and K must therefore be similar to, or exceed, these values.

6.5 Applications: elemental analysis

The interpretation of data from 'bulk' analyses of mineral separates or fragments is often influenced, not by growth zoning or heterogeneity within a given mineral, but by the presence of other mineral phases. While the O^- beam currents required for trace element analysis give relatively large beam diameters ($10-30 \mu m$), there are many applications where this is not a significant limitation. The method of analysis will depend on a number of factors, including the instrument characteristics (especially the physical size, which influences transmission losses at high mass resolution) and the computer control program. Most trace element analyses to date have been made with relatively small instruments where transmission losses for high resolution are generally comparable with losses which occur using energy filtering. Energy filtering can be used in conjunction with low mass resolution; this approach limits the effect of factors such as sample charging and magnetic field drift on secondary ion intensities. Energy filtering may also be preferred because it reduces the sensitivity of the ion yields to matrix and instrumental conditions. Many light and trace element applications have been discussed in previous reviews (Shimizu, Semet and Allègre, 1978; Shimizu and Hart, 1982; Reed, 1989) and in the following discussion, emphasis will be placed, where possible, on recent work.

6.5.1 Light element analysis

The water content of minerals and glasses and the presence of other volatile species, e.g. F and CO_2, can have important implications for the formation of melts and their subsequent evolution. Volcanic glasses, in particular, frequently lose volatiles during eruption, so that the analysed water content may not reflect the original composition. Analysis of trapped melt inclusions in quartz, feldspar or olivine can give a much better indication of pre-eruptive melt chemistry than fresh glass, especially for the volatile species H_2O and CO_2 (Kovalenko, Hervig and Sheridan, 1988; Hervig and Dunbar, 1992; Webster and Duffield, 1991). Other trace elements, including ore-forming elements such as Mo, Sn and W, can also be measured by ion microprobe in the same volume of material. This information can be combined with independent electron microprobe analyses of the major elements and volatiles (F and Cl), thus giving a comprehensive picture of the major and trace element chemistry of the melts. In general, these analyses demonstrate that the magmas readily lose volatiles and alkalis but that other trace elements, including ore-forming elements such as Sn, are retained.

The concentrations of volatiles in experimental run products can readily be determined by the ion microprobe. Pan, Holloway and Hervig (1991) have measured the pressure and temperature dependence of carbon dioxide solubility in tholeiitic basalt melts by measuring the C^-/Si^- ratios of experimentally produced glasses. These studies confirmed that carbon was present as carbonate in the melt, but demonstrated that the CO_2 solubility is temperature independent. Similarly, Webster (1992(a),(b)) determined the water content of granitic glasses as part of a study of water solubility and chlorine partitioning in Cl-rich granites.

6.5.2 Trace element analysis

Trace element concentrations have been measured in quartz grains from mylonites (Hervig and Peacock, 1989). Variations were observed in Li concentrations (<1 to 25 ppm), but not Na or Al. A single grain was shown to be zoned: Li was lower in the outer $100 \mu m$ and was assumed to have been lost during metamorphism. The observed elemental profile implied diffusion coefficients of 7–10 orders of magnitude less than expected from consideration of laboratory-determined values, taking into account the temperature and duration of the metamorphism. The correlation between cathodoluminescence and trace elements in quartz has been studied by Perny et al. (1992). Changes in trace element concentration of three orders of magnitude were observed on a scale of a few tens of micrometres. Li and Al were found to be well correlated and zones of blue luminescence showed the highest Al and Li contents.

Oscillatory zoning of trace elements in augite (Shimizu, 1990) has also been demonstrated, with very large changes in concentration over small distances being found. The amplitude of the oscillations varied significantly between the different trace elements studied (Sc, Ti, Cr, V, Sr and Zr) and compatible (e.g. Cr) and incompatible (e.g. Ti) elements behaved antithetically. The trace elements Sc, Cr and V were all shown to suffer from minor molecular interferences, but these were insignificant at the concentration levels in these samples (>80 ppm) and could be ignored. Ba and Sr distributions in plagioclase in calc-alkaline magmas were studied by Blundy and Shimizu (1991). Observed variations were attributed to changes in magma chemistry rather than changes in temperature and pressure.

Trace element measurement in carbonates suffers distinct problems, owing to the presence of inclusions and the frequently strong zoning. Veizer *et al.* (1987) demonstrated variations in Sr, Mn, Fe and Mg in calcites and dolomites. Silicate inclusions can give high apparent trace element contents, caused not only by the presence of the trace element within the silicate but also by molecular interferences. Thus, Na and Si were also measured in this study, but these elements could only be used as a check for the presence of silicate inclusions. Swart (1990) showed that energy filtering can be used for the trace element analysis of calcite and dolomite and noted that the ion yield differences between the two matrices observed by Veizer *et al.* (1987) are then eliminated. In a study of zoned limestone cements, Mason (1987) confirmed the importance of Mn as an activator and Fe as a suppressor of cathodoluminescence. Potential molecular interferences in the carbonate mass spectrum were assessed systematically.

6.5.3 Rare-earth element analysis

Rare-earth elements (REE) are extensively used in the study of many geological processes, from high-temperature melt formation through to low-temperature aqueous geochemistry, and there is considerable interest in the REE concentrations and variations in many major and accessory minerals. Only moderate mass resolution is required for the analysis of light REE since many potential molecular interferences consist of combinations of three, or more, relatively low mass ($m < 56$) elements. Similarly, energy filtering reduces molecular interferences due to the major rock-forming elements to extremely low levels (generally $\ll 0.1$ ppm). However, oxides of Ba and the light REE overlap the heavy REE peaks and can only be resolved with very high mass resolution ($M/\Delta M \geq 8000$). Even then, some unresolved interferences may still occur, especially under poor vacuum conditions (Maas *et al.*, 1992).

The similarity in the chemistry of the REE permits systematic study of both molecular interferences and ion yields (Zinner and Crozaz, 1986(a), (b); Crozaz and Zinner, 1985; Reed, 1983, 1984, 1986; Mitchell and Reed, 1988; Hinton, 1990). Reed (1983) showed that large differences exist between the low-energy ion yields of the REE; most notably, Lu^+ yield is some 3.2 times lower than La^+ and Eu^+ is 2.1 times higher. However, Zinner and Crozaz (1986(b)) have shown that energy filtering not only reduces the MO/M ratios but also the differences in ion yields between elements, as is also shown in Figure 6.14 (R.W. Hinton, unpubl.).

Corrections for REE oxide overlap can be made in a number of ways. If

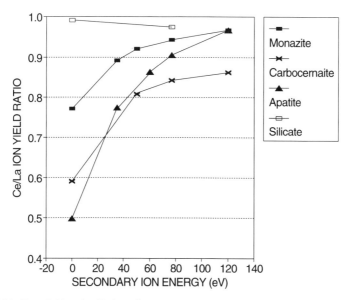

Figure 6.14 Ion yield ratio (Ce^+/La^+) dependence on secondary ion energy (40 eV energy window) for monazite, carbocernaite, apatite and silicate glass.

all the REE masses are recorded, it is possible to use matrix algebra to correct for the overlap of the individual light REE oxides on the heavy REE (Crozaz and Zinner, 1985; Zinner and Crozaz, 1986(b)). However, the overlap of GdO on Yb cannot be calculated in this way owing to the similarity of the isotopic abundance patterns of Gd and Yb, nor can the overlap of the mono-isotopic TbO on Lu be determined. The MO/M ratios correlate only approximately with the oxide bond strength. However, the fit can be improved if corrections are made to allow for the differences in the ionization potentials between elements and oxides (Reed, 1983). This relationship between MO/M and modified oxide bond strength permits an estimate to be made of the GdO/Gd and TbO/Tb ratios.

Corrections can also be made for oxide overlap by assuming that, for a given matrix, the REEO/REE ratios bear a constant relationship to each other so that, if one ratio is determined accurately, this can be used to calculate all the others. Thus, Hinton et al. (1988) showed that for a given matrix, the CeO/Ce, LaO/La, PrO/Pr and SmO/Sm ratios correlate with NdO/Nd over a wide range of secondary ion energies. Similar curves have been generated for monazite over the energy range 35–120 eV (Figure 6.15; R.W. Hinton, unpubl.).

The mass spectrum becomes significantly more complicated in fluorine-bearing minerals (e.g. apatite and biotite), owing to the presence of REE-F species. If the fluorine concentration is high, the REE signals cannot be

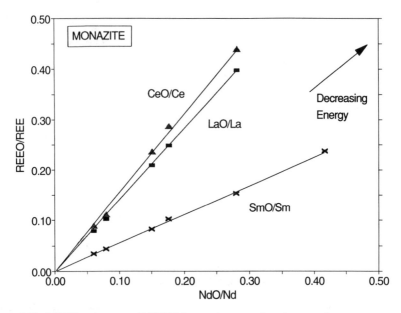

Figure 6.15 MO/M ratio versus NdO/Nd for varying secondary ion energies.

calculated until a correction is made for the REE-F species. A study of the presence of both oxide and fluoride molecular overlaps in apatite has been made using the Durango apatite standard, which has a REE pattern significantly biased towards the light REE (Zinner and Crozaz, 1986(b)). The CeO and CeF peaks are relatively large compared to the elemental peaks of Gd and Tb which they overlap and can be calculated accurately by making measurements with variable degrees of energy filtering.

Table 6.2 gives the REE ion yields and MO/M ratios for minerals and glasses determined by a number of laboratories by conventional energy filtering techniques. The ratios, when normalized to Nd, illustrate the consistent behaviour of the REE in the ionization of both the elements and the oxides (Figure 6.16). Given the consistency of these ion yield patterns in comparison with the errors inherent in the analyses of individual REE in many 'standard materials', the use of ion yields based on well-characterized glasses (e.g. NBS 610), adjusted by a common factor for all elements, may be a better guide to absolute concentrations than working curves based on a calibration standard having a few poorly determined concentrations.

(a) REE applications

The ion microprobe has been applied from an early date to REE determination in a variety of phases including silicates, oxides and phosphates. Although much of this work was on phases in which REE were reasonably abundant (>1 ppm), analyses are now being extended to silicate minerals

Table 6.2 REE ion yields relative to Ca and MO/M ratios for various phases

	Apatite [1]	Hibonite [2]	Silicate [3]	NBS silicate [4]	REE silicate [5]	Carbocernaite [5]	Garnet [6]
M/Ca							
La	0.74	0.68	0.75	0.70	0.68	0.62	–
Ce	0.67	0.63	0.66	0.68	0.60	0.52	–
Pr	0.82	0.75	0.75	0.76	0.68	0.66	–
Nd	0.85	0.77	0.82	0.79	0.73	0.68	–
Sm	1.05	0.93	0.91	0.85	0.82	–	–
Eu	1.23	1.09	1.03	0.90	0.97	–	–
Gd	1.00	0.88	0.87	0.84	0.76	–	–
Tb	0.68	0.60	0.79	0.77	0.76	–	–
Dy	0.79	0.69	0.77	0.77	0.71	–	–
Ho	0.71	0.62	0.77	0.78	0.75	–	–
Er	0.69	0.61	0.72	0.75	0.73	–	–
Tm	0.69	0.61	0.75	0.74	0.73	–	–
Yb	0.59	0.60	0.74	0.76	0.72	–	–
Lu	0.53	0.50	0.57	0.63	–	–	–
MO/M							
La	0.211	0.133	0.158	0.193	0.191	0.208	0.167
Ce	0.306	0.163	0.195	0.203	0.227	0.253	0.194
Pr	0.238	0.126	0.141	0.155	0.170	0.185	0.135
Nd	0.172	0.089	0.121	0.130	0.130	0.144	0.111
Sm	0.079	0.042	0.065	–	0.074	–	0.061
Eu	0.059	0.031	0.048	–	0.059	–	0.052
Gd	0.155	0.082	0.103	0.117	0.117	0.130	0.107
Tb	0.140	0.074	0.096	–	0.115	0.120	0.105
Dy	0.097	–	0.065	0.092	0.084	–	0.076
Ho	0.093	–	0.064	–	0.082	–	0.082
Er	0.095	–	0.066	0.078	0.081	–	0.083
Tm	0.077	–	0.052	0.054	0.065	–	0.070
Yb	0.062	–	0.047	0.042	0.061	–	0.059
Lu	0.098	–	0.067	0.068	–	–	0.115

[1] Zinner and Crozaz (1986b).
[2] Drake and Weill REE glass. Zinner and Crozaz (1986(a)).
[3] Madagascar Hibonite Fahey et al. (1987).
[4] NBS-610 glass. Hinton (1990).
[5] Carbocernaite REE glass. This work.
[6] Synthetic garnet glass. J. Craven (pers. comm.).

and glasses down to the 10 ppb level (Lundberg et al., 1988). Precisions, based on counting statistics, are generally ±1–10 ppb, although calculated errors of only ±0.2 ppb have been quoted for La in pyroxene (Lundberg, Crozaz and McSween, 1990).

The REE distributions in garnet and pyroxene are now being actively studied by a number of groups and measurements frequently include analysis of other trace elements by ion microprobe, including Sc, V, Zr and Nb. Previous experimental studies of partitioning have often been made on only

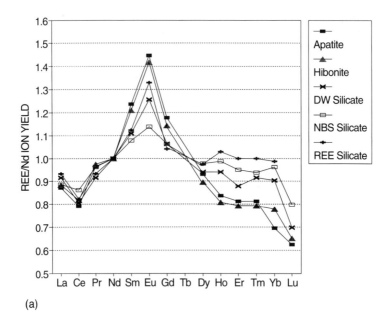

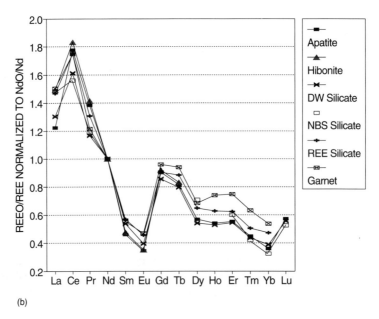

Figure 6.16 (a) Nd normalized ion yields of REE for apatite, hibonite (Fahey *et al.*, 1987) and silicate glasses. (b) REEO/REE ratios for apatite, hibonite (Fahey *et al.*, 1987) and silicate glasses normalized to the NdO/Nd ratio.

a limited number of elements and, in these circumstances, it is difficult to derive an internally consistent set of partition coefficients. Furthermore, where partitioning is extreme, the possibility that contamination or inclusions contribute to errors in the calculated coefficients is great. Examples of ion microprobe studies in this field include Caporuscio and Smyth (1990), Johnson, Dick and Shimizu (1990), Mazzuchelli et al. (1992), Sisson (1991) and Sisson and Bacon (1992) and Hart and Dunn (1993). Also, Hickmott and Spear (1992) studied trace element and REE zoning in metamorphic garnets. Harris, Gravestock and Inger (1992) measured trace element distributions in garnet and coexisting phases from anatectic assemblages. Shimizu and Richardson (1987) reported REE and trace element concentrations in garnets from diamond inclusions.

Trace elements have been measured in feldspars; the presence of Ba might be expected to lead to significant problems of BaO overlap on Eu. Fortunately Eu is often enriched in feldspars, so the correction is usually relatively minor. BaO is more likely to create problems in minerals such as amphiboles, where Ba is relatively high and Eu low. The chondrite-normalized REE concentrations of sanidine were found to decrease by almost three orders of magnitude from La to Sm (Hinton and Upton, 1991). The heavy REE were not measured in this case, owing to their low concentration and the large overlap of light REE interferences; however, the measured Y concentration indicated that the REE concentration continued to fall for the heavy REE.

A number of accessory minerals which are commonly present in rocks have relatively high REE concentrations. Both allanite and monazite, where REE are the dominant major elements, are usually extremely rich in the light REE and chondrite-normalized concentrations fall by three, or more, orders of magnitude from La to Lu (e.g. Reed, 1985). In these circumstances, even small errors in the oxide correction will lead to significant apparent changes in the heavy REE pattern. Apatite is a common accessory phase in terrestrial rocks, though many of the analyses to date have been on meteoritic material (Crozaz and McKay, 1990, Crozaz et al., 1989). Similar phosphates, e.g. merrillite (Reed, Smith and Long, 1983, Reed and Smith, 1985), stansfieldite and whitlockite (Davis and Olsen, 1991, Olsen et al., 1991), can be measured using the same methods as for apatite. The smooth chondrite-normalized curves through both the light and heavy REE testify to the effective removal of both the oxide and fluoride interferences. Large variations have been observed within and between individual apatite grains from a pegmatite (Jolliff et al., 1989), including Eu anomalies changing from positive to negative across a single crystal.

Zircon has been analysed extensively by ion microprobe for U/Pb isotopes, but only relatively recently for its trace element abundances. The REE pattern is invariably heavy REE-rich and the overlap correction for the light REE on the heavy REE is relatively minor. Analyses of terrestrial zircons all give large positive Ce anomalies (Zinner and Crozaz, 1986(b); Hinton and Upton, 1991; Maas *et al.*, 1992). Although the analyses of Maas *et al.* (1992) were made at high mass resolution, Pr could not be measured in some zircons, owing to the overlap of CeH. Somewhat unusually, the light REE were also measured as oxides, since (under the low ion energy conditions used) these were about three times more intense than the elemental peaks. Hinton and Upton (1991) have shown that the large positive Ce anomalies are caused by the presence of only small amounts of Ce^{4+} and are not associated with highly oxidizing conditions.

6.5.4 Imaging

Improvements in both detectors and computer hardware now permit sophisticated digital ion microprobe imaging. Digital images can be made by recording the fluorescent screen image with a CCD camera or by replacing the fluorescent screen with a position sensitive detector (e.g. resistive anode encoder, or RAE). Rastered ion images are generated by synchronizing the storage of the secondary ion counts from the electron multiplier with the primary beam raster. Ion images can readily be manipulated by software available for microcomputers and even three-dimensional images are possible.

Imaging has not been applied extensively to natural samples, but mostly to experimental run products, especially in diffusion studies (Elphick *et al.*, 1991). As the conducting coat is removed over relatively large areas, high-resolution imaging requires the use of electron flooding to prevent sample charging. Imaging was used to study fluid–mineral interactions by Elphick *et al.* (1991). Elphick, Dennis and Graham (1986) and Graham and Elphick (1991) also used imaging in diffusion experiments (section 6.5.5). Walker (1990) imaged ^{18}O which was exchanged with ^{16}O in alkali feldspars and showed a correlation with micropermeability.

To achieve the highest spatial resolutions in ion microscope images ($<5\,\mu m$), considerable restriction must be placed on the energy and angular spread of the secondary ions. The loss of intensity with increasing spatial resolution is much steeper for the ion microscope than the decrease in primary beam current with primary beam diameter for the ion microprobe. Thus, although ion microscope imaging is more efficient for low resolution images, rastered ion imaging is the most efficient method for high-resolution images. Slodzian (1988) noted that the crossover between the efficiencies of the two techniques is at a resolution of about $2.5\,\mu m$.

6.5.5 Depth profiling

The removal of material as a result of sputtering has the positive advantage that chemical and isotopic variations can be measured with depth into a crystal. In depth profile measurements, the beam is either rastered or defocused to produce a uniform beam density and generate a flat-bottomed crater (Colby, 1975). The fundamental processes which affect the depth resolution obtainable in depth profiling are outlined by Armour *et al.* (1988). Collisions with the matrix atoms lead to a collision cascade (Figure 6.1) and, as well as ejecting ions from the surface, internal relocation also takes place. This causes a net movement of atoms towards the surface as the implanted ions are accommodated. Some matrix atoms are also forced to greater depth through direct momentum transfer from the incoming primary beam. Since half to two-thirds of the sputtered surface consists of implanted primary beam atoms, much of the original atomic bonding structure must be lost. The depth over which mixing of atoms occurs is related to the primary ion energy, angle of impact, etc.: for most depth profile measurements of minerals (with a 10–30 keV primary beam energy and less than 45° incidence angle) it is approximately $0.02\,\mu m$.

Many synthetic materials on which depth profiling is carried out have the inherent advantages that, as well as being conducting or semiconducting, the surface is initially flat and contains no flaws. In contrast, geological samples are likely to be non-conducting and contain defects or grain boundaries. Crystal surfaces are rarely perfectly flat and even when flat surfaces are specially prepared for experimental diffusion studies, they are often degraded during the experiments (Elphick *et al.*, 1991). The ultimate depth resolution is therefore degraded, not only by sputtering effects, but also by the roughness of the original surface and by crystal imperfections. Diffusion profiles are also affected by any change in sputter rate with depth; however, Giletti, Semet and Yund (1978) demonstrated that the ratio between the implanted and matrix O quickly stabilizes and therefore the sputter rate is constant with depth (to a depth of at least $10\,\mu m$). It is thus possible to determine depth profiles of $^{18}O/^{16}O$ ratios, and thereby determine O diffusion rates, using a $^{16}O^-$ primary beam.

Depth profile measurements are most frequently made in diffusion studies. Laboratory studies are limited by the time scale, which must balance the selection of geologically reasonable temperatures, pressures, water contents, etc. against the diffusion distances which can be measured experimentally. If smaller distances (depths into a crystal) can be measured accurately, the experiments can be extended towards more geologically relevant conditions. Experiments frequently require the determination of the depth of movement of a trace element or isotopic tracer into the flat surface of a crystal. In this type of experiment, very high concentrations of an element or isotope may be present on the sample surface, while the concentration decreases rapidly with depth. As the layers are sputtered away, the contrast in concentration between the edges and the bottom of the crater increases and they may eventually differ by many orders of magnitude. The secondary ions collected must therefore be restricted to those generated in the centre of the sputtered area, by either electronic or mechanical gating of the secondary ion signal.

However, while greatly reducing crater edge effects, these measures cannot prevent contamination owing to the transfer of ions from the crater walls onto the crater floor.

The detection limit for depth profiling ultimately depends on the number of atoms sputtered, the ionization efficiency of the element and the transmission losses. Williams (1985) calculated that part-per-billion detection limits can only be achieved with a depth resolution of $0.02\,\mu m$, if a relatively large area ($>70\,\mu m^2$) is sputtered and ionization efficiencies are high ($>1\%$).

The diffusion rate of oxygen is very important for understanding isotopic variations in high temperature metamorphic rocks. Diffusion rates have been determined by ion microprobe analysis, under varying conditions of P, P_{H_2O} and T, for a variety of minerals. The relatively high efficiency of negative ion formation for O, coupled with the high concentrations of this element, enable the diffusion to be measured using virtually any source of primary ions. Elphick, Dennis and Graham (1986) used an Ar^+ beam to depth profile grossular, quartz, rutile and albite. Charging of the sample was prevented by use of electron flooding of the sputtered area. The shape of the diffusion profile of the tracer isotope (^{18}O), normalized to the total oxygen ($^{18}O + ^{16}O$), can be used to identify whether diffusion alone is operating. Evidence for the presence of a surface layer was identified in the experimental diffusion of albite (Elphick, Dennis and Graham, 1986). In contrast, Farver and Yund (1990, 1991) studied oxygen diffusion in alkali feldspar and quartz by this method, using a $^{16}O^-$ primary beam and demonstrated correlation of diffusion rates with water fugacity but not oxygen fugacity, hydrogen fugacity or hydrogen concentration.

Cs^+ is potentially the best primary beam species for oxygen isotope depth profiles, given that its sputter yield is high, it can produce a well-focused spot and it enhances the O^- ion yield. The charging inherent with its use can be compensated by flooding with electrons. Elphick and Graham (1988) measured oxygen diffusion in quartz and anorthite using a Cs^+ primary ion beam and low-energy electron charge neutralization. Finely focused beams can also be used for imaging the distribution of isotopes or elements in diffusion run products (Graham and Elphick, 1990, 1991; Elphick et al., 1991). While only approximate diffusion rates can be obtained from the images, they can be used to demonstrate whether diffusion is the only exchange process occurring. Graham and Elphick (1990) and Elphick et al. (1991) used ion imaging to show that in some cases solution/precipitation was occurring rather than diffusion. Both mechanisms were observed in the same crystal and appeared to be related to the crystal structure.

Depth profiling can be used to study the chemical changes in the surfaces of minerals during weathering. The specimen isolation (SI) technique and depth profiling have been used to study the weathering of feldspar in both natural surfaces and experimental run products (Muir and Nesbitt, 1991, 1992, and references therein). The SI technique has the advantage that no conductive gold or carbon coat is required and therefore the surface can be analysed in a pristine condition. The Si-normalized depth profiles for Ca and Al indicate that Ca is removed in advance of Al during leaching and may be lost by exchange with protons prior to the break-up of the Si–Al framework.

An interesting application of the depth profiling technique is the analysis of fluid inclusions. Recent work by Diamond et al. (1990) attempted quantification of the observed secondary ion yields for the alkali metals, Na and K, and for Ca in such inclusions. Importantly, this work demonstrated that individual inclusions could be analysed in minerals containing more than one generation of inclusion. Sputtering using a 200–500 nA O^-, 12.5 keV, primary beam focused to 75–100 μm diameter permitted excavation of pits at a rate of 0.6–0.9 μm/min^{-1}.

6.6 Applications: isotopic analysis

The application of the ion microprobe to the isotopic analysis of geological and meteoritic materials is extensively covered in proceedings of workshops (Shanks and Criss, 1989; Anon, 1992) and especially the reviews by Zinner (1989) and Reed (1989). Although significant progress has been made in this field, the precision achieved in isotope ratio measurements rarely approaches that possible by conventional gas- and solid-source mass spectrometry. In general, ion microprobe analyses are at least an order of magnitude less precise than conventional methods. Precisions of ± 0.5–1‰ may be expected for most elements, with the exception of H, where a more typical value is ± 10‰. However, where isotopic variations occur on a scale smaller than the minimum sample size required for conventional techniques, the ability to determine such variations may outweigh the sacrifice in precision. Furthermore, the summation of repeat analyses can improve the overall precision to a level that approaches that of conventional analyses, while still using substantially less material.

Isotope ratio measurement differs from elemental analysis, in that there is no choice as to which peaks are recorded and the abundances of the isotopes measured may differ by many orders of magnitude. Often the minor isotope has a significant molecular interference which cannot be totally eliminated by energy filtering. Furthermore, when the minor isotopes are heavier by one mass unit than the major isotope reference mass, there is a high possibility of a significant molecular hydride overlap. Such overlaps cannot be removed by conventional energy filtering techniques (section 6.3.5). Isotope ratio measurements are, therefore, more commonly made at high mass resolution rather than by energy filtering, with instrumental conditions tailored to those required for accurate determination of the minor isotope.

In geological processes, the isotopic fractionation at high temperature of major elements such as Si, Mg, Ca, etc. between minerals, or between minerals and melts, is very low. Probably only low-temperature water–solid interactions and very high-temperature evaporation/condensation reactions lead to isotopic fractionation in the major element cations sufficiently large to be measurable by ion microprobe techniques. For lower mass elements (O, C, B and H), mass fractionations can be sufficiently large to permit the detection of variations by ion microprobe analysis. However, larger matrix and instrument-related fractionations also occur for these elements. For H, in particular, the large relative difference in mass between the isotopes makes instrumental fractionation both large and more difficult to predict.

Isotope fractionation due to differential dead-time losses is discussed in section 6.4.7. The largest mass fractionation effects are those generated in the ionization process, and the other fractionation effects can be combined with these to give an overall matrix-dependent fractionation factor. Isotope ratios measured by SIMS always show a fractionation in favour of the light isotopes. For constant energy, velocity decreases as mass increases; therefore, it is the heavier, slower isotopes which are preferentially lost. The degree of fractionation between elements and, for the same element, between matrices, has also been shown to be correlated with the width of the energy distribution (Lorin, Havette and Slodzian, 1982). For example, Si has a wider energy distribution than Mg and has a significantly larger isotope mass fractionation (Si = 1.75‰/amu compared to Mg = 0.38‰/amu in olivine). Zinner (1989) also noted that fractionation is inversely related to the ionization efficiency; thus Mg has a higher ionization efficiency and a lower isotope fractionation than Si. The degree of fractionation is reduced if high-energy (faster) ions are measured. A linear relationship between velocity (energy) and measured mass fractionation has been demonstrated convincingly over the ion energy range 5–30 eV for B, Si and Ca (Gnaser and Hutcheon, 1987). As the velocity (energy) of the ions increases, the fractionation decreases, and approaches the absolute ratio determined by conventional mass spectrometry. In contrast, Deloule, Chaussidon and Allé (1992) demonstrated that, for hydrogen, the fractionation increases as the velocity increases. Furthermore, a linear relationship may not hold for all elements or for energies higher than 30 eV (Shimizu and Hart, 1982; Södervall et al. 1987; Engström et al., 1987).

While the degree of mass fractionation generally decreases with increasing mass (Shimizu and Hart, 1982), there are significant differences between elements of similar mass (e.g. Mg and Si, Mo and Zr). For some elements, the fractionation between matrices can be very large, e.g. 1.5–6% per amu for S (Eldridge et al., 1989) and close matching of the mineralogy (and chemistry) between standard and unknown is required. However, fractionation factors between matrices should be constant: thus, once they have been accurately established, a single standard may be used.

For stable isotope measurement, the mass fractionation is calculated by direct comparison between the standard and the unknown such that:

$$F = \left(\left(\frac{(m_2/m_1)_{\text{meas}}}{(m_2/m_1)_{\text{ref}}}\right) - 1\right) \times 1000 \qquad (6.5)$$

where F is the mass fractionation in parts per 1000 and m_1 and m_2 the individual isotopes. Here m_1 is often the lighter, more abundant, isotope. However, for stable isotope measurement the absolute fractionation factor can be neglected if a direct comparison can be made between the isotope ratios on a standard (of known isotopic composition) and those measured on an unknown of similar chemistry and mineralogy. Thus, if the isotope ratio measured on the standard, $(m_2/m_1)_{\text{std}}$, is substituted for $(m_2/m_1)_{\text{ref}}$, F becomes equivalent to δ (the difference in parts-per-thousand between the unknown and the internationally defined standard material (e.g. SMOW)). If an element has three or more isotopes, it is possible to determine the mass fractionation using one isotope pair and give an absolute ratio for all other isotopes. Thus, if the $^{87}Sr/^{86}Sr$ ratio is required, the instrumental fractionation can be measured and corrected for, using the $^{86}Sr/^{88}Sr$ ratio. Also, in the analysis of Ca isotope anomalies in meteorites, instrumental fractionation is measured using the ^{44}Ca and ^{40}Ca isotopes. The fractionation of ^{48}Ca relative to ^{40}Ca is not simply twice that of ^{44}Ca to ^{40}Ca, but is slightly less. The observed fractionation, per mass unit, decreases with increasing mass. Fahey et al. (1987) have shown that the instrumental mass fractionation (α) can be best corrected using a power law (frequently referred to as an exponential law):

$$\left(\frac{M_2}{M_1}\right)^\alpha = \left(\frac{(m_2/m_1)_{\text{meas}}}{(m_2/m_1)_{\text{ref}}}\right) \qquad (6.6)$$

where M_2 and M_1 are the isotope masses and m_2/m_1 are the measured isotope ratios (meas) or reference ratio value (ref). Once α is determined for one isotope pair, this correction can be used for all other isotopes (usually referenced to the most abundant isotope). A comparison between a linear correction and the exponential law is given in Figure 6.17. It should be noted that where fractionation is generated by natural processes, e.g. volatilization, the observed mass fractionation may follow a different law, e.g. the Rayleigh law (Ireland et al., (1992)).

6.6.1 Stable isotope measurement

Hydrogen occurs as two stable isotopes, H ($m = 1$) and D ($m = 2$). The abundance fraction of deuterium (D) is only 1.6×10^{-4}, hence it is very difficult to determine δD with high precision. However, hydrogen has the largest relative difference in mass between the isotopes and very large variations in D/H are observed in nature. According to McKeegan, Walker

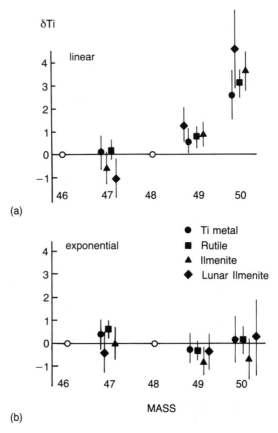

Figure 6.17 The titanium isotopic composition measured in three terrestrial and one lunar sample. Plotted are deviations (per mil) between ratios measured relative to ^{48}Ti corrected for mass fractionation based on the isotope pair ^{46}Ti and ^{48}Ti with (a) linear corrections and (b) assumed exponential mass fractionations. (From Fahey et al., 1987).

and Zinner (1985), approximately equal numbers of positive and negative secondary ions are produced. There is significant H_2^+ at mass 2, which requires moderate mass resolution (>1300), whereas H_2^- is negligible, permitting analysis of the negative ions at low mass resolution (McKeegan, Walker and Zinner, 1985). Although higher intensities are possible using a Cs^+ primary beam, charging problems generally inhibit its use on insulators. Deloule, France-Lanord and Albarède (1991(b)) and Deloule, Chaussidon and Allé (1992) demonstrated that H^+ intensities generated using an O^- primary beam are sufficiently high to permit the analysis of hydrogen isotopes in amphiboles to a precision of ±10‰. Although observed hydrogen isotope fractionation is very sensitive to variations in instrumental conditions and the vacuum (P_{H_2O}), these effects can be controlled. The H_2^+/H^+ ratio of hydrogen from minerals and that from adsorbed water differ by over an order of magnitude. Measurement of this ratio therefore permits the relative contributions of the two components to be monitored and limits to be set on the H_2^+/H^+ ratio to ensure that analysis is predominantly of structural water (Deloule, Chaussidon and Allé, 1992). Matrix-dependent isotope fractionation, although substantial, can be corrected by regressing the measured fractionation of standards against the major element concentrations of the minerals analysed. Deloule, FranceLanord and Albarède (1991(b)) demonstrated that Fe, Mn and Ti appear to have the greatest influence on the fractionation. In a study of hydrogen isotopes in amphiboles from mantle xenoliths, Deloule, Albarède and Sheppard (1991(a)) identified large variations within mantle-derived materials. The hydrogen isotopic composition of a single 2 mm pargasite grain was mapped and showed approximately concentric zonation, with variation of δD from $-60‰$ in the core to $-120‰$ in the rim.

Chaussidon and Albarède (1992) demonstrated that it is possible to measure the B isotopic composition of tourmalines, despite matrix-dependent variations in mass fractionation. Corrections can be applied, based on the mass/charge ratio and ionic intensities of the octahedral cations, in a manner similar to that used by Deloule, Albarède and Sheppard (1991(a)) for hydrogen isotopes. Large variations in the boron isotopic composition were observed (from -2 to $-30‰$). Two distinct groups were identified: a high Li and high $\delta^{11}B$ group which included B from a marine reservoir, and a low Li and lower $\delta^{11}B$ group derived from a continental reservoir.

Carbon is highly abundant in nature and is fractionated by a variety of physical, chemical and biological processes. The minor isotope ^{13}C is reasonably abundant, at approximately 1%; the accuracy of isotopic measurements is, therefore, limited by instrumental fractionation rather than counting precision. Carbon isotope measurements have been made in meteoritic material (McKeegan, Walker and Zinner, 1985; Zinner, Tang and Anders, 1987, 1989) on either conducting samples (graphite) or non-conducting samples pressed into gold foil. Precisions were approximately 5‰; higher precision was not required, owing to the extreme isotopic variations found (over 100%). Diamonds are sufficiently conducting that they can be analysed using a Cs^+ beam without charge build-up (Harte and Otter, 1992). The C isotopes are measured at $M/\Delta M$ of 4300 to permit complete separation of $^{12}CH^+$ from $^{13}C^+$ and a counting precision of 0.6‰ can be obtained. However, variations of up to 1.5‰ were observed over small areas of the diamond standard and this is a more realistic estimate of the reproducibility of the analytical technique.

Oxygen is one of the principal constituents of geological materials and knowledge of oxygen isotope variations within individual mineral grains would give considerable information on igneous, metamorphic and hydrothermal processes. The isotopic analysis of oxygen requires a Cs^+ primary beam; charging of non-conducting samples is therefore a problem. The ability to make precise and accurate ion microprobe measurements routinely and relatively quickly on silicates and carbonates (i.e. non-conductors) is an important goal which has yet to be fully realized. An initial study, made on meteoritic material, avoided charging by analysing very small crystals crushed into gold foil (McKeegan, 1987). Overall reproducibility of the mass fractionation was 2‰ (2σ) for individual grains but variations of up to 5–8‰ from the average were observed between individual analyses (McKeegan, 1987). This study also included measurement of ^{17}O and, although determination of absolute fractionation was relatively poor compared to conventional techniques, the $^{17}O/^{16}O$ ratio, once corrected for mass fractionation using the $^{18}O/^{16}O$ ratio, was very well determined. These authors demonstrated that the ^{17}O abundance as determined by E.O. Nier is incorrect and the true value for Standard Mean Ocean Water (SMOW) is 0.00038288 ± 0.00000028 (2σ). This measurement has been confirmed by Lorin (1992). Further studies have been made on meteoritic material by Lorin and co-workers (Lorin et al., 1990; Lorin, 1992), using minerals prepared as thin sections. Analysis of terrestrial standards showed a fractionation of 0.35‰/amu, which was not dependent on either the energy band width or matrix (Lorin et al., 1990).

Valley and Graham (1991, 1992) demonstrated that it is possible to measure O isotope variations of a conducting sample (magnetite) to 1‰ (1σ) precision. The combination of conventional and ion microprobe analyses showed that the grains were homogeneous to ±1‰, except at the crystal rim. The variation in surface topography and the presence of a nearby non-conductor (calcite) prevented analysis of the crystal rims by point analysis; however, depth profiling of a flat crystal face revealed depletion of $\delta^{18}O$ in the outer 10 μm of up to 9‰. Subsequent bulk analysis of small magnetite grains confirmed the presence of low $\delta^{18}O$ rims. Oxygen isotope variations ere also mapped across a magnetite grain, highlighting the heterogeneity of some grains.

As with the H isotopes, the S isotopes can be measured as either positive or negative secondary ions. Since many sulphides are conducting, analyses can be made using Ar^+ primary ions (Pimminger et al., 1984) or Cs^+ (Graham and Valley, 1992). However, most measurements have been made using O^- bombardment and positive secondaries, owing to instrument limitations (Eldridge et al., 1989, and references therein). Measurements must be made at moderate (3000–4500) mass resolution to separate O_2^+ (introduced from the primary beam) and hydrides (introduced from the vacuum). Operation at moderate resolution does not cause significant limitations on the count rates for single-collector instruments. Large matrix-dependent mass fractionation of the S isotopes and ion yield variations are found; great care must therefore be taken to match the chemistry of standard and unknown. Analysis of sulphide inclusions from diamonds (Chaussidon, Albarède and Sheppard, 1987; Eldridge *et al.*, 1991) demonstrated significant isotopic heterogeneity. This was used to infer the recycling of sedimentary materials into the diamond source region. Comparison between ion microprobe S isotope data and conventional analyses gave very similar ranges of values (Eldridge *et al.*, 1989).

6.6.2 Radiometric dating

The ion microprobe cannot make accurate measurements in cases where isobaric overlaps of a radioactive parent and daughter occur. Thus, ion microprobe isotopic analysis of Nd–Sm and Rb–Sr systems is not possible. In samples where the Sr/Rb ratios are very high and molecular interferences are low (e.g. aragonite or carbonate from carbonatite), low-resolution measurement of Sr isotopes has been possible (Exley, 1983). The precision obtainable for Sr isotopes (1‰) is sufficient to give the isotopic signature of the Sr source.

Even with no isobaric overlap, the concentrations of Sm are too low to give the precision necessary for dating in all but a few minerals (e.g. allanite and monazite). However, the measurement of the distribution of Nd and Sm and variations in the Nd/Sm ratios can be useful in the interpretation of conventional isotopic analysis of mineral separates. Similarly, distributions of Rb and Sr within and between individual mineral phases can help in the interpretation of Rb/Sr systematics.

6.6.3 Pb–Pb and U–Pb dating of zircon

Zircon is an extremely useful mineral for dating igneous and metamorphic events since it is capable of withstanding high-grade metamorphism, including anatexis. Furthermore, it survives weathering and transport and can also be found in sediments. Not only do the original grains survive, but later overgrowths may occur during igneous and metamorphic activity, and can also be dated. Conventional U–Pb analysis has developed from the study of multi-grain separates to the analysis of single grains and even grain fragments. The presence of areas within zircon crystals which have suffered element mobility, caused largely by radiation-induced metamictization, has led to special techniques to analyse separate areas of individual grains selected by either mechanical abrasion (Krogh, 1973) or stepped heating methods (Kober, Pidgeon and Lippolt, 1989). In ion microprobe analyses, areas of alteration and metamictization can be avoided. Furthermore, where different growth stages can be observed optically, individual areas from each stage of growth can be readily analysed.

In the analysis of zircon using the ion microprobe technique, operating conditions have also developed, from initial attempts using low mass resolution (Hinthorne et al., 1979) to recording the Pb isotopes alone at moderate resolution (Hinton and Long, 1979) and finally high-resolution U–Pb dating using the SHRIMP ion microprobe (Compston, Williams and Meyer, 1984). The best precisions that can be achieved in a single analysis using the ion microprobe, (i.e. 1‰ in ^{207}Pb/^{206}Pb and 2% in U–Pb (Williams and Claesson, 1987), are a factor of between two and ten times worse, respectively, than conventional analytical methods (Roddick, Loveridge and Parrish et al., 1987). However, if a number of analyses on a single grain are combined, precision can approach that of the single grain technique. The uncertainty in the U/Pb ratio becomes more significant when relatively young rocks are analysed, since the age is mainly dependent on this ratio (Compston et al., 1992). The mean ages of Palaeozoic volcanic zircons have been determined by this method to precisions of about 7 Ma (Compston et al., 1992), compared to precisions of 2 Ma for the conventional technique (Tucker et al., 1990). In both cases there was evidence that the magma contained some much older (inherited) grains, again highlighting the danger of multi-grain analysis.

Extensive measurements of Sri Lankan zircon reference samples by the SHRIMP ion microprobe have demonstrated that significant variations in the U^+ and Pb^+ ion yields occur, depending on the analytical conditions, such as beam density and sample charging. However, it has been shown that instrumentally derived changes in the measured U^+/Pb^+ ratio correlate with the UO^+/U^+ ratio and that an absolute U/Pb ratio can be calculated using UO^+/U^+ vs. U^+/Pb^+ curves generated from a Sri Lankan zircon reference sample (Compston, Williams and Meyer, 1984; Compston et al., 1992; Williams and Claesson, 1987). The observed correlation may be associated with matrix-dependent variations in ion yield but may also be related to the extreme differences in the energy distributions of Pb^+, U^+ and UO^+ (Reed, 1989). The energy distribution widths are in the order $U^+ \gg UO^+ \gg Pb^+$ and, as the average ion energy increases, the Pb^+ intensity decreases substantially more than that of UO^+, while U^+ remains virtually unchanged. Thus U^+/Pb^+ increases faster than UO^+/U^+. The matrix dependence is probably related to the strong formation of U–O bonding; increased oxygen will increase UO^+ but decrease U^+, at the same time as improving the efficiency of Pb^+ (M. Schumacher, personal communication). It is probable that the observed U^+/Pb^+ vs. UO^+/U^+ correlation is caused by a combination of both factors. Whatever the cause, the same curves are observed for both O^- and O_2^- primary beams (P. Kinny, personal communication). Since U^+ and Th^+ have similar mass, oxide formation and energy distributions, the ratio of these two elements is relatively unaffected by changes in matrix composition or charging effects and the ratio can be determined to at least 1%. While U^+/Pb^+ ratios can be determined to 2%, the absolute concentrations are known only to a precision which is limited by observed variations within the reference sample (approximately 20%; Williams and Claesson, 1987).

The correction for common lead can be made by direct analysis of ^{204}Pb, but this peak is often of very low intensity and has large statistical errors. Corrections can be applied using the measured ^{208}Pb/^{206}Pb and U/Th ratios. The ^{208}Pb/^{206}Pb ratio is almost entirely dependent on the U/Th ratio and is relatively insensitive to age. In closed systems there should be a linear relationship between the ^{208}Pb/^{206}Pb ratio and Th/U (Compston et al., 1986). Deviations from linear arrays where ^{208}Pb is in excess may be assumed to be caused by the presence of common lead and corrections can be made based on this assumption (Compston, Williams and Meyer, 1984; Compston et al., 1992). In relatively young rocks (<200 Ma) ^{207}Pb/^{206}Pb does not change rapidly with age and ^{207}Pb is also very sensitive to the presence of common lead. Ages can therefore be determined in young rocks using ^{238}U/^{206}Pb ratios, following a correction for common lead based on the ^{207}Pb peak (Coenraads, Sutherland and Kinny, 1990).

The ion microprobe is especially suited to the analysis of zircons from Precambrian metamorphic and sedimentary rocks, since these can contain variations in ages between populations, as well as between different areas of individual crystals (including overgrowths). Ion microprobe analyses have identified the oldest known terrestrial material, with an age of 4.2 Ga (Froude et al., 1983; Compston and Pidgeon, 1986). Ion microprobe measurements of Precambrian zircons from sedimentary rocks frequently give a small number of ages which are significantly older than those determined by the conventional analysis of single zircon grains. There has been some debate as to whether ion microprobe analyses represent real ages (Hinton and Long, 1979; Froude et al., 1983; Schärer and Allègre, 1985; Compston et al., 1985), especially for ages in excess of 4 Ga. Kober, Pidgeon and Lippolt (1989) used thermal ionization mass spectrometry to analyse zircons from ancient metaquartzites which had previously been identified by ion microprobe analysis as containing a few grains of greater than 4 Ga and confirmed that some ^{207}Pb/^{206}Pb ages were indeed in excess of 4 Ga. Further, Kinny, Compston and Williams (1991) studied the Lu/Hf isotope system using the ion microprobe for the same zircons and also confirmed the very old ages.

6.7 Future developments

It is clear from the applications given above that the ion microprobe is already extensively used in geological studies. There are many projects which can be undertaken using existing ion microprobe technology and it is likely that the present instruments will have a part to play for many years to come. The increasing use of ion microprobes with very large mass spectrometers, and hence high transmission at high mass resolution, will permit lower detection limits to be achieved and the analysis of extremely low abundances in either trace element or isotopic work. However, since the large high-transmission instruments already approach 100% transmission at >5000 mass resolution, further improvements will only be possible if the efficiency of secondary ionization can be increased, perhaps by laser ionization. Future isotopic research may, for example, include the study of the U and Th decay series. A major advance in stable isotope research will probably occur only when isotopic measurements can be made routinely with a multi-collector instrument. It is possible that the greatest improvements in precision will come not simply from increased count rates on minor species, but from the ability to control the fractionation which occurs in the collection of the secondary ions. Improved knowledge of the ionization process would reduce the reliance on standards and extend quantitative analysis to elements for which standards are difficult to obtain.

Acknowledgements

The author wishes to acknowledge support from the UK Natural Environment Research Council (Research Grants and Scientific Services Sections) for the Edinburgh ion microprobe facility.

References

Andersen, C.A. (1970) Analytical methods for the ion microprobe microanalyser. II. *Int. J. Mass Spectrom. Ion Phys.*, **3**, 413–28.

Andersen, C.A. and Hinthorne, J.R. (1973) Thermodynamic approach to the quantitative interpretation of sputtered ion spectra. *Anal. Chem.*, **45**, 1421–38.

Anon. (1992) Short reports (with discussion) of the Ion Microprobe Stable Isotope Workshop held at the University of Manchester on January 5, 1991. Chem. Geol. (Isotope Geosci.), **101**.

Armour, D.G., Wadsworth, M., Badheka, R. *et al.* (1988) Fundamental processes which affect the depth resolution obtainable in sputter depth profiling, in *Secondary Ion Mass Spectrometry, SIMS VI* (eds A. Benninghoven, A.M. Huber and H.W. Werner), Wiley, New York, pp. 399–407.

Blundy, J.D. and Shimizu, N. (1991) Trace element evidence for plagioclase recycling in calc-alkaline magmas. *Earth Planet. Sci. Lett.*, **102**, 178–97.

Caporuscio, F.A. and Smyth, J.R. (1990) Trace element crystal chemistry of mantle eclogites. *Contrib. Mineral. Petrol.*, **105**, 550–61.

Chaussidon, M. and Albarède, F. (1992) Secular boron isotope variations in the continental crust: an ion microprobe study. *Earth Planet. Sci. Lett.*, **108**, 229–41.

Chaussidon, M., Albarède, F. and Sheppard, S.M.F. (1987) Sulfur isotope heterogeneity in the mantle from ion microprobe measurements of sulfide inclusions in diamonds. *Nature*, **330**, 242–4.

Coenraads, R.R., Sutherland, F.L. and Kinny, P.D. (1990) The origin of sapphires: U–Pb dating of zircon inclusions sheds new light. *Mineral. Mag.*, **54**, 113–22.

Colby, J.W. (1975) Ion microprobe mass analysis, in *Practical Scanning Electron Microscopy* (eds J.I. Goldstein and H. Yakowitz), Plenum, New York, pp. 529–72.

REFERENCES

Colton, R.J., Ross, M.M. and Kidwell, D.A. (1986) Secondary ion mass spectrometry: polyatomic and molecular ion emission. *Nucl. Instr. Meth. Phys. Res.*, **B13**, 259–77.

Compston, W. and Pidgeon, R.T. (1986) Jack Hills, evidence of more very old detrital zircons in Western Australia. *Nature*, **321**, 766–9.

Compston, W., Williams, I.S. and Meyer, C. (1984) U/Pb geochronology of zircons from Breccia 73217 using a sensitive high mass-resolution ion microprobe. *Proc. 14th Lunar Planetary Sci. Conf., J. Geophys. Res.*, **89**, B525–34.

Compston, W., Kinny, P.D., Williams, I.S. and Foster, J.J. (1986) The age and Pb loss behaviour of zircons from the Isua supracrustal belt as determined by ion microprobe. *Earth Planet. Sci. Lett.*, **80**, 71–81.

Compston, W., Froude, D.O., Ireland, T.R. *et al.* (1985) The age of (a tiny part of) the Australian continent. *Nature*, **317**, 559–60.

Compston, W., Williams, I.S., Kirschvink, J.L. *et al.* (1992) Zircon U–Pb ages for the early Cambrian time-scale. *J. Geol. Soc. London*, **149**, 171–84.

Crozaz, G. and McKay, G. (1990) Rare earth elements in Angra dos Reis and Lewis Cliff 86010, two meteorites with similar but distinct evolutions. *Earth Planet. Sci. Lett.*, **97**, 369–81.

Crozaz, G. and Zinner, E. (1985) Ion microprobe determination of the rare earth concentration of individual meteoritic phosphate grains. *Earth Planet. Sci. Lett.*, **73**, 41–52.

Crozaz, G., Pellas, P., Bourot-Denise, M. *et al.* (1989) Plutonium, uranium and rare earths in the phosphates of ordinary chondrites – quest for a chronometer. *Earth Planet. Sci. Lett.*, **93**, 157–69.

Davis, A.M. and Olsen, E.J. (1991) Phosphates in pallasitic meteorites as probes of mantle processes in small planetary bodies. *Nature*. **353**, 637–40.

Deloule, E., Albarède, F. and Sheppard, S.M.F. (1991(a)) Hydrogen isotope hetorogeneities in the mantle from ion microprobe analysis of amphiboles from ultramafic rocks. *Earth Planet. Sci. Lett.*, **105**, 543–53.

Deloule, E., Allegre, C. and Doe, B. (1986) Lead and sulfur isotope microstratigraphy of Mississippi Valley-type Deposits. *Econ. Geol.*, **81**, 1307–21.

Deloule, E., Chaussidon, M. and Allé, P. (1992) Instrument limitations for isotope measurements with a Cameca-ims-3f ion microprobe: example of H, B, S and Sr. *Chem. Geol.*, **101**, 187–92.

Deloule, E., France-Lanord, C. and Albarède, F. (1991(b)) D/H analysis of minerals by ion probe, in *Stable Isotope Geochemistry: A tribute to Sam Epstein* (eds H.P. Taylor, J.R. O'Neil and I.R. Kaplan), *Geochem. Soc. Spec. Publ.* **3**, 53–62.

Diamond, L.W., Marshall, D.D., Jackman, J.A. and Skippen, G.B. (1990) Elemental analysis of fluid inclusions in minerals by secondary ion mass spectrometry (SIMS): applications to cation ratios in fluid inclusions in an Archean mesothermal gold-quartz vein. *Geochim. Cosmochim. Acta*, **54**, 545–52.

Elphick, S.C. and Graham, C.M. (1988) The effect of hydrogen on oxygen diffusion in quartz: evidence for fast proton transients? *Nature*, **335**, 243–5.

Elphick, S.C., Dennis, P.F. and Graham, C.M. (1986) An experimental study of the diffusion of oxygen in quartz and albite using an overgrowth technique. *Contrib. Mineral. Petrol.*, **92**, 322–30.

Elphick, S.C., Graham, C.M., Walker, F.D.L. and Holness, M.B. (1991) The application of SIMS ion imaging techniques in the experimental study of fluid-mineral interactions. *Mineral. Mag.*, **55**, 347–56.

Eldridge, C.S., Compston, W., Williams, I.S. and Walshe, J.L. (1989) Sulfur isotopic analyses on the SHRIMP ion microprobe. *US Geol. Surv. Bull.*, **1890**, 163–74.

Eldridge, C.S., Compston, W., Williams, I.S. *et al.* (1991) Isotope evidence for the involvement of recycled sediments in diamond formation. *Nature*, **353**, 649–53.

Engström, E., Lodding, A., Odelius, H. and Södervall, U. (1987) SIMS yields from glasses; secondary ion energy dependence and mass fractionation. *Mikrochim. Acta*, **I**, 387–400.

Exley, R.A. (1983) Evaluation and application of the ion microprobe in the strontium isotope geochemistry of carbonates. *Earth Planet. Sci. Lett.*, **65**, 303–10.

Fahey, A.J., Goswami, J.N., McKeegan, K.D. and Zinner, E. (1987) ^{26}Al, ^{244}Pu, ^{50}Ti, REE, and trace element abundances in hibonite grains from CM and CV meteorites. *Geochim. Cosmochim. Acta*, **51**, 329–50.

Farver, J.R. and Yund, R.A. (1990) The effect of hydrogen, oxygen, and water fugacity on oxygen diffusion in alkali feldspar. *Geochim. Cosmochim. Acta*, **54**, 2953–64.

Farver, J.R. and Yund, R.A. (1991) Oxygen diffusion in quartz: dependence on temperature and water fugacity. *Chem. Geol.*, **90**, 55–70.

Froude, D.O., Ireland, T.R., Kinny, P.D. *et al.* (1983) Ion microprobe identification of 4100–4200 Myr-old terrestrial zircons. *Nature*, **304**, 616–8.

Giletti, B.J., Semet, M.P. and Yund, R.A. (1978) Studies in diffusion – III. Oxygen in feldspars: an ion microprobe determination. *Geochim. Cosmochim. Acta*, **42**, 45–57.

Gnaser, H. and Hutcheon, I.D. (1987) Velocity-dependent isotope fractionation in secondary-ion emission. *Phys. Rev. B*, **35**, 877–89.

Graham, C.M. and Elphick, S.C. (1990) A re-examination of the role of hydrogen in Al–Si interdiffusion in feldspars. *Contrib. Mineral. Petrol.*, **104**, 481–91.

Graham, C.M. and Elphick, S.C. (1991) Some experimental constraints on the role of hydrogen in oxygen and hydrogen diffusion and Al–Si interdiffusion in silicates, in *Diffusion, Atomic Ordering and Mass Transport* (ed. J. Ganguly), *Advances in Physical Chemistry*, **8**, Springer-Verlag, New York, pp. 248–85.

Graham, C.M. and Valley, J.W. (1992) Sulphur isotope analysis of pyrites. *Chem. Geol.*, **101**, 169–72.

Harris, N.B.W., Gravestock, P. and Inger, S. (1992) Ion-microprobe determinations of trace-element concentrations in garnets from anatectic assemblages. *Chem. Geol.*, **100**, 41–49.

Hart, S.R. and Dunn, T. (1993) Experimental cpx/melt partitioning of 24 trace elements. *Contrib. Mineral. Petrol.*, **113**, 1–8.

Harte, B. and Otter, M. (1992) Carbon isotope measurement on diamonds. *Chem. Geol.*, **101**, 177–83.

Hayes, J.M. and Schoeller, D.A. (1977) High precision pulse counting: limitations and optimal conditions. *Anal. Chem.*, **49**, 306–11.

Hervig, R.L. and Dunbar, N.W. (1992) Cause of chemical zoning in the Bishop (California) and Bandalier (New Mexico) magma chambers. *Earth Planet. Sci. Lett.*, **111**, 97–108.

Hervig, R.L. and Peacock, S.M. (1989) Implications of trace element zoning in deformed quartz from the Santa Catalina mylonite zone. *J. Geol.*, **89**, 343–50.

Hervig, R.L. and Williams, P. (1986) Non-oxygen negative ion beams for oxygen isotopic analysis in insulators, in *Secondary Ion Mass Spectrometry*, *SIMS V*, (eds A. Benninghoven, R.J. Colton, D.S. Simmons and H.W. Werner), Springer-Verlag, New York. pp. 152–4.

Hervig, R.L., Thomas, R.M. and Williams, P. (1989) Charge neutralization and oxygen isotopic analysis of insulators with the ion microprobe. *US Geol. Surv. Bull.*, **1890**, 137–42.

Hickmott, D. and Spear, F.S. (1992) Major- and trace-element zoning in garnets from calcareous pelites in the NW Shelburne Falls Quadrangle, Massachusetts: garnet growth histories in retrograde rocks. *J. Petrol.*, **33**, 965–1005.

Hinthorne, J.R., Andersen, C.A., Conrad, R.L. and Lovering, J.F. (1979) Single-grain ^{207}Pb/^{206}Pb and U/Pb age determination with a 10 micron spatial resolution using the ion microprobe mass analyser (IMMA). *Chem. Geol.*, **25**, 271–303.

Hinton R.W. (1990) Ion microprobe trace-element analysis of silicates: measurement of multi-element glasses. *Chem. Geol.*, **83**, 11–25.

REFERENCES

Hinton, R.W. and Long, J.V.P. (1979) High resolution ion microprobe measurements of lead isotopes: variations within single zircons from Lac Seul, Northwestern Ontario. *Earth Planet. Sci. Lett.*, **45**, 309–25.

Hinton, R.W. and Upton, B.J.G. (1991) the chemistry of zircon: variations within and between large crystals from syenite and alkali basalt xenoliths. *Geochim. Cosmochim. Acta*, **55**, 3287–302.

Hinton, R.W., Davis, A.M., Scatena-Wachel, D.E. *et al.* (1988) A chemical and isotopic study of hibonite-rich refractory inclusions in primitive meteorites. *Geochim. Cosmochim. Acta*, **52**, 2573–98.

Ireland, T.R., Zinner, E., Fahey, A.J. and Esat, T.M. (1992) Evidence for distillation in the formation of HAL and related hibonite inclusions. *Geochim. Cosmochim. Acta*, **56**, 2503–20.

Jolliff, B.L., Papike, J.J. Shearer, C.K. and Shimizu, N. (1989) Inter- and intra-crystal REE variations in apatite from the Bob Ingersoll pegmatite, Black Hills, South Dakota. *Geochim. Cosmochim. Acta*, **53**, 429–41.

Johnson, K.T.M., Dick, H.J. and Shimizu, N. (1990) Melting in the oceanic mantle: an ion microprobe study of diopsides in abyssal peridotites. *J. Geophys. Res.*, **95**, 2661–78.

Jull, A.J.T., Wilson, G.C., Long, J.V.P. *et al.* (1980) Sputtering rates of minerals and implications for abundances of solar elements in Lunar samples. *Nucl. Instr. Meth.*, **168**, 357–65.

Kinny, P.D., Compston, W. and Williams, I.S. (1991) A reconnaissance ion-probe study of hafnium isotopes in zircons. *Geochim. Cosmochim. Acta*, **55**, 849–59.

Kober, B., Pidgeon, R.T. and Lippolt, H.J. (1989) Single-zircon dating by stepwise Pb-evaporation constrains the Archean history of detrital zircons from the Jack Hills, Western Australia. *Earth Planet. Sci. Lett.*, **91**, 286–96.

Kovalenko, V.I., Hervig, R.L. and Sheridan, M.F. (1988) Ion microprobe analyses of trace elements in anorthoclase, hedenbergite, aenigmatite, quartz, apatite and glass in pantellerite: evidence for high water contents in pantellerite melt. *Am. Mineral.*, **73**, 1038–45.

Krogh, T.E. (1973) A low-contamination method for hydrothermal decomposition of zircon and extraction of U and Pb for isotopic age determinations. *Geochim. Cosmochim. Acta*, **37**, 485–94.

Lau, W.M., McIntyre, N.S., Metson, J.B. *et al.* (1985) Stabilization of charge on electrically insulating surfaces during SIMS experiments – experimental and theoretical studies of the specimen isolation method. *Surf. Interface Anal.*, **7**, 275–81.

Levi-Setti, R., Crow, G. and Wang, Y.L. (1985) Progress in high resolution scanning microscopy and secondary ion mass spectrometry imaging microanalysis. *Scann. Elec. Microsc.*, 535–51.

Long, J.V.P. and Hinton, R.W. (1984) The intensity of metal hydride peaks in secondary positive-ion spectra from silicates. *Int. J. Mass Spectrom. Ion Proc.*, **55**, 307–18.

Lorin, J.C. (1992) Oxygen isotope analysis on the Cameca ims-300. *Chem. Geol.*, **101**, 193–5.

Lorin, J.C., Havette, A. and Slodzian, G. (1982) Isotope effect in secondary ion emission, in *Secondary Ion Mass Spectrometry, SIMS III* (eds A. Benninghoven, J. Giber, J. Laszlo *et al.*). Springer-Verlag, Berlin, pp. 140–50.

Lorin, J.C., Slodzian, J.C., Dennebouy, R. and Chaintreau, M. (1990) SIMS measurement of oxygen isotope-ratios in meteorites and primitive solar system matter, in *Secondary Ion Mass Spectrometry, SIMS VII* (eds A. Benninghoven, K.D. McKeegan, H.A. Storms and H.W. Werner), Wiley-Interscience, New York. pp. 377–80.

Lundberg, L.L., Crozaz, G. and McSween, H.Y. (1990) Rare earth elements in minerals of the ALHA 77005 shergottite and implications for its parent magma and crystallization history. *Geochim. Cosmochim. Acta*, **54**, 2535–47.

Lundberg, L.L., Crozaz, G., McKay, G. and Zinner, E. (1988) Rare earth element carriers in the Shergotty meteorite and implications for its chronology. *Geochim. Cosmochim. Acta*, **52**, 2147–63.

Maas, R., Kinny, P.D., Williams, I.S. *et al*. (1992) The Earth's oldest known crust: a geochronological and geochemical study of 3900–4200 Ma old detrital zircons from Mt Narryer and Jack Hills, Western Australia. *Geochim. Cosmochim. Acta*, **56**, 1281–1300.

MacRae, N.D., Botazzi, P., Ottolini, L. and Vannucci, R. (1993) Quantitative REE analysis of silicates by SIMS: conventional energy filtering vs specimen isolation mode. *Chem. Geol.*, **103**, 45–54.

Mason, R.A. (1987) Ion microprobe analysis of trace elements in calcite with an application to the cathodoluminescence zonation of limestone cements from the lower carboniferous of South Wales, UK. *Chem. Geol.*, **64**, 209–24.

Mazzuchelli, M., Rivalenti, G., Vannucci, R. *et al*. (1992) Trace element distribution between clinopyroxene and garnet in gabbroic rocks of the deep crust: an ion microprobe study. *Geochim. Cosmochim. Acta*, **56**, 2371–85.

McKeegan, K.D. (1987) Oxygen isotopic abundances in refractory stratospheric dust particles: proof of extra-terrestrial origins. *Science*, **237**, 1468–71.

McKeegan, K.D., Walker, R.M. and Zinner, E. (1985) Ion microprobe isotopic measurements of individual interplanetary dust particles. *Geochim. Cosmochim. Acta*, **49**, 1971–87.

Metson, J.B., Bancroft, G.M. and Nesbitt, H.W. (1985) Analysis of minerals using specimen isolated secondary ion mass spectrometry. *Scann. Electr. Microsc.*, **1985**, 595–603.

Migeon, H.N., Schumacher, M., Le Goux, J.J. and Rasser, B. (1989) Three-dimensional analysis of trace elements with the Cameca IMS 4F. *Fresenius Z. Anal. Chem.*, **333**, 333–4.

Mitchell, R.H. and Reed, S.J.B (1988) Ion microprobe determination of rare earth elements in perovskite from kimberlites and alnoites. *Mineral. Mag.*, **52**, 331–9.

Muir, I.J. and Nesbitt, H.W. (1991) Effects of aqueous cations on the dissolution of labradorite feldspar. *Geochim. Cosmochim. Acta*, **55**, 3181–9.

Muir, I.J. and Nesbitt, H.W. (1992) Controls on differential leaching of calcium and aluminum from labradorite in dilute electrolyte solutions. *Geochim. Cosmochim. Acta*, **56**, 3979–85.

Muir, I.J., Bancroft, G.M., and Metson, J.B. (1987) A comparison of conventional and specimen isolation filtering techniques for the SIMS analyses of geologic materials. *Int. J. Mass Spectrom. Ion Proc.*, **75**, 159–70.

Muir, I.J., Bancroft, G.M., Shotyk, W. and Nesbitt, H.W. (1990) A SIMS and XPS study of dissolving plagioclase. *Geochim. Cosmochim. Acta*, **54**, 2247–56.

Olsen, E., Schwade, J., Davis, A.M. *et al*. (1991) Watson: a new link in the IIE iron chain. *Lunar Planet. Sci.*, **XXII**, 999–1000.

Pan, V., Holloway, J.R. and Hervig, R.L. (1991) The pressure and temperature dependence of carbon dioxide solubility in tholeiitic basalt melts. *Geochim. Cosmochim. Acta*, **55**, 1587–95.

Perny, B., Eberhardt, P., Ramseyer, K. *et al*. (1992) Microdistributions of Al, Li, and Na in α quartz: possible causes and correlation with short-lived cathodoluminescence. *Am. Mineral.*, **77**, 534–44.

Pimminger, M., Grasserbauer, M., Schroll, E. and Cerny, I. (1984) Microanalysis of galena by secondary ion mass spectrometry for the determination of sulfur isotopes. *Anal. Chem.*, **56**, 407–11.

Ray, G. and Hart, S.R. (1982) Quantitative analysis of silicates by ion microprobe. *Int. J. Mass Spectrom. Ion Phys.*, **44**, 231–55.

Reed, S.J.B. (1983) Secondary ion yields of rare earths. *Int. J. Mass Spectrom. Ion Proc.*, **54**, 31–40.

REFERENCES

Reed, S.J.B. (1984) Secondary-ion mass spectrometry/ion probe analysis for rare earths. *Scann. Electr. Microsc.*, **1984**, 529-35.

Reed, S.J.B. (1985) Ion probe determination of rare earths in allanite. *Chem. Geol.*, **48**, 137-43.

Reed, S.J.B. (1986) Ion microprobe determination of rare earth elements in accessory minerals. *Mineral. Mag.*, **50**, 3-15.

Reed, S.J.B. (1989) Ion microprobe analysis - a review of geological applications. *Mineral. Mag.*, **53**, 3-24.

Reed, S.J.B., and Smith, D.G.W. (1985) Ion probe determination of rare earth elements in merrillite and apatite in chondrites. *Earth Planet. Sci. Lett.*, **72**, 238-44.

Reed, S.J.B., Smith, D.G.W. and Long, J.V.P. (1983) Rare earth elements in chondritic phosphates - implications for ^{244}Pu chronology. *Nature*, **306**, 172-3.

Roddick, J.C., Loveridge, W.D. and Parrish, R.R. (1987) Precise U/Pb dating of zircon at the sub-nanogram Pb level. *Chem. Geol.*, **66**, 111-21.

Rudat, M.A. and Morrison, G.H. (1978) Detector discrimination in SIMS: ion-to-electron converter yield factors for positive ions. *Int. J. Mass Spectrom. Ion Phys.*, **27**, 249-61.

Rudat, M.A. and Morrison, G.H. (1979) Energy spectra of ions sputtered from elements by O^{2+}: a comprehensive study. *Surf. Sci.*, **82**, 549-76.

Schärer, U. and Allègre, C.J. (1985) Determination of the age of the Australian continent by single-grain zircon analysis of Mt Narryer metaquartzite. *Nature*, **315**, 52-5.

Shanks, W.C. III and Criss, R.E. (eds) (1989) New frontiers in stable isotopic research: laser probes, ion probes, and small-scale analysis. *US Geol. Surv. Bull.*, **1890**.

Shaw, D.M., Higgins, M.D., Hinton, R.W. et al. (1988(a)) Boron in chondritic meteorites. *Geochim. Cosmochim. Acta*, **52**, 2311-9.

Shaw, D.M., Higgins, M.D., Truscott, M.G. and Middleton, T.A. (1988(b)) Boron contamination in polished thin sections of meteorites: implications for other trace element studies by alpha track images or ion microprobe. *Am. Mineral.*, **73**, 894-900.

Shimizu, N. (1990) The oscillatory trace element zoning of augite phenocrysts. *Earth Sci. Rev.*, **29**, 27-37.

Shimizu, N. and Hart, S.R. (1982) Applications of the ion microprobe to geochemistry and cosmochemistry. *Ann. Rev. Earth Planet. Sci.*, **10**, 483-526.

Shimizu, N. and Richardson, S.H. (1987) Trace element abundance patterns of garnet inclusions in peridotite-suite diamonds. *Geochim. Cosmochim. Acta*, **51**, 755-8.

Shimizu, N., Semet, M.P. and Allègre, C.J. (1978) Geochemical applications of quantitative ion-microprobe analysis. *Geochim. Cosmochim. Acta*, **42**, 1321-34.

Sisson, T.W. (1991) Pyroxene-high silica rhyolite trace element partition coefficients measured by ion microprobe. *Geochim. Cosmochim. Acta*, **55**, 1575-85.

Sisson, T.W., and Bacon, C.R. (1992) Garnet/high-silica rhyolite trace element partition coefficients measured by ion microprobe. *Geochim. Cosmochim. Acta*, **56**, 2133-6.

Slodzian, G. (1980) Microanalyzers using secondary ion emission. *Adv. Electronics Electron Phys. Suppl.* **13B**, 1-44.

Slodzian, G. (1988) Introduction to fundamentals in direct and scanning ion microscopy, in *Secondary Ion Mass Spectrometry, SIMS VI* (eds A. Benninghoven, A.M. Huber and H.W. Werner), John Wiley, Chichester, pp. 3-12.

Slodzian, G., Lorin, J.C. and Havette, A. (1980) Isotopic effect on the ionization probabilities in secondary ion emission. *J. Physique Lett.*, **41**, L555-8.

Slodzian, G., Daigne, B., Girard, F. and Hillion, F. (1991) A thermal ionization source for a Cs^+ ion probe. Abstracts of Eighth Int. Conf. Secondary Ion Mass Spectrometry, SIMS VIII (Amsterdam), p. 132.

Södervall, U., Odelius, H., Lodding, A. and Engström E.U. (1987) Mass fractionation and energy distribution of sputtered monatomic positive ions. *Scann. Microsc.* **1**, 471-7.

Steele, I.M., Hervig, R.L., Hutcheon, I.D. and Smith, J.V. (1981) Ion microprobe techniques and analyses of olivine and low-Ca pyroxene. *Am. Mineral.*, **66**, 526–46.

Stevie, F.A., Kahora, P.M., Singh, S. and Kroko, L. (1988) Atomic and molecular relative secondary ion yields of 46 elements in Si for O_2^+ and Cs^+ bombardment, in *Secondary Ion Mass Spectrometry, SIMS VI* (eds A. Benninghoven, A.M. Huber and H.W. Werner), Wiley, New York, pp. 319–22.

Storms, H.A., Brown, K.F. and Stein, J.D. (1977) Evaluation of a cesium positive ion source for secondary ion mass spectrometry. *Anal. Chem.*, **49**, 2023–30.

Swart, P.K. (1990) Calibration of the ion microprobe for the quantitative determination of strontium, iron, manganese, and magnesium in carbonate minerals. *Anal. Chem.*, **62**, 722–8.

Thompson, M.W. (1987) The velocity distribution of sputtered atoms. *Nucl. Inst. Meth.*, **B18**, 411–29.

Tucker, R.D., Krogh, T.E., Ross, R.J. and Williams, S.H. (1990) Time-scale calibrated by high-precision U–Pb zircon dating of interstratified volcanic ashes in the Ordovician and lower Silurian stratotypes of Britain. *Earth Planet. Sci. Lett.*, **100**, 51–8.

Valley, J.W. and Graham, C.M. (1991) Ion microprobe analysis of oxygen isotope ratios in granulite facies magnetites: diffusive exchange as a guide to cooling history. *Contrib. Mineral. Petrol.*, **109**, 38–52.

Valley, J.W. and Graham, C.M. (1992) Oxygen isotope measurement of magnetites. *Chem. Geol.*, **101**, 173–6.

Veizer, J., Hinton, R.W., Clayton, R.N. and Lerman, A. (1987) Chemical diagenesis of carbonates in thin-sections: ion microprobe as a trace element tool. *Chem. Geol.*, **64**, 225–37.

Walker, F.D.L. (1990) Ion microprobe study of intergrain micropermeability in alkali feldspars. *Contrib. Mineral. Petrol.*, **106**, 124–8.

Webster, J.D. (1992(a)) Fluid-melt interactions involving Cl-rich granites: Experimental study from 2 to 8 kbar. *Geochim. Cosmochim. Acta*, **56**, 659–78.

Webster, J.D. (1992(b)) Water solubility and chlorine partitioning in Cl-rich granitic systems: effects of melt composition at 2 kbar and 800°C. *Geochim. Cosmochim. Acta*, **56**, 679–87.

Webster, J.D. and Duffield, W.A. (1991) Volatiles and lithophile elements in the Taylor Creek Rhyolite: constraints from glass inclusion analysis. *Am. Mineral.*, **76**, 1628–45.

Williams, P. (1985) Limits of quantitative microanalysis using secondary ion mass spectrometry. *Scann. Electr. Microsc.*, **1985**, 553–61.

Williams, I.S. and Claesson, S. (1987) Isotopic evidence for the Precambrian provenance and Caledonian metamorphism of high grade paragneisses from the Seve Nappes, Scandinavian Caledonides. II. Ion microprobe zircon U–Th–Pb. *Contrib. Mineral. Petrol.*, **97**, 205–17.

Wilson, R.G. and Novak, S.W. (1988) Systematics of SIMS relative sensitivity factors versus electron affinity and ionization potential for Si, Ge, GaAs, GaP, InP and HgCdTe determined from implant calibration standards for about 50 elements, in *Secondary Ion Mass Spectrometry, SIMS VI* (eds A. Benninghoven, A.M. Huber and H.W. Werner), Wiley, New York. pp. 57–61.

Wilson, R.G., Novak, S.W., Smith, S.P. *et al.* (1988) Relative sputter rates and ion yields of semiconductors, metals and insulators under oxygen and cesium ion bombardment, in *Secondary Ion Mass Spectrometry, SIMS VI* (eds A. Benninghoven, A.M. Huber and H.W. Werner), Wiley, New York, pp. 133–4.

Yu, M.L. (1987) A bond breaking model for secondary ion emission. *Nucl. Instr. Meth.*, **B18**, 542–8.

Yurimoto, H., Kurosawa, M. and Sueno, S. (1989) Hydrogen analysis in quartz and quartz glasses by secondary ion mass spectrometry. *Geochim. Cosmochim. Acta*, **53**, 751–5.

REFERENCES

Zinner, E. (1989) Isotopic measurements with the ion microprobe. *US Geol. Surv. Bull.*, **1890**, 145–62.

Zinner, E. and Crozaz, G. (1986(a)) Ion microprobe determination of the abundances of all the rare earth elements in single mineral grains, in *Secondary Ion Mass Spectrometry, SIMS V* (eds A. Benninghoven, R.J. Colton, D.S. Simmons and H.W. Werner), Springer-Verlag, New York, pp. 444–6.

Zinner, E. and Crozaz, G. (1986(b)) A method for the quantitative measurement of rare earth elements in the ion microprobe. *Int. J. Mass Spectrom. Ion Proc.*, **69**, 17–38.

Zinner, E., Fahey, A.J. and McKeegan, K.D. (1986) Characterisation of electron multipliers by charge distributions, in *Secondary Ion Mass Spectrometry, SIMS V* (eds A. Benninghoven, R.J. Colton, D.S. Simmons and H.W. Werner), Springer-Verlag, New York, pp. 170–2.

Zinner, E., Tang, M. and Anders, E. (1987) Large isotopic anomalies of Si, C, N, and noble gases in interstellar silicon carbide from the Murray meteorite. *Nature*, **330**, 730–2.

Zinner, E., Tang, M, and Anders, E. (1989) Interstellar SiC in the Murchison and Murray meteorites: Isotopic composition of Ne, Xe, Si, C, and N. *Geochim. Cosmochim. Acta*, **53**, 3273–90.

CHAPTER SEVEN
Mineral microanalysis by laserprobe inductively coupled plasma mass spectrometry

William T. Perkins and Nicholas J.G. Pearce

7.1 Introduction

Laser ablation inductively coupled plasma mass spectrometry (ICP-MS) has the potential to determine almost all the elements in the periodic table at the part-per-million level in solid samples with a spatial resolution of around 20 μm. To gain an insight into this powerful technique it is necessary to understand the principles of operation of the inductively coupled plasma mass spectrometer and the operation of lasers. In this chapter, a detailed explanation of the ICP-MS instrument and the development of a laser system will be followed by a review of the history and current applications of laser ablation inductively coupled plasma mass spectrometry (LA-ICP-MS).

7.2 Inductively coupled plasma mass spectrometry

ICP-MS has emerged as a powerful technique for geochemical analysis during the last decade. The instrument combines very low limits of detection with a rapid analysis capability. In the following sections a description of the component parts of the instrument is presented so that the reader may understand the operation of the instrument. Those interested in the development of the technique and a more detailed description of the instrumentation are referred to the excellent book by Jarvis, Gray and Houk (1992).

7.2.1 The inductively coupled plasma

The inductively coupled plasma is the ion source used in ICP-MS instrumentation. It consists of a toroidal argon plasma, generated by a radio frequency field and propagated at the end of a plasma torch. A typical plasma torch as used in a commercial instrument is illustrated in Figure 7.1. The torch is composed of three concentric fused silica tubes which are generally held in a rigid arrangement, although some operators use a

Microprobe Techniques in the Earth Sciences. Edited by P.J. Potts, J.F.W. Bowles, S.J.B. Reed and M.R. Cave. Published in 1995 by Chapman & Hall, London. ISBN 0 412 55100 4

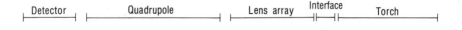
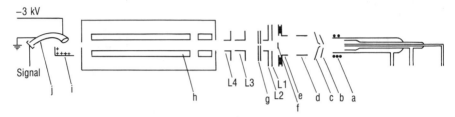

Figure 7.1 Schematic diagram of the ICP-MS instrument. The diagram shows each of the sections of the instrument described in the text. Individual components are labelled as follows: a, RF load coil; b, sample cone; c, skimmer cone; d, extractor lens; e, collector lens; f, photon stop; g, differential aperture; h, quadrupole rods; i, ion deflector; j, electron channel multiplier. L1, L2, L3 and L4 indicate the positions of the four main ion lenses.

demountable torch system (Jarvis, Gray and Houk, 1992). Argon gas (better than 99.996% pure) is introduced tangentially into the outer two tubes whilst the central channel has a linear flow. The inner flow is usually referred to as the nebulizer, carrier or injector flow and is responsible for the transfer of sample to the plasma. The intermediate, auxiliary or plasma gas is set at a low flow rate in most systems. The outer or coolant gas is required to keep the high temperature plasma away from the torch. Typical flow rates for laser ablation operation are given in Table 7.1.

An intense radio frequency (RF) field is used to generate and sustain the plasma. This field is produced by a load coil of between 2 and 4 turns, which encircles the end of the plasma torch (Figure 7.1). The coil, which is made from copper tube through which cooling water circulates, is the antenna for a high-power RF generator which is either crystal controlled at 27.12 MHz or free running at 40 MHz. An intense electromagnetic field (the induction region) is created at the end of the torch inside the coil. To produce a plasma it is necessary to 'seed' the argon gas with charged particles, which is achieved by passing a high-voltage spark along the inner wall of the

Table 7.1 Typical operating conditions

Gas	Flow rates (l min^{-1})
Coolant gas	12
Plasma gas	0.5
Carrier gas	0.9

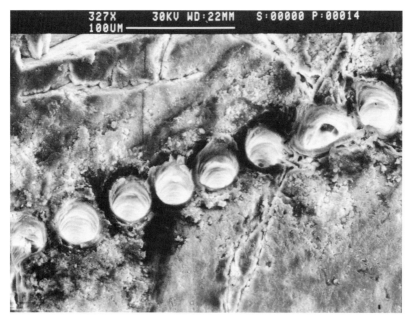

Plate 7.1 SEM image of evenly spaced Nd–YAG laser ablation craters in an olivine from Stapafell Quarry, Iceland. Note how scattering of incident laser light by fractures in the crystal causes irregular ablation craters.

Plate 7.2 SEM image of closely spaced Nd–YAG laser ablation craters in olivine. The smooth-sided craters remain as discrete, cylindrical pits and are surmounted by a small rim of quenched, ejected material. Note how craters are separated from each other by walls only a few micrometres thick.

outermost silica tube of the torch. Argon ions and electrons are produced along this discharge and as these particles are swept through the inside of the coil, they enter the induction region and become coupled to the oscillating field. The intense RF field causes non-elastic collisions between electrons and argon atoms which give rise to more argon ions. This process results in the production of an argon gas in which a significant proportion of the atoms are ionized; i.e. an argon plasma. The plasma is maintained as long as the RF field provides the induction. A temperature of about 10 000 K is attained in the core of the plasma. The inner nebulizer gas flow punches a hole through the plasma and the sample is heated by radiation and conduction from the surrounding annular plasma. It is estimated that the temperature in this central channel rises to between 5000 and 7000 K at the mouth of the torch. The function of the argon plasma in an ICP-MS instrument is to excite positive ions from sample atoms. Given the estimated temperature of the central gas channel and values for the first ionization energy, it is possible to use the Saha equation to estimate the degree to which atoms will be ionized. The values for the first ionization energy and percentage of ionization for those elements for which data are available are given in Table 7.2. Elements with low first ionization energies are almost completely ionized, and those which have first ionization energies below 10 eV give rise to more than 50% ionization in the plasma. Thus the argon ICP is an efficient ionization source for mass spectrometry. To take advantage of this ion source, it is necessary to transfer a fraction of the plasma gas and sample ions from the atmospheric plasma into a mass spectrometer maintained at relatively high vacuum. This is achieved by means of an ion extraction interface.

7.2.2 Ion extraction interface

The interface consists of two water-cooled metal cones, through which plasma gas and ions are extracted into the instrument, through chambers of increasingly high vacuum (Figure 7.1). The plasma impinges on a sample cone, made of nickel or sometimes platinum, which has a hole of about 1 mm diameter passing through the centre. The volume behind the sample cone, referred to as the expansion chamber, is evacuated by a rotary vacuum pump which maintains a pressure of approximately 10 mbar. Hot argon gas and sample ions are accelerated through the sample cone orifice to produce a supersonic jet of gas in the expansion chamber. A second nickel cone, the skimmer, is designed to pierce the back of the supersonic jet and extract a small proportion of the plasma gas and ions into the main body of the mass spectrometer. The region behind the skimmer is maintained at a vacuum of $\sim 10^{-3}$ mbar, by either oil vapour diffusion pumps or turbomolecular pumps. Once inside the main chamber of the instrument, the residual gas is separated from the positive ions and pumped away. The

Table 7.2 First ionization energy and % ionization in an argon ICP

Element	First ionization energy (eV)	Ionization (%)
Li	5.392	100
Be	9.322	75
B	8.298	58
F	17.422	9×10^{-4}
Na	5.139	100
Mg	7.646	98
Al	5.986	98
Si	8.151	85
P	10.486	33
Cl	12.964	0.9
K	4.341	100
Ca	6.113	99 (1)
Sc	6.54	100
Ti	6.82	99
V	6.74	99
Cr	6.766	98
Mn	7.435	95
Fe	7.870	96
Co	7.86	93
Ni	7.635	91
Cu	7.726	90
Zn	9.394	75
Ga	5.999	98
Ge	7.899	90
As	9.81	52
Se	9.752	33
Br	11.814	5
Rb	4.177	100
Sr	5.695	96 (4)
Y	6.38	98
Zr	6.84	99
Nb	6.88	98
Mo	7.099	98
Ru	7.37	96
Rh	7.46	94
Pd	8.34	93
Ag	7.576	93
Cd	8.993	65
In	5.786	99
Sn	7.344	96
Sb	8.641	78
Te	9.009	66
I	10.451	29
Cs	3.849	100
Ba	5.212	91 (9)
La	5.577	90 (10)
Ce	5.47	96 (4)
Pr	5.42	90 (10)

Table 7.2 *Continued*

Element	First ionization energy (eV)	Ionization (%)
Nd	5.49	99 (unknown)
Sm	5.63	97 (3)
Eu	5.67	100 (unknown)
Gd	6.14	93 (7)
Tb	5.85	99 (unknown)
Dy	5.93	100 (unknown)
Ho	6.02	n.d.
Er	6.10	99 (unknown)
Tm	6.18	91 (9)
Yb	6.254	92 (8)
Lu	5.426	n.d.
Hf	7.0	96
Ta	7.89	95
W	7.98	94
Re	7.88	93
Os	8.7	78
Ir	9.1	n.d.
Pt	9.0	62
Au	9.225	51
Hg	10.437	38
Tl	6.108	100
Pb	7.416	97
Bi	7.289	92
Th	~6.8	100
U	~6.0	100

Figures in parenthesis are for doubly charged ions; n.d. no data available. Ionization energy taken from *Handbook of Physics and Chemistry*, CRC Press, Florida; % ionization from Jarvis, Gray and Houk (1992).

remaining ions are extracted and focused into the mass spectrometer by a series of ion lenses.

7.2.3 *Ion lenses*

The function of the ion lenses is to produce a beam of ions focused into the mass analyser. A typical arrangement of ion lenses is shown in Figure 7.1. The lenses are cylindrical or disc-shaped electrodes to which a voltage is applied. In a typical instrument the ions are removed from the region behind the skimmer cone by the 'extraction' lens and pass through a series of lenses referred to in Figure 7.1 as the collector, L1, L2, L3 and L4 respectively, before entering the mass analyser. The lens region of the instrument also contains a photon stop which prevents light from the plasma passing through the mass analyser and falling onto the detector, which, being light sensitive, would otherwise show an enhanced background signal. Ions which travel at the same velocity but have different masses will have different kinetic energies and their paths through the lenses will vary according to their mass/charge ratio. In any system, the voltages set on the lens array will therefore be a compromise, usually chosen to maximize the transmission of an ion near the middle of the mass range of interest. When data are required across the full mass range, (i.e. ^{6}Li to ^{238}U), an element such as ^{115}In or ^{103}Rh would be chosen to optimize lens settings such that there is reasonable transmission of all the mass/charge values of interest. Within the lens system there is usually an orifice, known as the differential aperture, which further separates the instrument into different vacuum chambers, the region behind the skimmer being held at $\sim 10^{-3}$ mbar, whereas the mass analyser is held at $\sim 10^{-6}$ mbar. On leaving the ion lenses, the ion beam enters the mass analyser.

7.2.4 Mass analyser

In most commercial ICP-MS instruments the type of mass analyser used is a quadrupole, although a limited number of more expensive instruments make use of a high-resolution electrostatic and magnetic sector mass spectrometer. The quadrupole consists of four metal rods held such that they form a circle (Figure 7.1). Opposite pairs of rods are electrically connected and superimposed DC and RF voltages are applied to each pair of rods. The applied DC voltage is positive for one pair of rods and negative for the other whilst the RF voltages applied to each pair are equal in amplitude but 180° out of phase. Ions entering the quadrupole interact with the applied fields and are made to oscillate. For a given DC and RF voltage, only one mass/charge (m/z) value can be transmitted through the quadrupole whilst the others will spiral into the rods, where they are neutralized. By controlling the DC and RF voltages applied to the rods it is possible to change the m/z value transmitted by the quadrupole. The quadrupole can be operated in several modes:

1. With fixed DC and RF only one m/z value is transmitted, a mode of operation called single ion monitoring;
2. when the DC and RF are changed to discrete values under computer control it is possible to move rapidly between preselected m/z values, a mode of operation called peak jumping;
3. when the DC and RF are changed continuously under computer control, the quadrupole sweeps across a predetermined range of m/z values, a mode of operation called multi-channel scanning.

Ions which posses an appropriate m/z value for the selected DC and RF fields are transmitted through the quadrupole and are detected by the ion detection system.

7.2.5 Ion detection

The most commonly used type of ion detector in ICP-MS instruments is the Channeltron (or single channel) electron multiplier illustrated in Figure 7.1. This detector consists of a horn-shaped glass tube coated with a semiconducting material. The mouth of the tube is held at a high negative potential ($-3\,\text{kV}$) to attract the positive ions which leave the quadrupole. The end of the horn is electrically grounded so that a potential gradient exists along the length of the tube. A positive ion colliding with the wall of the multiplier produces one or more secondary electrons. These electrons are accelerated along the tube and when they collide with the walls further electrons are released. As a result of this internal multiplication effect, a single ion can give rise to a large population (10^8) of electrons at the grounded end of the tube. The signal pulse produced by this cloud of electrons is fed to a wide-band amplifier and then to the multichannel scaler. Because the scanning of the quadrupole is controlled by the DC and RF voltages applied to the rods, it is possible to synchronize the accumulation of ion counts in discrete channels of the analyser and thus produce a mass spectrum (i.e. in multichannel scanning mode). The proportion of ions sampled from the plasma that reach the detector is very small (1 in 10^6-10^8) and the high gain of the ion detection system ($\sim 10^6$) is essential if low limits of detection are to be achieved.

7.2.6 Sample introduction to the ICP-MS

In the majority of cases ICP-MS instruments are used to analyse samples prepared in the form of a solution. Sample solutions are converted to an aerosol which is entrained in the carrier gas flow by use of a nebulizer. Any large droplets are removed from the gas flow in a spray chamber. Ideally, droplets reaching the plasma should have a diameter of less than $10\,\mu\text{m}$. On reaching the plasma, the droplets undergo rapid desolvation, volatilization, atomization and ionization, the temperature in the plasma core ensuring efficient production of sample ions (Table 7.2).

7.2.7 Laser ablation

An alternative method of sample introduction involves the use of a laser to remove (ablate) a portion of material from a solid sample, so generating microparticulate matter which is entrained in the carrier argon stream and injected into the plasma. Lasers used in this type of work are generally optically pumped or pulsed lasers. In early work ruby lasers were used, but these have now been replaced in most laboratories by the more versatile Nd-YAG laser. The Nd-YAG laser comprises a rod of yttrium aluminium garnet ($Y_3Al_5O_{15}$) doped with approximately $3\,\text{wt}\%$ Nd_2O_3.

7.2.8 Laser operation

The Nd-YAG rod used in a typical 500 mJ laser system is 6.35 mm in diameter and 80 mm long. The laser rod has a xenon flashlamp which runs parallel with it and the two are enclosed in a light-tight enclosure, bounded at either end by mirrors (Figure 7.2). The xenon flashlamp is connected to a flashlamp driver circuit, which consists of a discharge capacitor connected to an ignition transformer. The discharge capacitor is charged to -1000 V DC and, when the flashlamp is ignited, the output from the ignition transformer may reach 20 000 V.

The light-tight cavity is constructed as a resonator such that the distance (d) between the two end mirrors is dictated by the equation:

$$d = \frac{n\lambda}{2} \qquad (7.1)$$

where λ = the characteristic wavelength and n = an integer.

When a discharge occurs in the flashlamp, the light pulse passes through the laser rod and is reflected from the two mirrors. The light interacts with the Nd in the laser rod, raising it to an excited state; interaction is greatest for a beam travelling parallel with the laser resonator. Because of the optical construction of the resonator, any beam of light not parallel with the optical axis or with a wavelength other than that dictated by equation 7.1 is destroyed by interference.

A schematic energy-level diagram for Nd is presented in Figure 7.3. Nd has three excited states and the laser transition takes place between the two metastable levels E_2 and E_1, whereas the optical pumping is responsible for raising the atom from the E_0 (ground state) to the E_3 (pumping level). The purpose of this optical pumping is to produce an overpopulation of excited-state Nd as a precursor to laser operation. When such an overpopulation occurs, the system becomes unstable and a photon, produced by fluore-

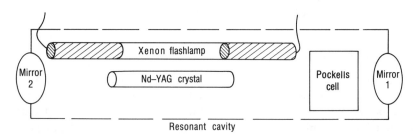

Figure 7.2 Optical diagram of the laser cavity. The Nd–YAG laser rod lies between two mirrors (1 and 2) within a light tight resonant cavity. The xenon flashlamp is used as an optical pump. The Pockels cell is used to alter the quality, or Q, of the cavity giving either normal mode (fixed-Q) or high-intensity (Q-switched) mode.

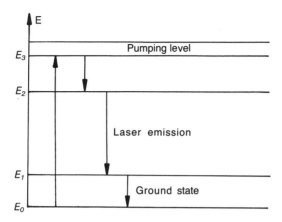

Figure 7.3 Energy level diagram for neodymium. The optical pumping from the xenon flashlamp raises the atoms from the ground state to the E_3, pumping level. Laser emission occurs during transition from the E_2 to the E_1 levels.

scence, will trigger or stimulate the emission of other photons. Because of the design of the resonator, light produced from such transitions is amplified until it reaches a threshold and emission can occur. This process is repeated for the duration of the flashlamp discharge and leads to a total pulse length of the order of 150 µs, characterized by many spikes, called relaxation oscillations, of a few microseconds duration. The wavelength of laser transition in the Nd-YAG system is 1064 nm (i.e. in the infrared). In this mode, the laser is said to be free running or in fixed-Q operation, 'Q' indicating the 'quality' of the resonator cavity. An alternative mode of operation alters the behaviour, or quality, of the resonator by placing an optical switch between the laser rod and the one of the mirrors. This can be achieved using an electro-optical device such as Pockels cell, a fast switch based on polarization (Figure 7.2). A polarizer is placed in the resonant cavity and when a voltage is applied to the Pockels cell, this device splits the polarized light into two different vibration directions having different velocities. The effect is destructive interference of the laser pulse. Whilst the voltage is applied, laser operation is suppressed resulting in a large overpopulation of excited-state Nd. When the voltage is removed, the Pockels cell no longer has an effect on the polarized light and the result is a single large output pulse of around 15 ns duration. This type of operation is the Q-switched mode. The cycle of optical pumping and laser output is variable from less than 1 Hz up to 20 Hz.

The energy output from the laser system can be controlled by the use of the Pockels cell and also by varying the voltage generated in the discharge capacitor. Table 7.3 gives the measured energy of a typical system used in

Table 7.3 Measured energy of typical laser ablation system

Voltage (V)	Fixed-Q output (mJ)	Q-switched output (mJ)
600	95	83
700	198	160
750	257	210
800	330	237
850	385	270
900	46	n.u.
950	514	n.u.

n.u. = not used.

ICP-MS applications for both fixed-Q and Q-switched output at different voltage settings with a 10 Hz cycle.

The interaction of the laser output with solid samples varies with the mode of operation. In fixed-Q mode operation, the relatively long laser pulse train causes a considerable heating of the target material. In Q-switched mode, very high power densities are produced, leading to the generation of a microplasma at or near the sample surface which is responsible for the ablation. In this mode of operation the laser can ablate material from samples which are transparent to the fixed-Q laser, including many minerals. Although fixed-Q operation gives rise to smaller craters on the sample surface, the ability of the Q-switched laser to ablate from transparent samples outweighs any advantage in working with fixed-Q output.

7.3 Laser ablation and ICP-MS

The ability of the laser to produce microparticulate material from solid samples makes it an ideal choice for sample introduction into an ICP. The plasma temperature is sufficiently high to volatilize solid particles, providing these have a relatively small diameter, and then to produce an optical emission signal (ICP-OES) or ionization (ICP-MS). The application of laser sampling has been comprehensively reviewed by Moenke-Blankenburg (1989) and the reader is referred to this work for details of non-ICP-MS applications. The potential of the laser system for ICP-MS analysis was first reported by Gray (1985) in a paper describing the use of a ruby laser. The samples analysed by Gray (*op. cit.*) included some British Geological Survey rock and mineral samples. Arrowsmith (1987) was the first to report the use of a Nd-YAG laser with an ICP-MS for the analysis of both microprobe reference materials and Cu standards. In these early applications the laser systems produced craters of about 150 μm diameter suitable for use in bulk sampling applications. Perkins, Fuge and Pearce (1991) and Fuge *et al.* (1993) used a Nd-YAG laser system to demonstrate the distribution of trace elements in carbonate shells, although the resolution achieved was limited.

The laser output is normally arranged as a horizontal beam which can be directed onto the sample surface using infrared reflecting surfaces. The sample is enclosed in a circular cell made of fused silica of typical internal diameter 53 mm, through which argon gas is passed. This argon transports the microparticulate material, produced during ablation, to the plasma. The sample cell is located on an x–y–z stage which can be driven by stepper motors. The optical system used in laser ablation systems is shown in Figure 7.4(a).

The spatial resolution of the laser system can be improved by reducing

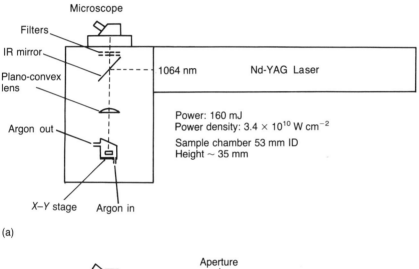

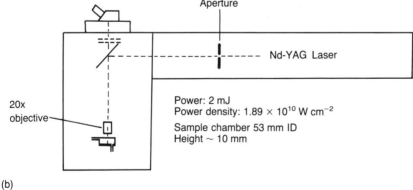

Figure 7.4 Typical laser ablation system used for ICP-MS. (a) The standard system with a large sample chamber and plano-convex focusing lens. This system will give ablation pits of about 150 μm diameter. (b) The modifications to the laser system for small spot applications. An aperture is inserted into the laser path and the plano-convex lens is replaced with a compound 20× long-working-distance objective.

the diameter of the laser output and replacing the simple plano-convex focusing lens with a compound objective (Figure 7.4(b)). This technique has been described as 'laserprobe microanalysis ICP-MS' (LPMA-ICP-MS) by Pearce *et al.* (1992(b)). Jackson *et al.* (1992) used the term 'laser ablation microprobe' (LAM) ICP-MS. Reduction of the beam diameter is achieved by inserting an aperture in the light path outside the resonant cavity. The compound objective used in this application must have a relatively long working distance because of the distance which must be left between the cell and the sample surface to prevent the microplasma, generated during Q-switched operation, from ablating material from the underside of the cell lid. The size of the craters produced using these modified systems depends on the type of material being ablated, its crystallographic orientation and the conditions of laser operation. Craters are smaller when the laser is operated in fixed-Q mode but, for the reasons summarized above, most applications have used Q-switched operation since ablation is less susceptible to variations in sample type. The crater diameter achieved with the Q-switched laserprobe thus varies between 20 and $50 \mu m$ in silicates and carbonates. The chemical composition of the sample has a marked effect on the absorption of light at the wavelength of Nd-YAG laser (1064 nm). Rossman (1988) presented absorption spectra for a variety of mineral types from which it is clear that many minerals are almost transparent at this wavelength (1064 nm), whilst the absorption observed for certain mineral types is attributed to Fe^{2+} spin-allowed bands. In addition to this chemical control of the mineral absorption, Hazen, Bell and Mao (1977) presented spectra for a lunar olivine showing that the crystallographic orientation has a marked effect on absorption at 1064 nm, which varied by a factor of three between the γ and the α and β vibration directions. These features have a marked effect on the efficiency with which the laser energy couples to the sample material, producing microparticulate material.

7.4 Analytical rationale

In ICP-MS analysis it is not possible to calibrate results directly by comparing the raw count data for an analyte with the signal from a calibration standard or reference material. This restriction particularly applies to laser ablation, where features such as the coupling efficiency of the laser to the sample surface and the method of sample preparation make a considerable difference to the amount of material removed during ablation.

The intensities of peaks are likely to vary dramatically between a reference material prepared as a compressed powder briquette and the cut or ground surface of a mineral. Material ablated from a mineral contains elemental abundances in the atomic proportions in which they occur in the mineral. Once the microparticulate material has been transported by the stream of Ar to the plasma, the number of ions which reach the detector (ignoring the possibilities of fractionation en route) depends on (i) the atomic proportions of the element in the source mineral, (ii) the amount of material removed during ablation (the ablation yield), (iii) the ionization potential of the element in the plasma and (iv) its isotopic abundance. To overcome differences in peak intensities caused by variations in ablation yields, it is necessary to normalize the analyte intensity to an internal standard. In the case of mineral analysis by LA-ICP-MS, the selection of a suitable internal standard is a major obstacle. Several methods have been employed in the selection of internal standards by ourselves (Pearce *et al.*, 1992(b)) and by others (e.g. Jackson *et al.*, 1992) and these various approaches are described in more detail below.

When the ratio of the analyte peak relative to that of an internal standard is calculated (i.e. analyte in unknown area counts per second (ACPS)/ internal standard ACPS), the concentration of the analyte can be calculated from the same ratio obtained from a calibration standard, provided that the internal standard is present at the same concentration in both the unknown and the calibration standard. For some minerals, reference materials exist (or can be produced) in which this assumption is valid and these minerals can be analysed directly (e.g. calcite, alkali feldspar, zircon, apatite and in some cases olivine; see below). For other minerals, where more complex chemical variations exist, it becomes necessary to employ other techniques to determine at least one element suitable for use as an internal standard. The most convenient supplementary technique for characterizing internal standard compositions is electron probe microanalysis (EPMA). Calibration lines can then be scaled according to the relative concentrations of the internal standard in the reference material and the unknown. This technique has been applied to a wide range of different minerals. The use of such an external analysis overcomes the need for reference materials of the same matrix type as the unknown, although discrepancies caused by matrix differences could then occur, a factor which may, however, be negligible (Perkins, Pearce and Jeffries, 1993, Jackson *et al.*, 1992).

In the following sections, a brief history of the application of laser ablation is discussed, followed by a review of the different methods of calibration and the choice of internal standards that have been employed. This section explains how various mineral types can be analysed, to provide an insight into the scope of this technique, as well as highlighting some of the potential problems. The examples given here will enable readers to select the most appropriate analytical strategy for their own application.

7.5 History of laser ablation analysis

Since the pioneering work of Gray (1985) and Arrowsmith (1987), a number of papers have addressed aspects of laser ablation ICP-MS analysis. In many of the contributions, workers have concentrated on the instrument response with a view to developing schemes of semiquantitative analysis (e.g. Hager, 1989; Denoyer 1991; Denoyer, Fredeen and Hager, 1991). This approach can be useful in a rapid survey where order-of-magnitude values are sufficient or where applications are restricted to bulk samples. However, it seems unlikely that levels of accuracy better than ±10% can be obtained by this method. There is now a consensus among most workers in this field (cf. Mochizuki *et al.*, 1988; Imai, 1990; Mochizuki, Sasashita and Iwata, 1990; van Heuzen, 1991; van Heuzen and Morsink, 1991; Perkins, Fuge and Pearce, 1991; Pearce *et al.*, 1992(b); Pearce, Perkins and Fuge (1992(a)); Perkins, Pearce and Fuge, 1992; Perkins, Pearce and Jeffries, 1993) that the prerequisites for high quality analytical data for bulk samples are

1. the availability of good calibration standards;
2. the use of an internal standard;
3. matrix matching, in pressed powder work;
4. the use of Q-switched laser operation.

Williams and Jarvis (1993) took a different approach and used the laser in a fixed-Q mode to determine major and trace elements in pressed powder briquettes. In this work, the importance of matrix matching between samples was emphasized whilst the authors relied on at least one trace element analysed by a supplementary technique as an internal standard.

7.6 Bulk sampling with LA-ICP-MS: an alternative to solution preparation

The laser ablation system offers an alternative to the preparation of samples in the form of a solution. The production of either pressed powder briquettes or fused glass disks has a number of advantages over the preparation of solutions for ICP-MS analysis. Imai (1990, 1992) has described the use of laser ablation for the analysis of bulk samples and Perkins, Pearce and Jeffries (1993) have presented trace element analyses for a wide range of silicate reference materials using laser ablation ICP-MS. More recently, however, attention has been focused on the use of the laser to provide spatially resolved elemental information (Pearce *et al.*, 1992(b); Pearce, Perkins and Fuge, 1992(a); Jackson *et al.*, 1992). In these applications a different approach to the calibration procedure has been applied, as reviewed in the following section.

7.7 Analytical methodology

7.7.1 Introduction

The instrumental characteristics listed above demonstrate that a range of options is available to the user in LA-ICP-MS. The ICP-MS can be operated in one of three modes: single ion monitoring, peak jumping or mass scanning, and the laser system can be operated with large or small crater diameters in either a fixed-Q or Q-switched mode. The application will, in most cases, dictate the appropriate instrumental operating conditions; however, it is necessary to consider carefully the calibration strategy. An increasing number of applications in LA-ICP-MS involve trace element microanalysis in minerals. Unfortunately, the number of calibration standards or reference materials available which are appropriate for calibration is limited. Furthermore, it is not possible to compare the raw element or isotope count data between standards and samples, for the reasons given above, so an internal standard must be selected.

Once the instrumental and calibration options have been selected the spectrometer can be optimized for the analytical procedure. In general, this tuning of the instrument is performed using a glass reference material (e.g. NIST SRM 610 or 612), which contains a range of elements from across the mass range (^{6}Li to ^{238}U). A typical analytical sequence might then be:

1. Monitoring of plasma gas blank;
2. analysis of calibration standard;
3. analysis of samples;
4. monitoring of calibration standard (as a drift monitor);
5. analysis of more samples or recalibration.

It is normal practice to monitor the calibration standard or a drift monitor every 30 minutes. In our experience, the instrumental drift is predictable and recalibration is only necessary once every 2–3 h.

7.7.2 Sample preparation

Analyses of minerals have been made from a range of mineral groups using samples prepared in a variety of ways. Jackson *et al.* (1992) used conventional 30 μm polished thin sections, Pearce *et al.* (1992(b)) used sawn and ground blocks as well as single grain mounts and Perkins *et al.* (unpublished data) have also used polished thin sections of 30, 45 and 60 μm thickness. Polished thin sections are extremely useful in that they allow petrographic descriptions of both the transparent and opaque minerals to be made, as well as analysis by a supplementary technique such as EPMA for the determination of an internal standard. The various methods of sample preparation have their own advantages and disadvantages.

(a) Uncovered or polished thin sections

These need only be polished if reflected light optics or EPMA analysis is required. They only allow a small number of laser shots to be placed into one point on the sample, each shot ablating 1–2 μm thickness of material from the sample surface. Dependent upon the diameter of the laser beam at the sample surface, the thickness of the sample can govern the effective spatial resolution achieved. To obtain the best spatial resolution the laser should not be allowed to bore deeper into the sample than the diameter of the crater – thus a 25–30 μm crater is well suited to a 30 μm thick thin section. It may be advantageous to use thicknesses greater than 30 μm so that the mineral grains are more coherent during ablation.

In the analysis of pale-coloured minerals (i.e. those with low transition metal abundances) or those minerals which do not absorb strongly at 1064 nm (in the infrared), much of the laser light passes through the sample and may be absorbed by the volatile mounting material. This can lead to catastrophic failure of the mineral undergoing ablation as chunks are blown from the surface by the expanding bubble of gas beneath (Jackson et al., 1992). Recently, a number of workers have modified the Nd-YAG laser by doubling or quadrupling the laser frequency to 532 and 266 nm respectively (Jackson et al., 1992 and Shibata, Yoshinaga and Morita, 1993). The use of the ultraviolet wavelength, 266 nm, may overcome some of these problems (see below and Section 8.4.2).

(b) Sawn blocks

It is not necessary to prepare a polished surface on the sample for LPMA-ICP-MS unless an auxiliary technique is to be used (e.g. EPMA). Simply sawing a flat surface on the sample is by far the easiest and quickest method of sample preparation, although it requires the use of incident light to locate the analysis points on a surface that is not strongly reflective. Blocks will not fail in the same manner as thin sections, although rough sawing of the sample may enhance brittle failure by weakening the surface.

(c) Polished blocks

As with sawn blocks, these will not fail easily during ablation (unlike thin sections). With a greater thickness of material to ablate, more shots can be placed in the same area, although this has implications for the ultimate spatial resolution achieved (see above). Polished blocks can be used for EPMA analyses and with reflected light microscopes.

(d) Mineral grains mounted in a binder

Separated mineral grains held in a binder, be it a cyanoacrylate glue (such as Superglue) or an epoxy resin, have been used in LPMA-ICP-MS analysis. In the case of grains mounted in Superglue or similar, little preparation is required but the grains are held relatively loosely, being blown out of the binder should the laser power be too high. If grains are mounted in resin blocks, they can be exposed at the surface of the block by sawing and can, if required, be polished. Grains held in resin blocks are more firmly secured than in Superglue and laser powers can be higher during analysis.

7.7.3 Calibration: quantitative analysis

To achieve a quantitative analysis of mineralogical samples it is necessary to calibrate the instrument with an appropriate standard and to select an internal standard which is present in both the samples and standards. A number of approaches have been taken to these analytical requirements. In the following section two examples are given to illustrate the different approaches to calibration strategy and the selection of appropriate internal standards.

(a) Calcite

Pearce, Perkins and Fuge (1992(a)) studied the variation in elemental concentrations in calcite (calcium carbonate) shells using spiked calcium carbonate, pressed powder calibration standards. Calcite is a chemically simple mineral, usually close to $CaCO_3$ in composition. Ca is present at one atom per formula unit (APFU) in pure calcite, which is equivalent to 40 wt% Ca. $CaCO_3$ occurs as the minerals calcite and aragonite in limestones, shelly materials and igneous rocks. In natural calcites, minor substitution of Mg, Fe, Mn and a range of trace elements for Ca can occur. However, the near constancy of Ca in these materials makes it an ideal internal standard during analysis and, allowing for some substitution, Ca can be assumed to be present at approximately 40 wt%. Analyte counts can be compared to those for a minor Ca isotope (e.g. ^{43}Ca) during LPMA-ICP-MS analysis without introducing substantial errors.

Pearce, Perkins and Fuge (1992(a)) produced a series of synthetic calcite standards by adding dilute solutions of metals in weak acid (atomic absorption single-element solutions) to a 10 g base of AnalaR $CaCO_3$ which was mixed thoroughly, dried and pressed into a powder briquette, giving a range of standards at 10, 20 and $30 \mu g\,g^{-1}$ addition. Ablation of these produced calibration lines with correlation coefficients of 0.99 or better, and the precision from an individual standard was nearly always better than ±4%, indicating homogeneity. Detection limits for this technique for a range of elements were better than 0.5 ppm.

In cases where elements other than Ca total more than 1 wt%, an iterative procedure can be adopted to re-estimate the Ca concentration and accordingly adjust the concentrations of all elements so that the cations total 1 APFU (for stoichiometric $CaCO_3$).

(b) Olivines

Olivines exhibit complete solid solution between fayalite (Fe_2SiO_4) and forsterite (Mg_2SiO_4) or tephroite (Mn_2SiO_4) (Deer, Howie and Zussman, 1992). In most geological environments, olivines show compositions between forsterite and fayalite, with Mn-substitution only becoming important in Fe-rich olivines. Olivines contain minor and trace elements, notably Ca, Ni, Co and Cr, which substitute for Fe or Mg and are present typically in the range 10–2000 ppm. Olivines often show compositional zoning with variation in Fe, Mg and trace element contents. In general, Ni and Cr show a similar behaviour to Mg and Co, and Mn follows Fe.

Most olivines are near-stoichiometric, with Si present at a level close to 1 APFU. The content of SiO_2, however, represented as a weight percentage, varies greatly between the end members forsterite ($SiO_2 \sim 41\,wt\%$) and fayalite ($SiO_2 \sim 30\,wt\%$), differences that are caused by changes in molecular weight of the end members due to substitution of Fe for Mg.

Calibration of LPMA-ICP-MS for the analysis of olivines treated as true unknowns is difficult, in part owing to a lack of single mineral reference materials of the correct composition. Available reference materials of a composition similar to natural olivines are whole-rock materials rich in olivine (e.g. peridotites and dunites), which also contain minor amounts of other minerals. They are all Mg-rich and similar to forsteritic olivines in composition. These rocks also all contain similar quantities of Ni, Cr and Co and thus are difficult to use to erect calibration lines, due to the clustering of data.

7.7.4 Major elements: direct analysis in terms of molecular formula

As we have seen, the analysis of a mineral by LPMA-ICP-MS requires the selection of an internal standard. In olivines, the three major components, Mg, Fe and Si, all vary widely in abundance. Optical properties could be used to determine the composition (from $2V$; Deer, Howie and Zussman, 1992) and the Mg content could be estimated. EPMA data could be used to determine the major elements for use as internal standards, and in general this would be the preferred method. It is however possible to determine the composition of an olivine by LPMA-ICP-MS because of the relatively simple and predictable variation in its chemistry. This approach may apply to other mineral groups of relatively simple chemistry.

Si is present at 1 APFU in almost all olivines. Mg varies from 0 to 2 APFU. In olivines, a calibration graph of Mg (ACPS)/Si (ACPS) against Mg (APFU) established from a range of compositions would give a straight line, but a similar graph of Mg (ACPS)/Si (ACPS) against Mg wt% would produce a curve. Thus, taking a single forsteritic standard, a calibration graph can be produced for the determination of Mg (APFU) in olivines. The residual Fe can be determined in a similar fashion or assumed (in order to make Mg + Fe = 2) and, as olivines are relatively simple minerals, the weight percentages of Mg, Si and Fe can be calculated. The errors introduced by this method will be relatively small, a result of the presence of minor components such as Mn, Ca, etc., and one of the major elements so determined can then be used as an internal standard in trace element analysis.

7.7.5 Trace elements: the problem of polymineralic standards

Since one element cannot be assumed to be present at a fixed weight fraction in olivines, an external analysis of an internal standard is necessary to compare against a reference material. This could be an EPMA analysis or a determination as described above. Once a suitable internal standard has been selected, trace element data can be produced, provided a suitable reference material is available (e.g. peridotite, dunite, etc.). These reference materials, however, illustrate one of the major problems of using a polymineralic substance for calibration, particularly when, in the natural rock, some elements are concentrated into minor phases. In ultrabasic rocks, this is particularly the case with Cr which is concentrated as discrete small grains of chromite ($FeCr_2O_4$), as well as within solid solution in olivines in these rocks. Despite the fine grinding of these rock samples, small dark grains of chromite are seen, and on the scale of LPMA-ICP-MS these powders lack homogeneity. Separate analyses may ablate different amounts of chromite, and thus dramatically change the instrument response for Cr. This problem was described by Pearce *et al.* (1992(b)) and is less acute for Ni, Co, Ca, Mn, etc., which are more evenly distributed throughout the rock.

7.7.6 Other mineral types

The two examples above illustrate different approaches to the choice of internal standard and calibration material. In some cases, stoichiometry allows an internal standard to be chosen without reference to an external analytical technique. Examples which have been studied to date include the following.

1. Zircon, where zirconium is present at 49.76 wt% and the minor isotope ^{96}Zr (2.8% isotopic abundance of natural zirconium) has been used as an internal standard. A single-mineral certified reference material is available for calibration (BCS-CRM 388).
2. Apatite, in which the calcium content can be used as an internal standard, although small variations occur depending on whether the apatite is fluorine-, chlorine- or hydroxyl-bearing. A single-mineral reference material is available for calibration (CTA-AC-1).
3. Alkali feldspars, in which there is solid solution between the two end members, $KAlSi_3O_8$ and $NaAlSi_3O_8$. Aluminium is present at, or close to, 1 APFU and, since the molecular weights of the two end members are very similar (556 and 524 respectively), Al can be assumed to be present at a near-constant weight percentage of 9.85 ± 0.37 wt%, and is thus suitable for use as an internal standard. Single-mineral reference materials are available for use as calibration standards (e.g. ANRT FK-N; GSJ JF-1, JF-2; NIST SRM 70a, 99a; BAS BCS 375, BCS 376).

7.7.7 Analysis using synthetic glass standards: the future

The limited availability of single-mineral concentrates has led several workers to use synthetic reference materials to analyse a variety of mineral species by LPMA-ICP-MS. To date, these studies have involved the use of spiked, silicate glasses such as those produced by the National Institute of Standards and Technology (NIST) in the United States. This approach promises to be the most important future application of LPMA-ICP-MS.

The technique depends on several factors. First, there must be an external analysis of the unknown (e.g. by EPMA), or the assumption must be made that one component is present at a composition which can be accurately estimated and can thus be used as an internal standard. The analysis of an internal standard in the unknown mineral is then used to correct for the differences in (analyte ACPS/internal standard ACPS) between the reference material and the unknown. The estimate of concentration must be sufficiently accurate for the application. Jackson et al. (1992) used external EPMA determinations of the unknown to define a minor isotope of one element for use as an internal standard for analysis against a synthetic reference material.

Secondly, there must not be any matrix effects relating to the differences in composition between the synthetic glass reference material and the unknown. This is as yet undefined, although the straight line calibrations obtained across a wide range of major and trace element compositions by Perkins, Fuge and Pearce (1991), Perkins, Pearce and Fuge (1992(a)), Perkins, Pearce and Jeffries (1993) suggest that matrix effects may be very limited, and can be corrected by the use of internal standards.

Thirdly, operating conditions must be consistent. Matrix effects and differences in ablation character, yield and any preferential volatilization effects can be minimized by consistency of operating conditions. Spectra from the unknown should be acquired under the same operating conditions as the reference material to avoid the above problems. Jackson et al. (1992) collected spectra for their unknowns at much lower laser powers than for their reference materials (due to difficulties in ablating sufficient material from the reference material) and used a different configuration of the laser optics. This may account for some differences in their LA analyses compared to solution determinations, although they achieved generally good agreements.

Fourthly, the reference material must be homogeneous. This is perhaps one of the major concerns with this type of analysis, where the spatial resolution is so small. In the use of pressed powder monomineralic reference materials (e.g. feldspar, zircon) we have seen that, due to the fine grinding of the sample and thorough mixture, standards can be produced which are effectively homogeneous. Single-mineral grains used as reference material standards are unlikely to be homogeneous at the trace element concentrations of interest, even though the major element composition may show no variation across the grain. Trace elements are likely to become strongly zoned across minerals which may remain relatively unzoned in terms of their major element compositions (e.g. olivines analysed by Pearce *et al.* (1992(b)) and garnets analysed by Jackson *et al.* (1992)). Jackson *et al.* (1992), Perkins, Pearce and Jeffries (1993 and unpublished data) have all used the NIST reference materials. These samples are silicate glasses (with a matrix composition of 72% SiO_2, 12% CaO, 14% Na_2O and 2% Al_2O_3), spiked to a nominal concentration of selected trace elements. The Certificate of Analysis supplied with the NIST materials states that 'the certified values given are for the entire wafer (not fragments thereof)' and as such, care must be exercised in the assumption that trace elements are homogeneously distributed at the resolution of the laser microprobe. Recently, Jeffries *et al.* (personal communication) have tested the homogeneity of the NIST glass standards NIST SRM 612 (nominal 50 ppm) and NIST SRM 610 (nominal 500 ppm). They conclude that, within the reproducibility of the LPMA-ICP-MS, the NIST glass standards are homogeneous but that instrumental drift may be significant over long periods of operation. They recommend that regular monitoring of the calibration standards is essential if instrumental drift is to be compensated.

In the only published account to date, Jackson *et al.* (1992) have described the analysis of a variety of minerals using synthetic glass standards. These have included apatite, garnet, sphene (titanite), uraninite, zircon and pyroxene, using a variety of internal standards appropriate to the mineral in question, including ^{42}Ca, ^{96}Zr, ^{235}U and ^{29}Si, all minor isotopes of major elements in the mineral. Internal standard concentrations were determined by EPMA analysis of the minerals and they achieved sub-part-per-million detection limits for many elements by LPMA-ICP-MS. Their data were tested for accuracy against solution analyses of mineral separates and gave acceptable results. For example, REE concentrations determined from five analyses across a garnet showed how these elements were strongly zoned, whilst the major elements remained almost constant across the crystal.

A similar approach has been employed by Perkins *et al.* (unpublished data) for the analysis of large (5 mm diameter) zoned apatites from a magnetite/apatite deposit in Chile. The concentration of Ca was used as an internal standard and all spectra were reduced using the calcium data for the NIST glasses. The results revealed complex chemical zonation with rare-earth element concentrations, as well as the shape of the REE pattern, changing across the grains.

7.7.8 *Quantitative calibration: a summary of calibration strategies*

To summarize, the main conclusions derived from calibration stategies of LPMA-ICP-MS analysis reviewed above:

1. Calibration lines are linear over a considerable range of concentration and can be erected from a single, high-concentration reference material and the origin.
2. Matrix effects are apparently relatively minor and reference materials and unknowns need not be of the same composition.
3. Analyses should be performed using the same operating conditions to minimize any possibilities of fractionation.
4. Minerals can be analysed directly as unknowns where reference materials of the same composition exist (e.g. calcite, zircon, alkali feldspar). In these cases, one element in the reference material which is used as an internal standard is present at the same concentration in the unknown.
5. Where reference materials of the same composition as the unknown are not available, or the unknown does not contain one component whose concentration can be assumed, an external analysis of the unknown is required. This requirement is probably best achieved by EPMA, although any other accurate method (such as optical properties) could be employed. The external analysis is then used to correct for the differences in the concentration of the internal standard.
6. Calibration. The LPMA-ICP-MS can be calibrated for the analysis of an unknown element using equation 7.2 below, assuming that the calibration line is linear and passes through the origin.

$$C\,el_{unk} = \frac{C\,el_{rm}}{(ACPS\,el/ACPS\,i.s.)_{rm}} \times \frac{ACPS\,el}{ACPS\,i.s._{unk}} \times \frac{C\,i.s._{unk}}{C\,i.s._{rm}} \quad (7.2)$$

where C = concentration; el = analyte element; i.s. = element selected as internal standard; rm = in the reference material; unk = in the unknown sample; ACPS = area counts per second.

7.7.9 Calibration: semiquantitative analysis

In some applications, semiquantitative analysis can be an alternative to the fully quantitative methods described above. A useful feature of ICP-MS is that the technique can be used to determine virtually all elements using only a single internal standard. This calibration strategy has been applied to the analysis of bulk samples where laser ablation has advantages over the production of solutions for ICP-MS analysis, for example in avoiding loss of volatile elements in the high-temperature digestions commonly employed in geochemical sample preparation. Data for virtually every element are obtained by setting the instrument to scan the whole mass range rapidly. This technique may detect the presence of unexpected elements which could be overlooked if a peakjumping analytical routine were employed.

The response of any element depends, to a large extent, on ionization energy as well as several other factors, most of which can be regarded as constant (including plasma temperature and isotopic abundance). Elements that are completely ionized in the plasma will, for the same concentration in a material, show higher sensitivity than elements which are poorly ionized. The instrument response for each element also varies according to the mass of the element – a result of the differential transmission of elements of different masses through the ion lenses and quadrupole.

A mass spectrum from a material that contains the same concentration of each element will thus show a series of peaks of different intensities which depend upon (i) isotopic abundance, (ii) mass transition factors and (iii) the degree to which the element is ionized.

Theoretically, the transmission of an element through the instrument should vary with mass according to a smooth curve, the shape of which will depend on instrumental factors. This curve would represent the theoretical response of the instrument, provided all other factors (such as the degree of ionization) are equal for each element. The actual response for an element can be scaled so that it fits the theoretical response curve by the application of the so-called Saha factor. The curve thus produced is called the 'instrument response curve' and represents the theoretical response for the same concentration of all elements in a material to be used for calibration.

In practice, the instrument response curve is generated from a calibration standard containing between six and eight elements spanning the mass range to which the curve is fitted mathematically. In semiquantitative analysis, the detection limits are relatively high because the whole mass range is being scanned and thus the dwell time on each isotope peak is short (see below). Results are inevitably less precise, although semiquantitative analysis can produce extremely useful data and involves less off-line data reduction than fully quantitative analysis.

7.8 Detection limits

Detection limits of an analytical technique can be calculated (at 3 standard deviations of the background) from

$$\text{LLD} = 3.(2\ B)^{1/2} \cdot \frac{C}{I} \tag{7.3}$$

where LLD = lower limit of detection; B = background intensity for analyte; I = peak intensity for analyte; C = concentration of analyte in reference material.

ICP-MS background intensities are generally extremely low, and the high sensitivity of the instrument for most elements leads to very low limits of detection. ACPS values for a blank are usually obtained as a plasma blank without the laser being fired. Because blank values are very low across the whole mass range, the LLD is effectively inversely proportional to the intensity of the analyte peak (I) so that factors which affect the number of ions reaching the detector will control the absolute LLD. These include the ablation volume, operating conditions and the relative isotopic abundance.

7.8.1 Ablation volume

A larger diameter laser beam will ablate more material and thus give a lower detection limit, as will deeper ablation craters. There is thus a compromise between LLD and spatial resolution; LLD is inversely proportional to (ablation volume)$^{1/3}$, or, for the same depth of ablation, LLD is inversely proportional to (ablation area)$^{1/2}$. Thus, as the size of the ablation crater decreases, the LLD for LPMA-ICP-MS approaches the detection limits of other techniques, such as EPMA, at analysis spot diameters of a few micrometres. Nonetheless, LPMA-ICP-MS still retains the distinct advantage of speed of analysis over EPMA at trace element levels.

7.8.2 Operating conditions

Careful selection of the instrument operating conditions will produce the optimum LLD for a particular application. Consider the analysis of a mineral prepared as a thin section, where the volume of material that can be ablated is finite. If the instrument is scanned across a large m/z range (each scan taking a relatively long time), LLD values will be relatively high; if the scan is limited to a smaller m/z range, LLD values will be lower as more counts for each analyte will reach the detector. Better detection limits are achieved by rapidly peak jumping across a small number of isotopes; the ultimate detection limit of the technique is achieved when only counts from the analyte and the internal standard are collected. The operating conditions can thus be selected to suit the requirements of the application, bearing in mind that the thickness of the sample will govern the acquisition time, and thus be the ultimate control on detection limit. Table 7.4 presents the theoretical LLD calculated from the NIST SRM 612 reference material at

Table 7.4 Detection limits (in ppm) from NIST SRM 612 glass

	(1)	(2)†	(3)‡
Li	0.2130	0.2004	0.0190
Be	0.4325	n.d.	0.1525
Ti	0.0351	n.d.	0.0342
V	0.2302	n.d.	0.0560
Cr	0.3131	n.d.	0.3035
Mn	0.2114	n.d.	0.0956
Co	0.1607	0.1346	0.0754
Ni	1.7790	n.d.	0.3637

DETECTION LIMITS

Table 7.4 *Continued*

	(1)	(2)	(3)
Cu	1.1426	n.d.	0.1913
Zn	0.9792	n.d.	0.2737
Ga	0.2228	n.d.	0.0307
As	1.3906	n.d.	0.3586
Rb	0.2864	n.d.	0.1336
Sr	0.2484	n.d.	0.0559
Y	0.3213	n.d.	0.1225
Zr	0.6408	0.3421	0.1606
Nb	0.3792	n.d.	0.1368
Sn	0.3101	n.d.	0.0791
Sb	0.4739	n.d.	0.1007
Cs	0.1304	n.d.	0.0618
Ba	0.9170	n.d.	0.1876
La	0.1216	0.0973	0.0198
Ce	0.1233	0.1113	0.0280
Pr	0.0948	n.d.	0.0213
Nd	0.7240	n.d.	0.1120
Sm	0.9664	n.d.	0.0845
Eu	0.1742	n.d.	0.0314
Gd	0.4743	n.d.	0.0263
Tb	0.0696	n.d.	0.0086
Dy	0.7412	n.d.	0.0730
Ho	0.0960	n.d.	0.0072
Er	0.1845	n.d.	0.0395
Tm	0.1048	n.d.	0.0086
Yb	0.4521	n.d.	0.0597
Lu	0.0714	n.d.	0.0189
Hf	0.3772	0.3138	0.0452
Ta	0.0926	n.d.	0.0080
W	0.3154	n.d.	0.1684
Pb	0.3050	0.1382	0.0116
Th	0.1730	n.d.	0.0197
U	0.0961	0.0683	0.0167

* n.d. = not determined in the short/selected scan.
(1) 40 μm diameter craters produced from a laser pulsed at 7 Hz. The full mass scan conditions were: mass range 6.02–239.1 with certain regions of the mass range skipped to avoid high matrix/gas peaks 11.5–23.5, 26.5–28.5, 29.5–41.5, 43.5–44.5 and 79.5–80.5 and other regions skipped because there is nothing of interest in the spectrum, i.e. 183.5–203.5 and 209.5–231.0. 2048 channels, 100 sweeps, 320 μs dwell time, pulse counting detector. Total scan time around 60 s.
(2) †Crater similar to (1) produced using the same laser conditions. In this run the skipped regions were enlarged to cover most of the mass range leaving only the elements of interest. Regions skipped were: 7.5–23.5, 24.5–58.5, 59.5–89.5, 90.5–138.5, 140.5–179.5, 180.5–207.5 and 208.5–237.5. 2048 channels, 400 sweeps, 160 μs dwell time, pulse counting detector.
(3) ‡150 μm diameter crater produced using large spot laser at 7 Hz. The other operating conditions were the same as (1).

two different scan conditions In the first (1), a full mass scan from ^{6}Li to ^{238}U was used, whilst in the second (2) a shorter scan was used. In a full mass scan the spectrometer and detection system will spend a smaller interval of time on each of the selected masses, which degrades the calculated detection limits. The table also contains the theoretical LLD calculated from the NIST SRM 612 glass using the large laser beam diameter (3), creating a crater of approximately 150 μm diameter. The much larger volume of material removed during ablation with the laser at 150 μm results in LLD values lower by more than an order of magnitude in many cases (Table 7.4).

7.8.3 Relative isotopic abundance

Under given analysis conditions, monoisotopic elements will have lower LLD values (theoretical) than polyisotopic elements. The LLD values for each isotope are lower for the more abundant isotopes as more ions are received by the detector (i.e. I is higher so LLD is lower). If we take, for example, Sr, which has four isotopes of mass (and relative isotopic abundance) 84 (0.56%), 86 (9.9%), 87 (7.0%) and 88 (82.6%), ^{88}Sr would be some eight times more sensitive than ^{86}Sr and some 150 times more sensitive than ^{84}Sr. Thus, where possible, taking into account the need to avoid spectral overlaps from polyatomic species, high-abundance isotopes should be selected for the analyte in preference to low-abundance isotopes.

Pearce et al. (1992(b)) suggested that detection limits of the order of 0.5–1 ppm were realistic, whilst Jackson et al. (1992), who describe the interplay of operating conditions and LLD values, cite detection limits for routine analysis at about 0.5 ppm. These figures may be compared with data listed in Table 7.4, which shows detection limits for a variety of elements in the NIST SRM 612 reference material, with spectra obtained at different operating conditions.

7.9 Applications

7.9.1 Quantitative analysis

It is important to realize that the technique of laser probe microanalysis ICP-MS is relatively new and the number of applications described to date is limited. Also important is the fact that instrumentation is being developed rapidly at the same time as new applications are being described. The following section presents a summary of the papers which have applied this technique to geochemical analysis.

LPMA-ICP-MS has been applied to the analysis of trace metals in modern shells by Fuge et al. (1993). Shelly creatures add layers to their shells as they grow and these layers reflect the changing environmental conditions in which the organism lived. Changes in Mg content have been related to

APPLICATIONS

changes in temperature of the water and Sr variations have been assigned to salinity changes. Fuge *et al.* (1993) used LA-ICP-MS and LPMA-ICP-MS to study chemical variation in shells from Cardigan Bay, off the coast of west Wales. Clear variations in Mg content were recorded in mussels and limpets, reflecting a seasonal change in temperature. In large bivalves such as *Arctica islandica*, large variations and spikes were recorded in the Pb content (Figure 7.5). Fuge *et al.* (1993) assigned these variations to relatively short-lived, single large pollution events related to former Pb-mining activity in mid-Wales, including large outbursts of highly contaminated water from disused mines.

Pearce *et al.* (1992(b)) analysed a series of olivines from Icelandic basalts by LPMA-ICP-MS, calibrating with the reference peridotite PCC-1 and using ^{29}Si as an internal standard. They achieved a spatial resolution of approximately 30 μm with a series of holes placed at about 50 μm spacing (Plates 7.1 and 7.2). They made no correction for the differences in Si

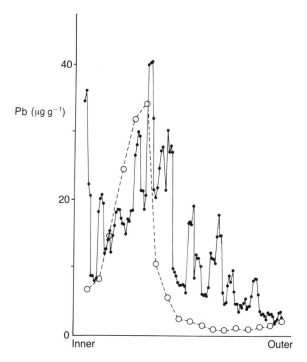

Figure 7.5 Lead profile through a valve from the bivalve *Arctica islandica* collected in Cardigan Bay, Wales. Large open symbols represent the data collected with the standard laser system producing craters of about 150 μm diameter. The smaller closed symbols represent data collected with the laserprobe system. The slight offset of the major lead peak is due to the fact that the two profiles were taken from different positions along the same valve.

between the reference material and the unknown as these concentrations were similar. The calibration line was produced by the random ablation of the reference material. They observed two distinct sets of olivines in these samples: (i) unzoned crystals up to 1 mm diameter from lava flows that had cooled relatively slowly and (ii) smaller, zoned olivines from rapidly quenched lavas.

7.9.2 Alkali feldspars

Pearce et al. (1992(b)) described the application of LPMA-ICP-MS techniques to the analysis of major and minor components in alkali feldspars. Single mineral reference materials exist for albitic and orthoclasic feldspars and thus the potential exists to compare natural reference materials with unknowns of a similar composition. Pearce et al. (1992(b)) erected calibration curves for the analysis of single feldspar grains, prepared by pressing individual grains into the surface of a bead of cyanoacrylate glue ('Superglue') on a glass slide. The glue was allowed to harden before ablation. Several determinations could be made from each grain when prepared in this manner. Low repetition rates for the ablation were used (in this case 1.7 Hz) at relatively low power to avoid dislodging the grains from the mount.

7.9.3 Zircon

Zircon ($ZrSiO_4$) is a refractory silicate with a wide range of industrial applications. It is a common accessory in igneous, metamorphic and sedimentary rocks (Deer, Howie and Zussman, 1992) and is an important host of Zr, Hf, REE, U and Th. Pb-isotope ratios from zircons are widely used to date rocks by the decay of U and Th to Pb (Faure, 1986).

Zircon concentrates are available commercially as reference materials, but only a few are issued with certificates of analysis for some of their components. Perkins, Pearce and Fuge (1992) have recently produced trace element data by laser ablation and solution ICP-MS analysis for all currently available zircon reference materials. Good agreement was obtained in values produced by both laser ablation and solution analysis, and in both cases good precision (better than ±10% on five repeats) and a high level of accuracy (against certified values) were achieved.

Perkins, Pearce and Fuge (1992) selected zircon reference materials containing varying amounts of trace elements and produced pressed powder calibration standards using BCS-388 (low concentrations of trace elements), SARM-13 (moderate levels) and BCS-204a (high levels). They collected three spectra from each standard, which were averaged to produce calibration lines for a wide range of trace elements using ^{96}Zr (2.8% of natural Zr) as an internal standard. A correction was made for the different concentration of Zr in BCS-204a which, unlike the others, is not a pure zircon concentrate.

In general, Perkins, Pearce and Fuge (1992) obtained excellent calibrations for a wide range of elements. In certain cases, the concentrations of some elements in the reference materials used by Perkins, Pearce and Fuge (1992) were similar (e.g. Eu and Hf) and this caused a clustering of the data which produced meaningless calibration lines. In these cases, the inclusion of the origin produced usable calibration lines, which were applied to natural (large) zircons from alkaline pegmatites using ^{96}Zr as an internal standard. Geologically meaningful data was obtained. Zircons are commonly strongly zoned with respect to REE, Hf, U and Th and thus Perkins, Pearce and Fuge (1992) were unable to test the data for reproducibility.

LPMA-ICP-MS can also offer isotopic information and this capability may have uses in determining the ages of individual zircon crystals in, or extracted from, rocks based on their Pb isotopic composition. Single-crystal dating by ion microprobe (e.g. Compston, Williams and Meyer, 1984) is now widely used and gives high precision at a resolution of approximately 30 μm, although instrument time is expensive. Although ICP-MS may not be able to better 1% precision on Pb isotope determinations (Date and Cheung, 1987), this capability may be sufficient to erect reconnaissance age dates from single zircon crystals. This is the subject of continued research at Aberystwyth.

7.9.4 Semiquantitative analysis

Semiquantitative analysis has been widely used in water analysis (e.g. Pearce, 1991; Fuge and Perkins, 1991) where an internal standard is added to the solution. However, semiquantitative methods have also been developed for the laser ablation analysis of solid samples. Pearce, Perkins and Fuge (1992(b)) described a semiquantitative technique for the analysis of carbonates by LA-ICP-MS. They produced a single calibration standard using a 10 g base of AnalaR CaCO$_3$ spiked to give known concentrations near to 100 ppm of Li, Mg, Mn, Sr, In, Ba, Pb and U. With a minor isotope of Ca (^{43}Ca) as the internal standard, this standard can be ablated to generate an instrument response curve. Unknowns were ablated in the same manner as the calibration standard, each analyte line being referenced to ^{43}Ca, scaled by the appropriate Saha factor and finally compared to the response curve to generate a concentration. To check the accuracy of this technique, Pearce, Perkins and Fuge (1992(b)) analysed two available limestone reference materials and showed that this technique is typically accurate to about ±10%, which is perfectly acceptable as a reconnaissance technique. Detection limits for trace elements, obtained by using an unmodified laser (150 μm diameter crater) and scanning the mass range from Mg to U, were of the order of 0.5 ppm.

A similar technique has been applied by Westgate et al. (1994) to the analysis of glass shards from recent (Quaternary) tephra from North America.

In this case, an instrument response curve was generated from the NIST SRM 612 glass standard, and an internal standard was selected from an external analysis (Ce from INAA). This application is strictly a bulk sampling technique; small samples (~0.01 g) prepared as piles of material are held on a mount in cyanoacrylate glue and about 20 glass shards are consumed in each ablation, but there are no reasons why this approach cannot be adapted for mineral microanalysis. A comparison of accepted and experimental results for selected trace elements in a basaltic volcanic glass, UTB-1, an in-house standard from the University of Toronto (Barnes and Gorton, 1984) is presented in Table 7.5. Determinations by semiquantitative LA-ICP-MS are compared in this table with accepted data obtained by X-ray fluorescence (XRF) and instrumental neutron activation analyses (INAA). It is clear that

Table 7.5 Comparison of semiquantitative LA-ICP-MS data with accepted values for UTB-1, a basaltic glass reference material from the University of Toronto (Westgate et al., 1994; Barnes and Gorton, 1984)

Element	Accepted (ppm)	LA-ICP-MS (ppm)	LA-ICP-MS/Accepted
Ga	22	15	0.68
Rb	33	29	0.88
Sr	307	361	1.18
Y	43	43	1.00
Zr	200	205	1.03
Nb	15.5	12.3	0.79
Cs	0.77	0.75	0.97
Ba	538	557	1.04
La	27	24	0.89
Ce	61	–	
Pr	7.3	8.5	1.16
Nd	32	34	1.06
Sm	8.0	7.5	0.94
Eu	2.40	2.63	1.10
Gd	7.7	8.5	1.10
Tb	1.25	1.50	1.20
Dy	7.9	8.42	1.07
Ho	1.6	1.68	1.05
Er	4.5	4.4	0.98
Tm	0.6	0.77	1.28
Yb	4.0	4.63	1.16
Lu	0.60	0.65	1.08
Hf	4.7	5.55	1.18
Ta	1.0	0.70	0.70
Th	4.3	4.3	1.00
U	1.2	1.27	1.06

Accepted results were obtained on bulk samples analysed by XRF and INAA.

there is good agreement between the two data sets, testifying to the accuracy of this technique. One advantage of the LA-ICP-MS technique is that it will allow much smaller samples of tephra to be analysed than was previously possible by other bulk trace element techniques such as INAA. Westgate *et al.* (1994) describe an application in which only very small samples of tephra from geographically widespread locations in America were available. These samples have been analysed by EPMA for major elements but could not be distinguished on this basis and were too small for INAA analysis. The LA-ICP-MS trace element analyses allow the tephra to be distinguished on element ratio diagrams. The technique is currently being developed for the analysis of individual shards using LPMA-ICP-MS and will represent a significant contribution to the study of distal tephra and tephra correlation.

7.9.5 Depth profiling

An interesting use of polished or sawn blocks is 'depth profiling'. Many LA-ICP-MS instruments have software available that will store the spectra acquired during a long analysis in relatively short 'time slices', the length of which can be defined by the operator. If these time slices can be correlated with the rate at which the laser bores into the sample, a record of the compositional variation with time (and thus depth) can be produced. This scheme assumes that the rate at which the sample is ablated is constant (which is likely to be the case), with actual penetration rates dependent upon laser power, repetition rate, coupling efficiency, etc. One problem with this approach is that material will be continually ablated from the walls of the deepening crater which will contribute to, and dilute, the signal from deeper material. This effect will become even more pronounced when the crater base is not flat. However, by using this method and selecting the operating conditions carefully, it is possible to achieve a high resolution profile through a mineral with depth slices of the order of $1-2\,\mu$m thick (see also Section 6.5.5).

7.10 Future trends

With modifications to the optics of commercial laser ablation systems, it is possible to achieve laser crater diameters of about $30\,\mu$m, with the rate of laser excavation of material from the bottom of the crater being the ultimate limit of resolution. However, a number of applications would be possible if a smaller crater diameter were available and to this end, other laser wavelengths have been investigated. It is known that the absorption of light by most mineral species increases strongly towards the ultraviolet and this phenomenon has led to an investigation of such wavelengths for laser ablation. Mermet (1991) discussed the application of an ultraviolet (UV) excimer laser for the analysis of metals and demonstrated significant advantages over

Nd-YAG lasers. The excimer laser is less dependent on the physical and chemical properties of the target material and does not show the significant element fractionation seen with the infrared source. The excimer laser has a less predictable beam profile, which leads to irregular cratering and may cause problems with mineralogical materials.

Using a different approach, some workers have experimented with quadrupling the frequency of the Nd-YAG laser to produce an ultraviolet laser (266 nm) with the advantages of the predictable quality of the Nd-YAG source (Jackson *et al.*, 1992; Chenery and Cook, 1993). The diameters of craters produced by these systems are about $5 \mu m$ but, because of the smaller amounts of material reaching the plasma, the limits of detection are poorer than with the infrared laser. Chenery and Cook (1993) quote limits of detection at the tens of ppm level. This detection limit can be improved using the newer plasma interface geometry now available. In addition to the modifications to interface geometry, instruments with improved first stage vacuum systems are becoming available. By pumping the expansion chamber harder, more material from the plasma is drawn into the instrument, and consequently sensitivity is improved. Preliminary results of this application have been presented for UV laser ablation ICP-MS by Pearce *et al.* (1994) for the analysis of single volcanic glass shards from tephra, using EPMA data for internal standardization, and by Gunther *et al.* (1994) for a range of mineral types, again with EPMA or stoichiometric constraints for internal standardization. Both authors report craters in the $10-40 \mu m$ range from frequency quadrupled Nd-YAG lasers. A recent comparison of UV and IR laser probe analyses is given by Jeffries *et al.* (1995(a)) who also present detection limit data from a normally aspirated ICP-MS at crater diameters from $5 \mu m-70 \mu m$. Vacuum system modifications however give theoretical detection limits up to an order of magnitude better than the normally aspirated ICP-MS. Foley (1993) described the application of a UV laser probe ICP-MS to the determination of mineral–melt and mineral–mineral distribution coefficients in experimental charges. Jeffries *et al.* (1995(b)) have presented distribution coefficients determined by IR laser probe ICP-MS for basalts and their phenocrysts from Icelandic and Hawaiian samples. This is a highly promising area of research which will provide important information for the modelling of petrogenetic processes.

Preliminary results from a laserprobe which uses an ICP source coupled to a double focusing mass spectrometer with multi-collectors (Walder *et al.*, 1993) show a significant improvement in both the limit of detection and the stability of the ion signal. Levels of precision obtained with this instrument are comparable to those obtained using a thermal ionization mass spectrometer and this performane promises to offer exciting new applications of laser ICP-MS. The greatest disadvantage with such a system will be its capital cost, which will be considerable.

7.11 Conclusions

Laser probe inductively coupled plasma mass spectrometry is a new analytical technique and the number of applications described to date is limited. The technique has limits of detection which are better than those of the EPMA but not as low as the ion microprobe. The advantages of the LPMA-ICP-MS are that it is rapid, has good limits of detection, does not require conducting samples and, with the development of the UV laser, has spatial resolution approaching that of the electron probe. Sample preparation is simple, requiring only an approximately flat surface. The combination of these factors makes this a promising technique for the determination of trace elements in minerals, shells and glasses. The necessity for an internal standard means that LPMA-ICP-MS will rely on another instrumental technique to provide analytical data for at least one element in the sample but, given the excellent dynamic range of the instrument, the internal standard can be a minor isotope of a major element, determined by EPMA, for example. Exciting new developments in instrumentation and applications are anticipated in this rapidly changing field in future years.

References

Arrowsmith, P. (1987) Laser ablation of solids for elemental analysis by inductively coupled plasma mass spectrometry. *Anal. Chem.*, **59**, 1437–44.

Barnes, S.J. and Gorton, M.P. (1984) Trace element analysis by neutron activation with a low flux reactor (Slowpoke-II): results for international reference rocks. *Geostandards Newsletter.* **VII**, 17–23.

Chenery, S.R.N. and Cook, J.M. (1993) Determination of rare earth elements in single mineral grains by laser ablation microprobe–inductively coupled plasma mass spectrometry – preliminary study. *J. Anal. Atom. Spectrom.*, **8**, 299–303.

Compston, W., Williams, I.S. and Meyer, C. (1984) U–Pb geochronology of zircons from lunar breccia 73217 using a sensitive high mass-resolution ion microprobe. *Proc. 14th Lunar Planet. Sci. Conf.*, Part 2, *J. Geophys. Res. Suppl.*, **89**, B525–34.

Date, A.R. and Cheung, Y.Y. (1987) Studies in the determination of lead isotope ratios by inductively coupled plasma mass spectrometry. *Analyst*, **106**, 1255–67.

Deer, W.A., Howie, R.A. and Zussman, J. (1992) *An Introduction to the Rock-forming Minerals*, 2nd edn, Longman, Harlow.

Denoyer, E.R. (1991) Analysis of powdered materials by laser sampling ICP-MS, In *Applications of Plasma Source Mass Spectrometry* (eds G. Holland and A.E. Eaton), Royal Society of Chemistry.

Denoyer, E.R., Fredeen, K.J. and Hager, J.W. (1991) Laser solid sampling for inductively coupled plasma mass spectrometry. *Anal. Chem.*, **63**, 445–57.

Faure, G. (1986) *Principles of Isotope Geology*, Wiley, New York.

Foley, S. (1993) Laser-ablation microprobe inductively coupled plasma mass spectrometry for the determination of trace element partitioning in high pressure experimental charges. Mineralogical Society Meeting, London, May 1993.

Fuge, R. and Perkins, W.T. (1991) Aluminium and heavy metals in potable waters of the north Ceredigion area, mid-Wales. *Environ. Geochem. Health*, **13**, 56–65.

Fuge, R., Palmer, T.J., Pearce, N.J.G. and Perkins, W.T., (1993) Minor and trace element chemistry of modern shells: a laser ablation inductively coupled plama mass spectrometry study. *Appl. Geochem. Suppl.*, **2**, 111–16.

Gray, A.L. (1985) Solid sample introduction by laser ablation for inductively coupled plasma source mass spectrometry. *Analyst*, **110**, 551–6.

Gunther, D., Forsythe, L.M., Jackson, S.E. and Longerich, H.P. (1994) Elemental analysis of minerals using laser ablation microprobe sample introductions system with a new ultra-high sensitivity inductively coupled plasma mass spectrometer. *Eos Trans. Am. Geophys. Union, 1994 Fall Meeting*, **75**, Supplement.

Hager, J.W. (1989) Relative elemental responses for laser ablation inductively coupled plasma mass spectrometry. *Anal. Chem.*, **61**, 1243–8.

Hazen, R.M., Bell, P.M. and Mao, H.K. (1977) Comparison of absorption spectra of lunar and terrestrial olivines. *Carnegie Inst. Washington Yearbook*, **77**, 853–5.

Imai, N. (1990) Quantitative analysis of original and powdered rocks and mineral inclusions by laser ablation inductively coupled plasma mass spectrometry. *Anal. Chim. Acta*, **235**, 381–91.

Imai, N. (1992) Microprobe analysis of geological materials by laser ablation inductively coupled plasma mass spectrometry. *Anal. Chim. Acta*, **269**, 263–8.

Jackson, S.E., Longerich, H.P., Dunning, G.R. and Fryer, B.J. (1992) The application of laser-ablation microprobe–inductively coupled plasma–mass spetrometry to *in situ* trace-element determinations in minerals. *Can. Mineral.*, **30**, 1049–64.

Jarvis, K.E., Gray, A.L. and Houk, R.S. (1992) *Handbook of Inductively Coupled Plasma Mass Spectrometry*, Blackie, Glasgow.

Jeffries, T.E., Perkins, W.T. and Pearce, N.J.G. (1995(a)) Comparisons of infrared and ultraviolet laser probe microanalysis inductively coupled plasma mass spectrometry (LPMA-ICP-MS) in mineral analysis. *Analyst*, (in press).

Jeffries, T.E., Perkins, W.T. and Pearce, N.J.G. (1995(b)) Measurements of trace elements in basalts and their phenocrysts by laser probe microanalysis inductively coupled plasma mass spectrometry (LPMA-ICP-MS). *Chem. Geol.*, (in press).

Mermet, J.M. (1991) Laser ablation of solids in inductively coupled plasma spectrochemical analysis. (Abstract). Fourth Surrey Conference on Plasma Source Mass Spectrometry.

Mochizuki, T., Sasashita, A. and Iwata, H. (1990) Laser ablation for direct elemental analysis of solid samples by ICP–atomic emission spectrometry and ICP–mass spectrometry. *NKK Tech. Rev.*, **58**, 19–27.

Mochizuki, T., Sasashita, A., Iwata, H. *et al.* (1988) Laser ablation for direct elemental analysis of solid samples by inductively coupled plasma mass spectrometry. *Anal. Sci.*, **5**, 311–17.

Moenke-Blankenburg, L. (1989) *Laser Microanalysis, Chemical Analysis*, **105**, Wiley, New York.

Pearce, F.M. (1991) The use of ICP-MS for the analysis of natural waters and an evaluation of sampling techniques. *Environ. Geochem. Health*, **13**, 50–5.

Pearce, N.J.G., Perkins, W.T. and Fuge, R. (1992(a)) Developments in the quantitative and semiquantitative determination of trace elements in carbonates by laser ablation inductively coupled plasma mass spectrometry. *J. Anal. Atom. Spectrom.*, **7**, 595–8.

Pearce, N.J.G., Perkins, W.T., Abell, I. *et al.* (1992(b)) Mineral microanalysis by laser ablation inductively coupled plasma mass spectrometry. *J. Anal. Atom. Spectrom.*, **7**, 53–7.

REFERENCES

Pearce, N.J.G., Westgate, J.A. and Perkins, W.T. (1994) Trace element analysis of single glass shards in volcanic deposits by laser ablation ICP-MS; application to tephrochronology. *Geol. Soc. Am. Annual Meeting 1994, Abstract with programs*, **26**, A-483.

Perkins, W.T., Fuge, R. and Pearce, N.J.G. (1991) Quantitative analysis of trace elements in carbonates using laser ablation inductively coupled plasma mass spectrometry. *J. Anal. Atom. Spectrom.*, **6**, 445–9.

Perkins, W.T., Pearce, N.J.G. and Fuge, R. (1992) Analysis of zircon by laser ablation and solution inductively coupled plasma mass spectrometry. *J. Anal. Atom. Spectrom.*, **7**, 611–16.

Perkins, W.T., Pearce, N.J.G. and Jeffries, T.E. (1993) Laser ablation inductively coupled plasma mass spectrometry: a new technique for the determination of trace and ultra-trace elements in silicates. *Geochim. Cosmochim. Acta*, **57**, 475–82.

Rossman, G.R. (1988) Optical spectroscopy. *Spectroscopic Methods in Mineralogy and Geology* (ed. F.C. Hawthorne), *Mineral. Soc. Am. Rev. Mineral.*, **18**.

Shibata, Y., Yoshinaga, J. and Morita, M. (1993) Micro-laser ablation system combined with inductively coupled plasma mass spectrometry for the determination of elemental composition in the micron range. *Anal. Sci.*, **9**, 129–31.

van Heuzen, A.A. (1991) Analysis of solids by laser ablation inductively coupled plasma mass spectrometry – 1. Matching with glass matrix. *Spectrochim. Acta*, **46B**, 1803–17.

van Heuzen, A.A. and Morsink, J.B.W. (1991) Analysis of solids by laser abalation inductively coupled plasma mass spectrometry – 2. Matching with pressed pellets. *Spectrochim. Acta*, **46B**, 1819–28.

Walder, A.J., Abell, I.D., Platzner, I. and Freeman, P.A. (1993) Lead isotope measurement of NIST 610 glass by laser ablation inductively coupled plasma spectrometry. *Spectrochimica Acta B*, **48B**, 397–402.

Westgate, J.A., Perkins, W.T., Fuge, R. *et al.* (1994) Trace-element analysis of volcanic glass shards by laser ablation inductively coupled plasma mass spectrometry: application to tephrochronological studies. *Applied Geochemistry*, **9**, 323–335.

Williams, J.G. and Jarvis, K.E. (1993) Preliminary assessment of laser ablation inductively coupled plasma mass spectrometry for quantitative multi-element determination in silicates. *J. Anal. Atom. Spectrom.*, **8**, 25–34.

CHAPTER EIGHT
Ar–Ar dating by laser microprobe
Simon Kelley

8.1 Introduction

The Ar–Ar dating technique was developed by Merrihue and Turner (1966) and proved to be an extremely powerful geochronological tool even before the advent of laser extraction techniques. The strength of Ar–Ar dating lay in the stepped heating procedure (Turner, Miller and Grasty, 1966), applied with great success to extraterrestrial samples. The technique involves sequentially increasing the temperature (stepped heating) of a vacuum furnace into which the sample has been loaded. As the extraction temperature is increased, deeper or more strongly held argon is released from the mineral lattice. The process can reveal not only an original cooling age for disturbed samples, but often (i) the age of the later event causing the disturbance, (ii) argon diffusion characteristics of the sample and (iii) the thermal history of the later event. Applications of the technique to terrestrial minerals have met with equal success, though it has become increasingly apparent that when the stepped heating technique is applied to hydrous minerals 'in vacuo', results reflect not only the distribution of argon within the grains but also the breakdown of the minerals by dehydration or decomposition during the procedure. This limits the usefulness of stepped heating.

In recent years there has been a resurgence of interest in the Ar–Ar dating technique, led by the development of laser-based extraction techniques. These can be used to investigate the distribution of argon in mineral grains at high spatial resolution without the problems of vacuum breakdown exhibited during stepped heating. The laser-based techniques allow the investigator to separate spatially distinct argon components in rocks and even individual mineral grains. The technique was originally developed to investigate the distribution of noble gases in meteorites and lunar rocks (Megrue, 1967, 1971). The first application of the technique to Ar–Ar dating by Megrue (1972, 1973) investigated the distribution of argon in a lunar breccia. Subsequently, in the late 1970s the technique was applied to a range of problems in lunar rocks (Plieninger and Schaeffer, 1976;

Microprobe Techniques in the Earth Sciences. Edited by P.J. Potts, J.F.W. Bowles, S.J.B. Reed and M.R. Cave. Published in 1995 by Chapman & Hall, London. ISBN 0 412 55100 4

Müller et al., 1977; Schaeffer, Müller and Grove, 1977; Eichhorn et al., 1979). In the early 1980s, the first work on terrestrial mineral analyses appeared (York et al., 1981; Maluski and Schaeffer, 1982; Sutter and Hartung, 1984) and since that time studies of terrestrial rocks and minerals have dominated the research effort, though the technique continues to be used in extraterrestrial studies (Glass, Hall and York, 1986; McConville, Kelley and Turner, 1988; Laurenzi, Turner and McConville, 1988). Currently, laser extraction techniques for argon take two basic forms; stepped heating or total fusion of separated grains (e.g. Lo Bello et al., 1987; Copeland and Harrison, 1990; Chesner et al., 1991; Walter et al., 1991; Wright, Layer and York, 1991; Kelley and Bluck, 1992) and *in situ* analysis of minerals and rocks (e.g. Burgess, Turner and Harris, 1989; Phillips, Onstott and Harris, 1989; Lee, Onstott and Hanes, 1990; Scaillet et al., 1990; Kelley and Turner, 1991; Onstott, Phillips and Pringle-Goodell, 1991; Burgess et al., 1992; Kelley and Bluck, 1992), both of which will be discussed here.

8.1.1 The relationship between K–Ar and Ar–Ar dating techniques

Both K–Ar and Ar–Ar dating techniques are based upon the decay of a naturally occurring isotope of potassium, ^{40}K to an isotope of argon, ^{40}Ar. In fact only 10.48% of ^{40}K decays to ^{40}Ar (by β^+ and electron capture to excited states, followed by γ decay to the ground state and by electron capture direct to the ground state) and 89.52% decays to ^{40}Ca (by β^- to the ground state). K–Ca dating is therefore also possible but rarely used since calcium is common in many rock-forming minerals and ^{40}Ca is the most abundant naturally occuring isotope (96.94%), making the small amounts of radiogenically produced ^{40}Ca very difficult to measure. Argon, in contrast, is always a rare trace element. Radiogenically produced ^{40}Ar generally exceeds the background ^{40}Ar levels, though this is not always the case. The essential difference between K–Ar and Ar–Ar dating techniques lies in the measurement of potassium. K–Ar dating relies on measuring potassium on a separate aliquot of the mineral or rock sample, whereas in Ar–Ar dating, as the name suggests, potassium is measured by the transmutation of ^{39}K to ^{39}Ar by neutron bombardment. In order to perform K–Ar analyses, wet chemistry, flame photometry (the standard technique for potassium measurement) and noble gas mass spectrometry are required. In order to perform Ar–Ar dating, only noble gas mass spectrometry is required though you must also have access to a powerful nuclear reactor!

The naturally occurring isotopes of argon are measured by mass spectrometry for K–Ar dating (^{36}Ar, ^{38}Ar and ^{40}Ar) and in fact ^{38}Ar is superfluous since the $^{36}Ar/^{38}Ar$ ratio is unchanged through time. The ^{36}Ar peak is measured in order to correct for atmospheric ^{40}Ar using the measured present-day atmospheric $^{40}Ar/^{36}Ar$ ratio (295.5; Steiger and Jäger,

1977). The Ar–Ar technique requires that reactor-produced isotopes (^{39}Ar and ^{37}Ar) are also measured in addition to the three stable isotopes (^{36}Ar, ^{38}Ar and ^{40}Ar). As we shall see later, there are reactions producing the stable argon isotopes as well during irradiation, and it is thus important to measure all argon masses by mass spectrometry.

The 'date' measured by both K–Ar and Ar–Ar techniques is the time since argon became trapped in the mineral or rock. This is generally termed 'closure' or 'blocking' and in individual minerals (Dodson, 1973) is controlled by composition, temperature, grain size and cooling rate. Argon loss can also occur as a result of alteration or weathering of minerals. In igneous rocks, the cooling rates are very fast and the K–Ar or Ar–Ar 'date' often coincides with eruption or intrusion to within the analytical errors. In metamorphic rocks the closure to argon loss is more complex and important owing to the slower cooling rates. Dodson (1973) defined an analytical solution for a mineral system with a cooling history linear in $1/T$ which, although it is an approximation, has proved to be an extremely powerful concept in the interpretation of K–Ar and Ar–Ar ages from metamorphic minerals. Full derivation and discussion of the 'closure temperature' concept appears in McDougall and Harrison (1988).

8.2 The ^{40}Ar–^{39}Ar dating technique

The aim of this section is to introduce briefly the Ar–Ar dating technique as an aid to those who are unfamiliar with it, before proceeding to the laser extraction techniques. For those who need to know more, *Geochronology and Thermochronology by the ^{40}Ar/^{39}Ar Method* by McDougall and Harrison (1988) gives the most comprehensive description. A shorter explanation is offered by the ^{40}Ar/^{39}Ar dating chapter in *Principles of Isotope Geology* by Faure (1986).

The Ar–Ar technique, first described by Merrihue and Turner (1966), is based on the formation of ^{39}Ar from ^{39}K by neutron bombardment in a nuclear reactor, by the reaction:

$$^{39}_{19}\text{K}(n,p)^{39}_{18}\text{Ar} \tag{8.1}$$

The ratio of ^{39}K to ^{40}K is constant in almost all natural rocks and thus the critical ^{40}Ar(radiogenic)/^{40}K ratio is proportional to the ratio of the two argon isotopes ^{40}Ar/^{39}Ar. Although ^{39}Ar is radioactive, decaying with a half life of 239 years, this effect is small for the period between irradiation and analysis (generally less than 6 months).

Mitchell (1968) showed that the number of ^{39}Ar atoms formed during irradiation is given by the equation:

$$^{39}\text{Ar} = {}^{39}\text{K}\,\Delta T \int \phi(\varepsilon)\sigma(\varepsilon)\mathrm{d}\varepsilon \tag{8.2}$$

where ^{39}K is the number of atoms, ΔT is the duration of the irradiation, $\phi(\varepsilon)$ is the neutron flux density at energy ε, and $\sigma(\varepsilon)$ is the neutron capture cross-section of ^{39}K for neutrons of energy ε for the neutron in/proton out reaction shown in equation 8.1.

The standard equation for the K–Ar decay scheme (full derivation is available in McDougall and Harrison, 1988) is:

$$t = \frac{1}{\lambda}\ln\left(1 + \frac{\lambda}{\lambda_e + \lambda'_e}\frac{^{40}\text{Ar}^*}{^{40}\text{K}}\right) \tag{8.3}$$

where t is the age of the sample, λ (Table 8.1) is the combined decay constant for ^{40}K; λ_e and λ'_e are the decay constants for electron capture to an excited state and to the ground state respectively (Table 8.1); ^{40}Ar$^*/^{40}$K is the ratio of radiogenic daughter product (shown conventionally as ^{40}Ar* to distinguish it from atmospheric ^{40}Ar) to the parent ^{40}K. Rearranging equation 8.3 in terms of ^{40}Ar* yields:

$$^{40}\text{Ar}^* = {^{40}\text{K}}\frac{\lambda_e + \lambda'_e}{\lambda}\left[(e^{\lambda t}) - 1\right] \tag{8.4}$$

Combining equations 8.2 and 8.4 for a sample of age t yields:

$$\frac{^{40}\text{Ar}^*}{^{39}\text{Ar}} = \frac{^{40}\text{K}}{^{39}\text{K}}\frac{\lambda_e + \lambda'_e}{\lambda}\frac{1}{\Delta T \int \phi(\varepsilon)\sigma(\varepsilon)\mathrm{d}\varepsilon}[(e^{\lambda t}) - 1] \tag{8.5}$$

This can be simplified by defining a dimensionless irradiation-related parameter, J, as follows:

$$J = \frac{^{39}\text{K}}{^{40}\text{K}}\frac{\lambda}{\lambda_e + \lambda'_e}\Delta T \int \phi(\varepsilon)\sigma(\varepsilon)\mathrm{d}\varepsilon \tag{8.6}$$

The J value is determined by using standard minerals of known age to monitor the neutron flux. Substituting equation 8.6 into equation 8.5 and rearranging, yields the standard Ar–Ar age equation:

Table 8.1 Constants for decay of ^{40}K

Decay	Decay factor	Value ($\times 10^{-10}\,\text{a}^{-1}$)
^{40}K→^{40}Ca by β^-	λ_β^-	4.962
^{40}K→^{40}Ar by electron capture and γ	λ_e	0.572
^{40}K→^{40}Ar by electron capture	λ'_e	0.0088
combined value	$\lambda = \lambda_\beta^- + \lambda_e + \lambda'_e$	5.543
present day ^{40}K/K		0.0001167

$$t = \frac{1}{\lambda}\ln\left(1 + J\frac{^{40}\text{Ar}^*}{^{39}\text{Ar}}\right) \qquad (8.7)$$

The ratio of the two isotopes of argon, naturally produced radiogenic ^{40}Ar and reactor-produced ^{39}Ar is thus proportional to the age of the sample. Under the simplest assumptions, the ^{40}Ar peak measured in the mass spectrometer has two components, radiogenic and atmospheric. The ^{40}Ar/^{36}Ar ratio of the atmosphere is 295.5 (Steiger and Jäger, 1977). Thus by assuming that all the non-radiogenic argon is atmospheric, the daughter/parent ratio (^{40}Ar*/^{39}Ar) can be determined from the equation:

$$\frac{^{40}\text{Ar}^*}{^{39}\text{Ar}} = \left(\frac{^{40}\text{Ar}}{^{39}\text{Ar}}\right)_m - 295.5\left(\frac{^{36}\text{Ar}}{^{39}\text{Ar}}\right)_m \qquad (8.8)$$

where subscript m denotes the measured ratio. The sample may contain a mixture of radiogenic argon with an additional component of argon with a non-atmospheric ^{40}Ar/^{36}Ar ratio. The sources of such non-atmospheric argon range from hydrothermal fluids (Kelley et al., 1986) to fluids from the deep crust or mantle (Burgess et al., 1992). The isochron or correlation diagram (a plot of ^{36}Ar/^{40}Ar vs. ^{39}Ar/^{40}Ar) can be used to surmount such problems (Figure 8.1). In such a plot, the slope is negative and the intercepts are proportional to the age (x-intercept) and the non-radiogenic component (y-intercept). Thus the isotope ratio of the non-radiogenic component can be measured directly rather than being assumed.

The irradiation procedure induces not only the $^{39}_{19}$K(n,p)$^{39}_{18}$Ar reaction but also a series of interfering reactions caused by neutron bombardment of potassium, calcium, chlorine and argon. The complete series is detailed in Table 8.2, but most have such low production levels that they can be ignored. The most important reactions are those involving calcium and potassium. The corrections are generally small, though they are critical both for samples less than 1 Ma old, when the interfering reactions producing ^{40}Ar from K are important, and for samples with Ca/K ratios >10, when reactions producing ^{36}Ar and ^{39}Ar from Ca become important. The magnitude of the interference reactions varies with the neutron flux energy spectrum. Measured interference factors for many of the world's reactors are listed in McDougall and Harrison (1988), and ranges for the three most important reactions are listed in Table 8.3.

The ^{42}Ca(n,α)^{39}Ar and ^{40}Ca(n,nα)^{36}Ar production ratios do not vary a great deal, because they are caused by fast neutrons and the energy spectrum of fast neutrons in most reactors is fairly similar. The far larger variation in the interference caused by the ^{40}K(n,p)^{40}Ar reaction is due to its higher sensitivity to the ratio of fast to thermal neutrons in the reactor. This ratio varies between reactors and also between different irradiation positions within a reactor, or even between samples shielded against slow neutrons

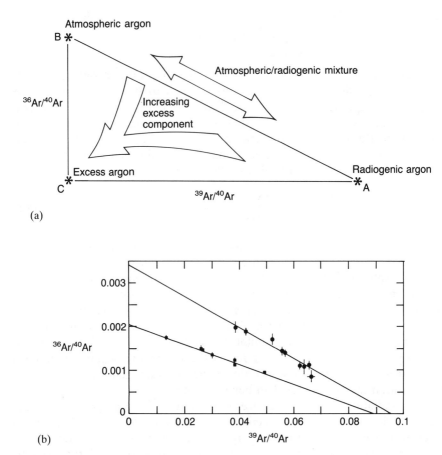

Figure 8.1 (a) A plot of $^{36}Ar/^{40}Ar$ vs. $^{39}Ar/^{40}Ar$ used to distinguish excess argon. The plot illustrates data as a mixture of several end member compositions, radiogenic argon (intercept (A) on the $^{39}Ar/^{40}Ar$ axis); atmospheric argon (intercept (B) on the $^{36}Ar/^{40}Ar$ axis) and excess argon ((C) at the origin). Simple cases exhibit mixtures of radiogenic and atmospheric end members, yielding a precise age from the $^{39}Ar/^{40}Ar$ intercept. More complex cases often involve mixtures between radiogenic argon and an end member with $^{36}Ar/^{40}Ar$ ratios less than atmospheric (0.003384), reflecting addition of ^{40}Ar. This may still be resolved to yield an 'age' if the non-radiogenic component is homogeneous and a line results. (b) An example of a correlation diagram showing two amphibole samples from the Tavsaneli zone, Turkey, with similar ages of 108.1 ± 3.8 Ma and 101.1 ± 3.8 Ma. The intercepts of the two lines with the $^{39}Ar/^{40}Ar$ axis are therefore similar (owing to the similar ages). However, one amphibole contains another component of argon with an atmospheric ratio indicated by the $^{36}Ar/^{40}Ar$ intercept close to 0.003384 ($^{40}Ar/^{36}Ar = 295.5$), while the other contains an argon component with a non-atmospheric ratio indicated by much lower intercept at 0.002058 ($^{40}Ar/^{36}Ar = 486$).

with, for example, cadmium foil and those unshielded. The correction factors are determined by irradiating pure salts of Ca and K (generally CaF_2 and K_2SO_4) with each irradiation. An additional correction for the decay of ^{37}Ar (produced by neutron bombardment of calcium) must always be made,

Table 8.2 Interfering reactions on Ca, K, Ar and Cl

Argon isotope	Ca	K	Ar	Cl
^{36}Ar	^{40}Ca(n,nα)^{36}Ar*		^{36}Ar(n,γ)^{37}Ar	^{35}Cl(n,γ)^{36}Cl$\rightarrow\beta^-\rightarrow^{36}$Ar
^{37}Ar	^{40}Ca(n,α)^{37}Ar*	^{39}K(n,nd)^{37}Ar		
^{38}Ar	^{42}Ca(n,nα)^{38}Ar	^{39}K(n,d)^{38}Ar	^{40}Ar(n,nd)^{38}Cl$\rightarrow\beta^-\rightarrow^{38}$Ar	^{37}Cl(n,γ)^{38}Cl$\rightarrow\beta^-\rightarrow^{38}$Ar
		^{41}K(n,α)^{38}Cl$\rightarrow\beta^-\rightarrow^{38}$Ar		
^{39}Ar	^{42}Ca(n,α)^{39}Ar*	^{39}K(n,p)^{39}Ar*†	^{38}Ar(n,γ)^{39}Ar	
	^{43}Ca(n,nα)^{39}Ar	^{40}K(n,d)^{39}Ar	^{40}Ar(n,d)^{39}Cl$\rightarrow\beta^-\rightarrow^{39}$Ar	
^{40}Ar	^{43}Ca(n,α)^{40}Ar	^{40}K(n,p)^{40}Ar*		
	^{44}Ca(n,nα)^{40}Ar	^{41}K(n,d)^{40}Ar		

The terminology (a,b) used here refers to nuclear reactions taking place during irradiation where a is the incident particle and b is the resulting emission. The terms are n = neutron, p = proton, d = deuteron, α = alpha particle, γ = gamma ray and β^+ = positron.
*Important interfering reactions; †Main ^{39}Ar producing reaction.

Table 8.3 Measured interference factors for the most important reactions

Reaction	Factor
$^{42}Ca(n,\alpha)^{39}Ar$	6×10^{-4} to 1×10^{-3}
$^{40}Ca(n,n\alpha)^{36}Ar$	2×10^{-4} to 3×10^{-4}
$^{40}K(n,p)^{40}Ar$	1×10^{-3} to 2×10^{-1}

since it has a half life of 35 days, which is very significant and means that all samples must be analysed within a year of irradiation and that the precision of the ages may be affected if they are not analysed within 6 months.

A final note on the technique is that to achieve optimum precision in the mass spectrometric measurements, the neutron flux (which affects the magnitude of the J value) must be carefully selected. There must be sufficient flux to produce measurable ^{39}Ar peaks and a $^{40}Ar^*/^{39}Ar$ ratio within the dynamic range of the mass spectrometer. The $^{40}Ar^*/^{39}Ar$ ratio is dependent upon the age of the sample, which is thus another parameter. Further, as the amounts of ^{39}Ar are increased by higher flux levels, so are the interfering reactions on Ca and K which degrade the precision with which the $^{40}Ar^*/^{39}Ar$ ratio may be determined. For each sample, therefore, there is an optimum flux level and given that many samples are irradiated together, each irradiation is a compromise; Turner (1971) calculated the fields for optimum J value and the corresponding integrated neutron flux, which were upgraded by McDougall and Harrison (1988) in the light of higher sensitivity mass spectrometers (Figure 8.2).

8.3 Estimation of errors

All Ar–Ar ages are calculated from a ratio of $^{40}Ar^*/^{39}Ar$ using a standard mineral of known age which is irradiated at the same time as the sample. There is a range of international reference materials, generally comprising hornblende, biotite, muscovite or sanidine. The most commonly used standards are the hornblendes Hb3gr, 1072 ± 11 Ma (Turner et al., 1971) and MMHb1, 520.4 ± 1.7 Ma (Samson and Alexander, 1987); biotite B4B, 17.3 ± 0.2 Ma and muscovite B4M, 18.6 ± 0.4 Ma (Flisch, 1982) and sanidine from the Fish Canyon Tuff, 27.9 ± 1.2 Ma (Cebula et al., 1986; 27.84 ± 0.3 Ma measured by Ar–Ar relative to an age for MMHb1 of 520.4 Ma).

The uncertainties quoted on the ages of the standards are about 1% (1σ), except MMHb1 where determinations were refined using data from many international K–Ar laboratories, to yield a pooled error of 0.3% (1σ). Uncertainties in reference material values define the best precision that can be achieved using the Ar–Ar technique since the uncertainty on the

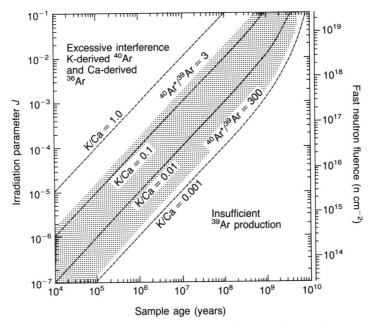

Figure 8.2 Optimization of the irradiation parameter *J*. The shaded area indicates the optimal zone for fast neutron fluence. Higher flux levels cause unacceptable interference effects of neutron interactions with calcium and potassium. Lower flux levels form insufficient ^{39}Ar, leading to very high ^{40}Ar/^{39}Ar ratios which may be difficult to measure with precision by mass spectrometry. (Based upon Turner, 1971, with modifications by McDougall and Harrison, 1988. Reproduced with permission).

reference material is combined with the errors on the *J* value and thus affects the final age.

The precision of the final ^{40}Ar*/^{39}Ar ratio ranges from as little 0.1% up to 1–2%. The size of the uncertainty is often controlled by the sample size, larger samples are less affected by the errors on the blanks. High errors may be acceptable if the aim of the study is to produce for example, an age profile with high spatial resolution.

The expression for the error in the calculated age, *t*, from Dalrymple et al. (1981) is:

$$\sigma_t^2 = \frac{J^2 \sigma_R^2 + R^2 \sigma_J^2}{\lambda^2 (1 + RJ)^2} \tag{8.9}$$

where σ_t is the error on the age, *J* is the value from equation 8.6, *R* is the ^{40}Ar*/^{39}Ar ratio, λ is the combined decay constant, σ_R is the error on the ^{40}Ar*/^{39}Ar ratio and σ_J is the error on the *J* value.

8.4 Instrumentation

There are currently relatively few purpose-built laser Ar–Ar systems in existence; most laboratories use laser extraction techniques as an added accessory to conventional stepped heating equipment. This situation is generally not a problem, though combining the two forms of sample extraction requires careful regulation since the large quantities of argon sometimes analysed in the conventional stepped heating technique may cause a memory effect in the mass spectrometer, deleterious to the analysis of small gas samples obtained by laser extraction.

Instrumentation for the Ar–Ar laser technique can be considered conveniently as two basic components: the laser extraction system and the gas handling/mass spectrometer system.

8.4.1 Gas handling and mass spectrometry

The quantity of gas available for analysis is the principal difference between mass spectrometric measurements made in conventional stepped heating and laser extraction. In general, smaller amounts of gas are analysed when laser extraction is used to achieve high spatial resolution. The need to analyse small amounts of gas while stepped heating can normally be avoided providing there is sufficient sample. Thus, by matching the sample size, the age of the sample, the number of steps in the heating procedure and the potassium content, the amount of gas at each step can be optimized and measured using a standard Faraday detector. The small quantities of gas used for high spatial resolution laser analysis generally require electron multiplier or Daly detectors.

It is not within the scope of this chapter to describe the detailed workings of noble gas mass spectrometers. Briefly, argon gas admitted into the mass spectrometer is ionized in an electron beam from an incandescent (thermally emitting) filament. Ions created within the electron beam are accelerated by applying a high potential (generally about 4 kV) to a series of ion extraction lenses. The ion beam is deflected into a circular path by a magnetic field created at the pole pieces of an electromagnet (the mass analyser) which separates the masses according to the standard equation:

$$\frac{m}{e} = \frac{H^2 r^2}{2V} \tag{8.10}$$

where m and e are the mass and charge of the ion respectively, H is the strength of the magnetic field, r is the radius of curvature of the ion path and V is the accelerating potential of the ion source. Fractionation of the measured masses, commonly known as discrimination, can occur in the ion source (McDougall and Harrison, 1988) though this is minimized by the design of modern ion sources so that the Faraday detector measures the atmospheric ^{40}Ar/^{36}Ar ratios as 296–300 (the accepted value for the atmosphere is 295.5; Steiger and Jäger, 1977). However, multiplier detectors generally cause greater degrees of fractionation, leading to measured ratios generally between 270 and 290. This fractionation or discrimination effect is corrected by analysing small aliquots of air or samples known to contain argon with an atmospheric ratio.

Most noble gas mass spectrometers use 60° or 90° sector magnets with radii of 12 cm or greater. The magnets are focused in two directions (taking the ions from a point at the exit slit of the ion source and focusing to a point at the collector slit). In addition, ion beams are designed to hit the magnetic field at a slight angle which further differentiates the masses. Resolution (a measure of the separation between peaks; see McDougall and Harrison, 1988, for full explanation) is generally greater than 100 to minimize mass overlap interferences.

The gas mixture released from the sample by the laser contains active gases such as H_2O, CO_2, CH_4, N_2 and H_2 together with the noble gases (principally radiogenic argon and helium). Various techniques are used to remove the active gases which would otherwise cause scattering effects in the mass spectrometer and ultimately oxidize the filament. Hot and cold 'getters' (Figure 8.3) made of fine-grained metal or alloys such as Ti or Zr-Al are used to remove the active gases. When held at temperatures of around 450°C in the vacuum chamber, the getters react with and absorb most active gases. Hydrogen is less soluble in the hot metals but has greater solubility at room temperature. Thus, many systems consist of two metal alloy getters, one at 450°C and one at room temperature. The getters are outgassed at around 800°C to renew their surface properties at least once a day. Some systems also include a cold trap cooled to about −95°C to assist in trapping CO_2 and H_2O, in order to prolong the life of the getters.

In most cases, the purified noble gases are allowed to equilibrate into the mass spectrometer (Figure 8.3), though they may be leaked in by partially opening the inlet valve and the making appropriate corrections for fractionation of the isotopes. The signs that the gas has not been sufficiently 'cleaned' are either a noisy signal indicated by erratically varying peak heights or 'curved build-up', where the peak heights describe a curve over time scales of a few minutes rather than the normal near linear build-up. Modern mass spectrometers normally have an additional getter at room temperature close to the source to maintain ultra-low partial pressures of the active gases during analysis. The mass spectrometer getters are outgassed only rarely, unless poisoned by 'dirty' samples.

When making a measurement, the ion beam is scanned over mass peaks from 36 to 40 by varying the magnet field (achieved by changing the current to the electromagnet), while maintaining a constant accelerating voltage in the ion source. Measurements are made by 'peak hopping', that is by altering the magnet field to preset values to focus each peak and trough or

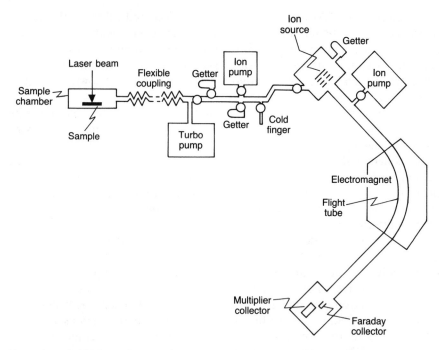

Figure 8.3 A schematic diagram of a gas extraction line and mass spectrometer system.

zero point in turn on the detector. All the argon isotope peaks, ^{36}Ar, ^{37}Ar, ^{38}Ar, ^{39}Ar and ^{40}Ar are generally measured 5–10 times during each analysis. The peak intensity and time of acquisition are recorded for each measurement and peak heights are extrapolated back to the inlet time to avoid errors caused by the adsorption and desorption of argon on the metal walls of the mass spectrometer, which significantly affects measured peak heights. Commonly, during blank analyses in the mass spectrometer, the ^{40}Ar peak (the most common isotope in the atmosphere) increases with time as argon diffuses out of the stainless steel walls of the mass spectrometer. During many sample analyses, however, the intensity of the ^{40}Ar peak will decrease as the pressure of argon in the mass spectrometer is such that the dynamic equilibrium is reversed and argon diffuses into the walls. Such memory effects are minimized by extrapolating the peak intensities back to the inlet time.

8.4.2 Lasers and laser/solid interaction

Since 1972, when the laser microprobe technique was first used to extract argon for Ar–Ar dating, virtually all types of high-powered lasers have been used for the technique. A pulsed ruby laser with a wavelength of 548 nm was the type first used to outgas samples of meteorites to analyse noble gases (Megrue, 1967). However, by 1971 a pulsed Nd-doped yttrium aluminium garnet (Nd-YAG) laser was also being used, producing a power density of between 10^{11} and 10^{13} W m^{-2} at the sample surface.

The general principles of laser operation can be found in texts such as Svelto (1989). Briefly, the light photons are generated by high-intensity lamps which cause stimulation of the active species in the rod (solid-state lasers such as Nd-YAG) or cavity (gas lasers such as argon ion) into excited modes. Lasing occurs mainly at the fundamental wavelength of the laser as the excited species emit photons with characteristic wavelengths and a full range of transverse and longitudinal modes (vibrating perpendicular or parallel to the laser rod respectively). The light is amplified by being repeatedly 'pumped' through the laser rod or cavity between two mirrors. Hence the term 'laser' (light amplified by stimulated emission of radiation).

A laser without apertures (termed 'multimode') will produce a wide beam of high power but most systems used in laser microprobe Ar–Ar dating operate TEM$_{00}$ (transverse electromagnetic mode 00, meaning a Gaussian distribution of light intensity across the beam), which is achieved by placing a small aperture in the laser cavity to restrict the transverse modes. TEM$_{00}$ mode produces a lower absolute power but high power densities, since it is essentially diffraction-limited, in other words the beam is less divergent and can be focused to a smaller spot. Lasing commences above a threshold flashlamp power and subsequent laser output power is controlled by the power of the flashlamps.

Lasers emit light in several different modes all of which may be used in Ar–Ar laser extraction. Pulsed laser output consists of a series of microsecond spikes which continue while the energy in the laser rod or cavity is above the lasing threshold. Such pulses have the appearance of a spiked hill and generally last a few hundred microseconds. 'Q-switching' is a technique for generating very short pulses of light only a few nanoseconds long by storing energy in the laser and releasing all the energy very rapidly. Consequently the powers of such pulses are extremely high. In general, Q-switching is not used for Ar–Ar laser microprobe dating, though a new application discussed below uses a Q-switched ultraviolet laser system. Continuous wave (CW) operation is the most common form used in Ar–Ar analysis, where a continuous beam of light is achieved by continuous pumping at high power. To achieve CW operation, however, the lasing rod or cavity must be resistant to thermal breakdown and CW operation is a relatively recent development in lasers.

The minimum spot size when the laser has been focused through a microscope is an important parameter when performing laser spot analyses. It can be calculated (Fallick *et al.*, 1992), by first calculating the spot area:

$$\text{area} = \frac{\pi d^2}{4} = \frac{\pi f^2 D^2}{4} \qquad (8.11)$$

where d is the spot diameter, D is the beam divergence (in radians) and f is the focal length of the microscope objective lens. The diffraction limited beam divergence for TEM_{00} mode is given by (Svelto, 1989):

$$D = \frac{1.22\lambda}{2b} \tag{8.12}$$

where λ is the wavelength of the laser light and b is the diameter of the beam incident on the microscope lens. From equations 8.11 and 8.12, it can be seen that the spot size is related to the wavelength of the laser. The average power density can thus easily be calculated from the laser output power and beam diameter, though more correctly, the power density has a Gaussian distribution if the laser is being operated in TEM_{00} mode. The relationship between laser output power and peak incident power can be calculated as follows (from Girard and Onstott, 1991):

$$P_0 = \int_0^\infty I(r) 2\pi r dr \tag{8.13}$$

$$P_0 = \int_0^\infty \exp^{(-2r^2/2W_0^2)} 2\pi r dr = I_0 \left(\frac{\pi W_0^2}{2}\right) \tag{8.14}$$

where P_0 is the laser output power, I is the incident power at radius r, I_0 is the peak incident power, and W_0 is the beam spot diameter.

The physical effects on a sample of different levels of irradiance (light density) can be characterized in three general ranges (Ready, 1971). Powers up to $10^9 \, W \, m^{-2}$ cause heating; between 10^9 and $10^{12} \, W \, m^{-2}$, melting and later vaporization occur, and above $10^{12} \, W \, m^{-2}$ the laser causes ablation, a term which indicates possible shock damage and the generation of a plasma at the point of impingement.

Lasers are required to perform two tasks in the Ar–Ar laser microprobe dating: (i) to heat reasonably large areas of samples slowly, for which a continuous laser producing up to about $10^9 \, W \, m^{-2}$ is preferable, and (ii) to extract argon from small areas of sample at high resolution without heating the surrounding areas, for which short high-energy pulses are required, with power densities greater than $10^9 \, W \, m^{-2}$ and preferably closer to $10^{12} \, W \, m^{-2}$. Pulsed laser systems have been used for high spatial resolution studies, including Nd-YAG, Nd-glass and ruby lasers (Pleininger and Schaeffer, 1976; Schaeffer, Müller and Grove, 1977; Laurenzi, Turner and McConville, 1988; McConville, Kelley and Turner, 1988). However, although pulsed lasers deliver high power densities causing instantaneous melting, there is a threshold below which they will not lase, a threshold which is still too high to cause gentle heating. The great majority of laboratories use only one laser system, which must be flexible enough to be able to both heat slowly and cause instantaneous melting, and thus pulsed lasers are less common than high-powered CW lasers. There is no 'first choice' system for this application, but currently two popular solutions are CW Nd-YAG and CW argon ion.

Both the Nd-YAG and argon ion lasers produce up to 10–20 W output power. Nd-YAG lasers produce an infrared beam with a wavelength of 1064 nm; argon ion lasers have several wavelengths in both visible and ultraviolet regions, the two main ones being 488 and 514 nm, in the blue/green visible region. Two advantages of the argon ion system are evident. Firstly, being visible, the argon ion lasers are inherently more safe and secondly, since the wavelength is shorter, the output beam can be focused to smaller spot sizes through a microscope system (equations 8.11 and 8.12). However, Nd-YAG lasers are generally easier to maintain and require less power and cooling than the argon ion systems for the same power output. Both Nd-YAG and argon ion laser beams can be passed through normal microscope optics and Kodial (specialist glass with the same thermal expansion characteristics as stainless steel) or silica laser port windows **with care**. The absorption characteristics of the laser beam by various minerals for both Nd-YAG and argon ion systems are similar since the absorption increases gradually from the near infrared into the visible (note, however, that biotite has significantly higher absorption for the argon ion laser), and increases rapidly in the ultraviolet (Figure 8.4).

Given that both the popular CW lasers perform to similar standards, though the argon ion system probably has the edge, how close do they come to the ideal system for this application? Both systems perform very well when used for stepped heating (Layer, Hall and York, 1987; Lee et al., 1991). By defocusing the beam to produce spots of 1–3 mm, very low power densities can be achieved (power densities as low as $10^5 \, W\,m^{-2}$ have been used). The laser power is generally absorbed in the top few tens of micrometres of the sample, but temperature variations across single grains caused by this effect are generally negligible (Hall, 1990). Stepped heating of clear minerals such as feldspars may not be as simple since a great deal of the light is transmitted and passes through the mineral, though by increasing the power it is generally possible to achieve melting. The earliest demonstration of the technique by Layer, Hall and York (1987), showed that 20–25 step heating schedules could be undertaken on single grains, providing high-resolution stepped heating profiles equal to those analysed by the conventional technique. Both Nd-YAG and argon ion are now used routinely for stepped heating analysis.

Consider now, the suitability of the CW lasers for producing laser spot ages. The highest powers will be achieved using the smallest spot size, which is generally 10–20 μm. A 10 μm spot and a maximum of 20 W power from the laser results in a power density of over $6 \times 10^{10} \, W\,m^{-2}$. This power density is adequate to melt most minerals virtually instantaneously and laser

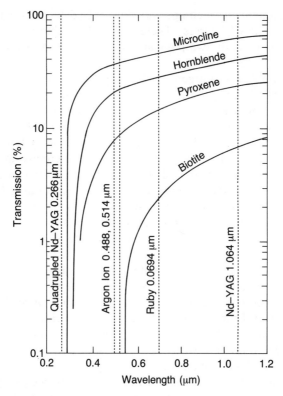

Figure 8.4 Diagram illustrating the absorption of light by various minerals used in laser microprobe Ar–Ar dating. The analyses were performed on a 30 μm thick rock section. Note the great variation of absorption between minerals but also with variation with changing wavelength and that all the minerals absorb strongly in the ultraviolet. (From McConville, 1985. Reproduced with permission).

pulses of a little as 1 ms can be produced using external shuttering systems. Such square wave laser pulses are therefore ideal for producing laser pits of the order of 50–100 μm in diameter (Figure 8.5(a)). However, in cases where the mineral absorption is low, such as feldspar, two problems arise. Although the laser power density may be well above that required to melt the mineral, as little as 0.2% may be absorbed at the surface (Girard and Onstott, 1991). In such a case, the sample in the centre of the focused laser beam may take several seconds to reach melting point even at maximum power. This effect will cause, at best, heating and argon loss of the surrounding areas and, at worst, melting of more susceptible minerals outside the laser spot. The second problem is caused by light transmitted through the surface and refracted or reflected within the mineral. This phenomenon can be a problem particularly in quartz and feldspar (Girard and Onstott,

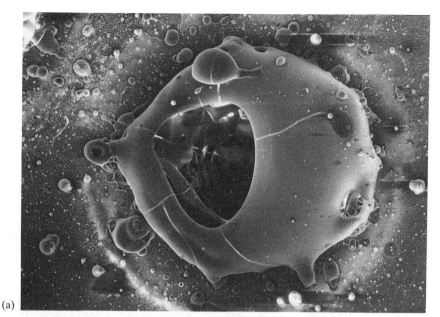

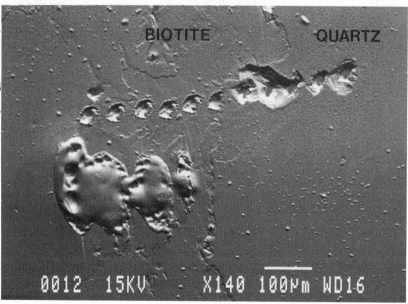

Figure 8.5 SEM photographs of laser pits used in Ar–Ar dating. (a) A 100 μm diameter laser pit in biotite drilled using a CW Nd-YAG laser at 10 W output for 10 ms. (b) A traverse from biotite, typically the least transparent mineral to visible and infrared lasers into quartz, typically the most transparent, without varying the laser power, which shows little variation in pit size using the ultraviolet laser (the upper traverse of pits). However, note that some of the ultraviolet laser pits in quartz resulted in fracturing at the top of the pit (the pits are several tens of micrometres deep). A similar traverse (the lower set of melt pits in biotite only) using a continuous infrared laser (Nd-YAG, 8 W continuous power for 10 ms) causes melting in the biotite but has no effect upon the quartz. In fact, the first noticeable effect on increasing the infrared laser power is that light refracted within the quartz causes melting in the adjacent biotite.

1991; Burgess *et al.*, 1992) or even in feldspathoids such as leucite. Experience shows that dark inclusions or adjacent minerals may melt at distances of up to 500 μm from the laser spot in feldspar. Great care must be taken, therefore, in interpreting high-resolution laser spot data for clear minerals.

In summary, CW lasers perform well for stepped heating and reasonably well for spot analysis. The problems occur in spot analysis when mineral absorption coefficients are low, for example in feldspars. Several solutions have been tried, such as coating the sample surface to increase absorption, but none are satisfactory. The best solution to the problem is to use two lasers, though there has to be some system for switching between the two lasers during analysis. Pulsed lasers perform better in spot analyses (McConville, Kelley and Turner, 1988), though in minerals with very low absorption coefficients pulsed Nd-YAG lasers cause shock in the mineral lattice and excavate craters while causing incomplete argon recovery.

Another solution to the problem is the use of a laser with a wavelength strongly absorbed by all natural mineral species. The closest approach to this ideal is an ultraviolet system such as a quadrupled Nd-YAG laser. The light produced has a wavelength of 266 nm which is strongly absorbed by all silicates (Figure 8.4). In addition, the short wavelength means that smaller spot sizes are achievable. Ultraviolet light absorption in quartz and feldspar is sufficiently high that small spots can be analysed in the same way as on darker minerals such as biotite (Figure 8.5(b)).

8.4.3 Microscopes and laser ports or sample holders

Lasers systems can be supplied with beam steering apparatus mounted on the laser (Figure 8.6(a)), comprising a beam steering mirror, focusing lens and binocular viewer. This is a convenient and common setup used for Ar–Ar laser microprobe work, though the optics are inflexible and tend to have long focal lengths and low magnification (typically ×5). An alternative, more flexible system utilizes a standard petrological or stereoscopic micro-

Figure 8.6 Three common laser/microscope system designs. (a) 'Machine head', bolted to the laser and consisting of a binocular microscope system. The laser beam passes into the body of the microscope and is reflected through 45° by an oxide coated lens (which reflects only the laser wavelength), and is focused through the lens which is also used to observe the sample. (b) Free-standing binocular microscope system looking through a reflecting mirror (coated to reflect only the laser wavelength) such that the laser does not pass through the microscope optics. Laser focusing is independent of the binocular system. (c) Petrological microscope system, where the laser passes through the microscope and is reflected through the objective lens by a small oxide-coated mirror (coated to reflect only the laser wavelength). The laser is focused through the same lens which is used to observe the sample. Working distances tend to be short (20–30 mm) and the lens turret allows changing of objective lenses.

INSTRUMENTATION

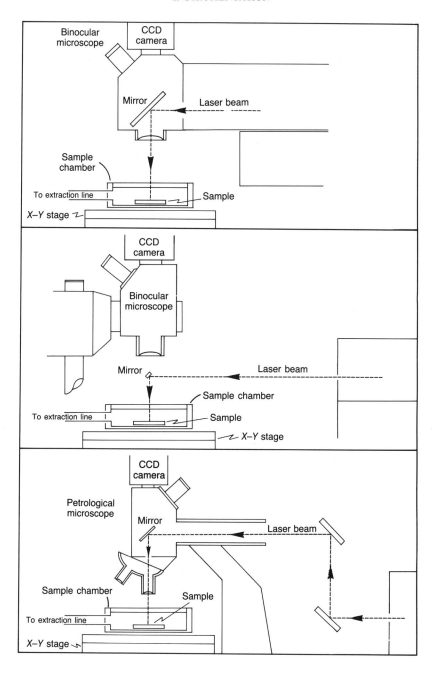

345

scope with the laser beam either directed below the microscope (Figure 8.6(b)) or passing through the microscope (Figure 8.6(c)). This setup provides better quality images of the samples and allows the magnification to be changed during observation (typically ×5 to ×40). There are additional problems of aligning the laser with the microscope but these systems have greater flexibility, even to the extent of allowing more than one laser to be used. Standard microscope objective lenses may be used, though they suffer degradation in the long term. However, objective lenses optimized for laser wavelengths tend to give inferior optical images, and the result is normally a compromise.

The optical quality of the image is further degraded by the laser port window. Windows mounted in ultra-high vacuum (UHV) flanges are available in a range of materials with variable optical properties, the most common being Kodial, fused silica (quartz) and sapphire. Fused silica has the highest transmission for visible and near infrared lasers though sapphire has better thermal properties. The main cause of transmission loss is reflection (generally about 10%), which can be ameliorated by appropriate coatings, though these are seldom applied.

The laser port or sample chamber (Figure 8.6) is generally constructed of stainless steel, and consists of two or three standard vacuum fittings, connected by a UHV flexible coupling to the gas cleanup section. The sample chamber is mounted on an $x-y$ stage (often remotely controlled) to allow the sample to be moved relative to the focused laser beam to select either the appropriate grain or position on the sample surface for analysis. Movement in the z-direction (focusing) is achieved either using the microscope focusing or by using an $x-y-z$ movement stage. There is an obvious advantage in being able to focus and refocus remotely, since the microscope will often be covered in order to enclose totally the laser beam path for safety reasons.

8.4.4 Final word

Most mass spectrometers which are used for laser analysis are computer controlled, so the magnet and data acquisition system will normally be automated. In an increasing number of systems, the valve operation is also automated and controlled by the computer. Further, the sample movement and focusing may be manual or remotely controlled and even the laser may be remotely controlled. For this reason, no attempt has been made here to describe a typical laser microprobe Ar-Ar system since the reader may be faced with anything from a fully manual system, with computer control of only data acquisition and requiring the analyst to move about the laboratory and work for the results, to a fully automated system which requires the analyst to remain seated throughout or even one with minimal operator intervention once a laser port has been loaded and pumped down.

8.5 Methodology

8.5.1 Sample preparation

Sample preparation is much the same as the conventional stepped heating technique and there is no need for chemical treatment of the sample, though extra care must be taken that the samples are clean, particularly of hydrocarbons. Mineral separations are generally achieved by hand picking clean grains from crushed samples. For laser stepped heating of individual grains, the mineral grains should be of sufficient size that they release measurable quantities of argon with each step. For laser spot analysis, the samples are prepared as single grains or freestanding polished slices of rock (again, any hydrocarbons will adversely affect the analyses). The samples may have cross-sectional areas of up to 10 mm^2, and are typically less than 1 mm thick. It is possible to create sections as thin as 30 μm thick on glass slides, which may then be carefully removed and might allow normal petrographical examination, though they are difficult to handle. However, such thin sections disintigrate due to thermal stresses during laser melting and the thinest sections used for Ar–Ar laser microprobe tend to be about 100 μm thick.

8.5.2 Laser spot analysis

Laser spot analyses are simple and require only that the laser port and getter system be closed off for a given time (generally about 5–10 min) while the laser is fired into the sample and getters (sometimes also with a cold trap; section 8.4.1) remove the active gas species. Blanks are performed in precisely the same manner and for the same period of time without firing the laser, to measure the argon buildup in the line while the pumps are closed. After the set time, the valve to the mass spectrometer is opened and gas equilibrates into the mass spectrometer. Isotope analysis begins at the instant the gas enters the mass spectrometer since the peak intensities change with time (section 8.4.1).

Spot analysis requires that the amount of sample extracted is matched to the amount of argon gas needed to achieve reasonable precision. Put simply, larger areas of mineral must be melted to date younger or low potassium samples, in the same way that larger amounts of material are required to date such samples by the conventional Ar–Ar stepped heating method. This requirement is the fundamental limit on laser spot analysis and it is a useful, even essential, exercise to calculate the size of laser pit required to release enough gas for a measurement, prior to the actual analysis of the sample. This calculation should be a balance of blank levels and sample parameters such as the age and potassium content. The other parameter which is inherent in laser spot dating is the volume of sample outside the nominal laser pit which may contribute argon to the analysis. In general, the area affected is less than 10% of the pit diameter outside the actual laser pit (McConville, 1985; Plieninger and Schaeffer, 1976). However, as described above, where clear minerals such as quartz and feldspar are involved, zones affected by the laser may be much larger. Girard and Onstott (1991) showed that when a Nd-YAG laser was used to date K-feldspar overgrowths, even though the laser pit did not impinge on the detrital core and thermal calculations showed that the core should not have been affected, very significant amounts of argon were released. They concluded that conductive heating through the core–overgrowth boundary was not responsible for partial degassing of the core. Rather, beam scattering along intra-grain structural defects and incipient melting of the core must have occurred. Great care must therefore be taken when interpreting laser spot data of such minerals.

8.5.3 Laser stepped heating

Stepped heating analysis of individual mineral grains was first demonstrated by Layer, Hall and York (1987) on grains of biotite and hornblende. Comparison of laser probe and conventional stepped heating (Ruffet, Feraud and Amouric, 1991) have demonstrated the ability of the technique to improve on conventional stepped heating results, though the problems of interpreting any stepped heating in terms of natural argon loss still remain (Lee et al., 1991). Laser stepped heating involves defocusing the laser to yield a beam wider than the grain being analysed in order to produce as even a heating as possible. The essential differences between conventional stepped heating and individual grain stepped heating are:

1. It is easier to separate and characterize single grains for laser stepped heating than characterize several tens or hundreds of grains for conventional stepped heating;
2. the temperature of each laser step can only be measured using an infrared pyrometer, since the conventional thermocouple cannot detect the temperature of such a small area;
3. the process of laser stepped heating is faster since the sample is heated for only 30 s to 2 min compared with 15–45 min for the conventional stepped heating procedure.

Despite the fact that the laser heats only the top surface of grains, thermal gradients across single grains caused by laser heating are small and the grains reach equilibrium rapidly, even for grains up to 1 mm in diameter (Hall, 1990). Temperature variations based upon thermal models are as little as 2°C for a 500 μm diameter grain and only 9°C for a 1 mm diameter grain (Hall, 1990).

Laser stepped heating involves firstly the heating, followed by an equal or greater time for (generally 2–5 min) 'gettering', during which the active gases are removed (see above), and finally, equilibration into the mass spectrometer. Isotope measurement is precisely the same as laser spot dating. Blanks will normally be performed between samples, sometimes as much as one blank per heating step. The sequential steps of increasing temperature are produced by simply increasing the power of the CW laser. Argon ion lasers often perform this task better since most are controlled in 'power mode' where a small photodiode in the output section is used in a feedback loop to smooth the output of the laser, allowing a simple method for the analyst to increase power linearly during stepped heating. Many Nd-YAG lasers do not have such a smoothing system since the output at any given power is stable. However, the relationship between flashlamp current and laser output power is not linear and a calibration is thus required.

Layer, Hall and York (1987) measured temperatures during laser stepped heating using a combined infrared detector and microscope system. The subject of infrared non-contact thermometry is not within the scope of this chapter; the important point to note is that the temperature measured by most detectors has to be corrected for the emissivity of the material being studied (emissivity is a measure of the difference between the sample and a theoretical material known as a 'black body'). Layer, Hall and York (1987) noted that hornblendes and biotites had emissivities of about 90% and 75% respectively, but that emissivity could vary by as much as 10% between grains of the same mineral. Minerals such as feldspar have much lower emissivities. The temperatures measured by such detectors are thus subject to errors of up to 10%. 'Ratio' or 'two-colour' detectors can be used to lessen the effects of emissivity, and high resolution infrared cameras, which have been developed to study heating in microchips, can measure accurately temperatures on any material though their cost is currently prohibitive.

8.6 Applications of the laser microprobe technique

The following examples describe ways in which the laser microprobe technique has been used to study geological problems. Firstly, some difficult or rare samples beyond the reach of 'bulk' techniques; secondly, examples of laser stepped heating and thirdly, examples of high spatial resolution argon diffusion studies in minerals.

8.6.1 Unique samples

One of the features of the laser microprobe system is its high spatial resolution, allowing the investigator to determine the distribution of components between or within grains. However, an additional feature is that blank levels are very low and reproducible since only the sample itself is heated, not large areas of metal furnace. This means that smaller samples, even single grains, can be analysed. Lo Bello *et al.* (1987) used this feature to great effect when they dated a Quaternary pumice flow near Neschers, France. While conventional stepped heating of bulk samples comprising separated K-feldspar yielded 'saddle'-shaped age spectra, with the youngest step around 18 Ma, laser fusions of individual grains in the size range 500–1000 μm yielded two populations of ages. Clear grains yielded a weighted mean of 0.58 ± 0.02 Ma, whereas cloudy grains yielded ages ranging from 233 to 334 Ma. Clearly, the cloudy grains were incorporated from the country rocks in the flow during the eruption. Lo Bello et al. (1987) demonstrated the advantage of being able to date individual grains in an application where there are two populations of ages in a single mineral species. A similar technique has been used more recently to obtain precise ages for pumice horizons bracketing the famous ancient human remains in the Olduvai Gorge of Tanzania (Walter et al., 1991).

Diamonds rarely contain inclusions but when they do, the inclusions rarely contain potassium. However, rare clinopyroxene inclusions contain around 100–1000 ppm potassium. Burgess, Turner and Harris (1989) and Phillips, Onstott and Harris (1989) utilized the laser probe to analyse very small samples by melting clinopyroxene inclusions in diamonds which had been carefully cleaved to reveal the inclusions. Ar–Ar ages close to that of the kimberlite intrusion generally resulted. Older ages occasionally measured in some inclusions may have been the result of excess argon or perhaps even long-term storage of radiogenically produced argon within the diamond inclusions in the mantle.

In another example from the extraterrestrial field of studies, a range of formation ages and cosmic ray exposure ages were extracted from diverse lithic fragment and adhering impact glass measuring only 2 mm × 1 mm (Laurenzi, Turner and McConville, 1988).

8.6.2 Laser microprobe stepped heating

While the laser stepped heating technique can reproduce the results of conventional stepped heating on smaller samples, innovative techniques are also appearing, such as the stepped heating of individual 'sized' grains to study slow cooling (Wright, Layer and York, 1991). These workers analysed 42 individual biotite grains from two granitoid rocks from the Red Lake greenstone belt of northwestern Ontario, Canada, by laser stepped heating. They were able to show that samples with radii above 225 μm yielded a narrow age range from 2610 Ma to 2645 Ma, whereas those with radii between 225 μm and 100 μm yielded ages from 2570 Ma to 2640 Ma (Figure 8.7). These results are suggestive of a critical radius for argon diffusion as hypothesized by Harrison, Duncan and McDougall (1985) (though they suggested a critical radius of 140 μm). Using the age–grain size effect for the smaller grain sizes, Wright, Layer and York were able to derive a continuous portion of the thermal history for the Red Lake batholiths. This

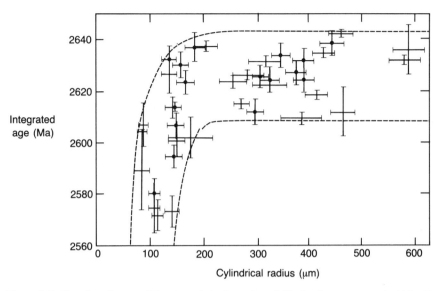

Figure 8.7 Results of stepped heating of single grains of biotite from two granitoid bodies from the Red Lake greenstone belt of north-western Ontario, Canada. The dates for grains larger than 225 µm yielded ages in a narrow range, whereas those with radii between 100 µm and 225 µm exhibited a strong correlation between age and grain size. Using this age/grain size effect, Wright, Layer and York (1991) were able to reconstruct a continuous portion of the thermal history of the Red Lake batholiths. The lack of age variation in larger grains was attributed to imperfections, fractures and cracks in larger grains which formed fast pathways for argon diffusion and defined a maximum effective grain radius of about 225 µm. (After Wright, Layer and York, 1991. Reproduced with permission).

study is a significant advance over the type of discontinuous thermal history derived from comparison of K–Ar ages and blocking temperatures of different minerals.

Comparisons of laser stepped heating and bulk stepped heating of the same sample illustrate a common problem with the bulk technique. Ruffet, Feraud and Amouric (1991) derived both bulk and single grain ages of biotites from the North Trégor Batholith (Amorican Massif, France). All stepped heating spectra showed disturbed patterns, related to minor alteration of the biotites to chlorite. However, the laser step heated grains showed consistently less disturbance than the bulk samples, not because of any special experimental advantage but because handpicking pure, unaltered single grains is easier than separating and purifying bulk samples.

8.6.3 High-resolution laser microprobe studies of argon diffusion

Although laser stepped heating can boast advantages over the conventional technique, it also suffers similar experimental artifacts. Lee *et al.* (1991)

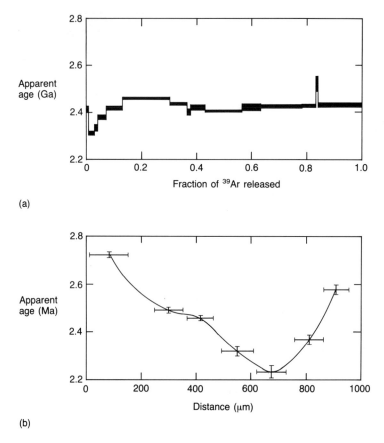

Figure 8.8 A comparison of laser stepped heating and laser spot analyses of hornblende grains from the same sample. Note that although the stepped heating yields an acceptable plateau (a), it has averaged the true range of ages shown by the spot analysis (b). The laser spot analyses illustrate the incursion of excess argon from the grain boundaries, yielding anomalously old apparent ages. (After Lee et al., 1991. Reproduced with permission).

reported a comparison of laser stepped heating and laser spot analysis (Figure 8.8), which showed a laser stepped heating plateau age of 2425 ± 11 Ma (Figure 8.8(a)). However, this result does not reflect the true distribution of ages (Figure 8.8(b)), which range from 2250 Ma to 2725 Ma, as can be revealed by high-resolution laser microprobe analysis.

'Chrontours', lines of equal age, have been one of the features of work produced using the laser microprobe to map argon distributions. Chrontours have been mapped in single grains of various minerals to produce two-dimensional maps of argon loss and gain. Phillips and Onstott (1988) used the resolution of the laser microprobe to show that argon diffusion in phlogopite is dominantly parallel to the cleavage in micas rather than

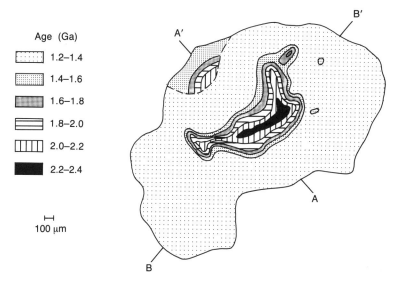

Figure 8.9 An apparent age contour (chrontour) map of a cleavage section from a phlogopite grain. The contours demonstrate the scale of argon diffusion in nature, though they do not correspond to theoretical diffusion models which predict bell-shaped profiles from the centre of the grain to the edge. (After Phillips and Onstott, 1988. Reproduced with permission).

perpendicular to it. The importance of this study was that it showed that relatively large-scale zoning of argon was present in nature (Figure 8.9), which commonly does not conform to the type of bell-shaped argon loss profile (loss at the grain margins but not at the centre) which might be predicted from theory. Scaillet et al. (1990) measured ages in large, deformed but unrecrystallized phengite grains and small recrystallized phengite grains from the Dora Maria nappe of the French Alps. Chrontours of a large phengite grain served to confirm that natural profiles are complex (Figure 8.10), and that the microstructure of the grains (fractures and dislocations in the phengite) strongly affected the resulting argon distribution.

Kelley and Turner (1991), showed that quantitative information can be extracted from laser microprobe traverses from individual hornblende grains. Samples from the Giants Range Granite in the contact metamorphic aureole of the Duluth Gabbro, Minnesota, showed that different grains in a single section lose argon during contact metamorphism in proportion to their grain size (Figure 8.11). The visible grain size does not, however, necessarily relate to the argon loss since biotite inclusions within the hornblende defined fast pathways for argon diffusion (argon drains). The cores of the grains, free of inclusions, acted as large single diffusion domains during the heating event; the outer rim, with biotite inclusions, acted as a large number of small grains (Figure 8.11). The resolution of the laser microprobe allowed

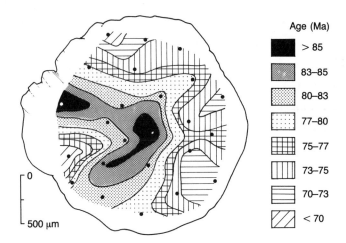

Figure 8.10 An apparent age contour (chrontour) map measured on an unrecrystallized phengite grain from the Dora Maira nappe of the Western Alps. The chrontours reflect diffusive argon loss not only via the grain boundaries but also fractures and dislocations caused by deformation, yielding a highly variable chrontour map. (After Scaillet et al., 1990. Reproduced with permission).

Kelley and Turner (1991) to use diffusion profiles from the hornblende cores to calculate integrated diffusive argon loss during the contact metamorphism for each grain. The amounts of argon lost were directly related to the measured size of the grain core (Figure 8.11) and allowed the derivation of a single time-integrated diffusion constant for hornablende in the sample.

The relationship between temperature and the diffusion constant for hornblende has been measured under laboratory conditions (Harrison, 1981; Baldwin, Harrison and Fitzgerald, 1990). The results for the laser microprobe study thus yielded a relationship between time and temperature for the contact metamorphism. In addition, an alternative measure of temperature from earlier stable isotope studies (Perry and Bonnichson, 1966) showed that the peak temperature at the margins of the intrusion were about 700°C. Thus, with a measure of the temperature, the length of time for which the rocks were hot could be calculated. Since the sample lay a few hundred metres from the contact, the temperatures would have been lower. So for example, if the temperature of the sample had been 600°C, the laser microprobe results indicate that the event must have lasted around 80 000 years, but if the temperature was 500°C, the event lasted more than 10 Ma. The study placed both time and temperature constraints upon the contact metamorphism and, given the temperature measurements, was able to demonstrate that a relatively long contact metamorphic event occurred.

Finally, though it is too large to cover in any detail here, the reader is directed to a laser microprobe study of natural- and laboratory- produced

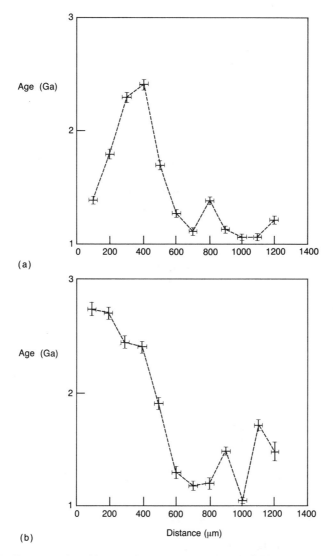

Figure 8.11 Two examples of laser probe age traverses from individual hornblende grains, in the Giants Range Granite, Minnesota. (a) Represents a small sub-grain, both sides of which exhibit age (argon loss) profiles over distances of about 300 μm. (b) Represents the edge of a larger grain exhibiting age profiles (argon loss) over distances of greater than 600 μm. The effective grain boundaries are defined not only by visible hornblende grain boundaries but also by biotite inclusions within the hornblende grains. Nevertheless, the diffusion profiles from the biotite-free sections of the grains all yield similar thermal histories. (After Kelley and Turner, 1991. Reproduced with permission).

diffusion which combines conventional stepped heating, laser spot analysis, electron microprobe and high-resolution transmission electron microscopy. Onstott, Phillips and Pringle-Goodell (1991), used all these techniques, measuring not only argon diffusion but also chlorine diffusion in natural phlogopite and hydrothermally treated biotite. They were able to show that argon diffusion and chlorine diffusion occurred on the scale of the whole grains in contrast to the smaller grain sizes (~150 μm) inferred by previous studies of hydrothermally treated biotite. The discrepancy was attributed to the numerous pores at layer terminations and the edges of dislocations detected by transmission electron microscopy, which appeared to be more common in the hydrothermally treated biotite. Such pores may permit rapid argon diffusion and may be the cause of enhanced argon diffusivity in micas seen in such chrontour maps as Figures 8.9 and 8.10.

References

Baldwin, S.L., Harrison, T.M. and Fitzgerald, J.D. (1990) Diffusion of ^{40}Ar in metamorphic hornblende. *Contrib. Mineral. Petrol.*, **105**, 691–703.

Burgess, R., Turner, G. and Harris, J.W. (1989) ^{40}Ar–^{39}Ar laser probe studies of clinopyroxene inclusions in eclogitic diamonds. *Geochim. Cosmochim. Acta*, **56**, 389–402.

Burgess, R., Kelley, S.P., Parsons, I. *et al.* (1992) ^{40}Ar–^{39}Ar analysis of perthite microtextures and fluid inclusions in alkali feldspars from the Klokken syenite, South Greenland. *Earth Planet. Sci. Lett.*, **109**, 147–67.

Cebula, G.T., Kunk, M.J., Mehnert, H.H. *et al.* (1986) The Fish Canyon Tuff, a potential standard for the ^{40}Ar–^{39}Ar and fission-track dating methods. *Terra Cognita*, **6**, 139.

Chesner, C.A., Rose, W.I., Denio, A. *et al.* (1991) Eruptive history of Earth's largest Quaternary caldera (Toba, Indonesia) clarified. *Geology*, **19**, 200–3.

Copeland, P. and Harrison, T.M. (1990) Episodic rapid uplift in the Himalaya revealed by ^{40}Ar/^{39}Ar analysis of detrital K-feldspar and muscovite, Bengal fan. *Geology*, **18**, 354–7.

Dalrymple, G.B., Alexander, E.C., Lanphere, M.A. and Kraker, G.P. (1981) Irradiation of samples for ^{40}Ar/^{39}Ar dating using the Geological Survey TRIGA reactor. *US Geol. Surv. Prof. Paper* **1176**.

Dodson, M.H. (1973) Closure temperature in cooling geochronological and petrological systems. *Contrib. Mineral. Petrol.*, **40**, 259–74.

Eichhorn, G., James, O.B., Scaeffer, O.A. and Müller, H.W. (1979) Laser ^{39}Ar–^{40}Ar dating of two clasts from consortium breccia 73215. *Proc. 9th Lunar Planet. Sci. Conf.*, pp. 855–76.

Fallick, A.E., McConville, P., Boyce, A.J. *et al.* (1992) Laser microprobe stable isotope measurements on geological materials: some experimental considerations (with special reference to δ^{34}S in sulphides). *Chem. Geol. (Isotope Geosci.)*, **101**, 53–61.

Faure, G. (1986) *Principles of Isotope Geology*, 2nd ed, Wiley, New York.

Flisch, M. (1982) Potassium–argon analysis, in *Numerical Dating in Stratigraphy* (ed. G.S. Odin), Wiley, Chichester, pp. 151–8.

Girard, J.-P. and Onstott, T.C. (1991) Application of ^{40}Ar/^{39}Ar laser-probe and step-heating techniques to the dating of diagenetic K-feldspar overgrowths. *Geochim. Cosmochim. Acta*, **55**, 3777–93.

REFERENCES

Glass, B.P., Hall, C.M. and York, D. (1986) $^{40}Ar/^{39}Ar$ laser-probe dating of North American tectite fragments from Barbados and the age of the Eocene–Oligocene boundary. *Chem. Geol.*, **59**, 181–6.

Hall, C.M. (1990) Calculation of expected thermal gradients in laser $^{40}Ar/^{39}Ar$ step-heating experiments. *Eos*, **71**, 653.

Harrison, T.M. (1981) Diffusion of ^{40}Ar in hornblende. *Contrib. Mineral. Petrol.*, **78**, 324–31.

Harrison, T.M., Duncan, I. and McDougall, I. (1985) Diffusion of ^{40}Ar in biotite: temperature, pressure and compositional effects. *Geochim. Cosmochim. Acta.*, **49**, 2461–8.

Kelley, S.P. and Bluck, B.J. (1992) Laser $^{40}Ar/^{39}Ar$ ages for individual detrital muscovite in the Southern Uplands of Scotland, UK. *Chem. Geol. (Isotope Geosci.)*, **101**, 143–56.

Kelley, S.P. and Turner, G. (1991) Laser probe $^{40}Ar-^{39}Ar$ measurements of loss profiles within individual hornblende grains from the Giants Range Granite, northern Minnesota, USA. *Earth Planet. Sci. Lett.*, **107**, 634–48.

Kelley, S.P., Turner, G., Butterfield, A.W. and Shepherd, T.J. (1986) The source and significance of argon isotopes in fluid inclusions from areas of mineralisation. *Earth Planet. Sci. Lett.*, **79**, 303–8.

Laurenzi, M.A., Turner, G. and McConville, P. (1988) Laser probe $^{40}Ar-^{39}Ar$ dating of impact melt glasses in Lunar breccia 15466. *Proc. 18th Lunar Planet. Sci. Conf.*, pp. 299–305.

Layer, P.W., Hall, C.M. and York, D. (1987) The derivation of $^{40}Ar/^{39}Ar$ age spectra of single grains of hornblende and biotite by laser step-heating. *Geophys. Res. Lett.*, **14**, 757–60.

Lee, J.K.W., Onstott, T.C. and Hanes, J.A. (1990) An $^{40}Ar/^{39}Ar$ investigation of the contact effects of a dyke intrusion, Kapuskasing Structural Zone, Ontario. *Contrib. Mineral. Petrol.*, **105**, 87–105.

Lee, J.K.W., Onstott, T.C., Cashman, R.J. *et al.* (1991) Incremental heating of hornblende *in vacuo*: implications for $^{40}Ar/^{39}Ar$ geochronology and the interpretation of thermal histories. *Geology*, **19**, 872–6.

Lo Bello, Ph., Feraud, G., Hall, C.M. *et al.* (1987) $^{40}Ar/^{39}Ar$ step-heating and laser fusion dating of a Quaternary pumice from Neshers, Massif Central, France: the defeat of xenocrystic contamination. *Chem. Geol. (Isotope Geosci.)*, **66**, 61–71.

Maluski, H. and Schaeffer, O.A. (1982) $^{39}Ar-^{40}Ar$ laser probe dating of terrestrial rocks. *Earth Planet. Sci. Lett.*, **59**, 21–7.

McConville, P. (1985) Development of a laser probe for argon isotope studies. PhD thesis, Sheffield University, UK.

McConville, P., Kelley, S.P. and Turner, G. (1988) Laser probe $^{40}Ar-^{39}Ar$ studies of the Peace River shocked L6 chondrite. *Geochim. Cosmochim. Acta.*, **52**, 2487–99.

McDougall, I. and Harrison, T.M. (1988) *Geochronology and Thermochronology by the $^{40}Ar/^{39}Ar$ Method*, Oxford University Press.

Megrue, G.H. (1967) Isotopic analysis of rare gas with a laser microprobe. *Science*, **157**, 1555–6.

Megrue, G.H. (1971) Distribution and origin of helium, neon, and argon isotopes in Apollo 12 samples measured by *in situ* analysis with a laser probe mass spectrometer. *J. Geophys. Res.*, **76**, 4956–4968.

Megrue, G.H. (1972) *In situ* $^{40}Ar/^{39}Ar$ ages of breccia 14301, and concentration gradients of helium, neon and argon isotopes in Apollo 15 samples, in *The Apollo 15 samples* (eds J.W. Chamberlain and C. Watkins), Lunar Science Institute, pp. 378–9.

Megrue, G.H. (1973) Spatial distribution of $^{40}Ar/^{39}Ar$ ages in Lunar breccia 14301. *J. Geophys. Res.*, **78**, 3216–21.

Merrihue, C.M. and Turner, G. (1966) Potassium–argon dating by activation with fast neutrons, *J. Geophys. Res.*, **71**, 2852–7.

Mitchell, J.G. (1968) The argon-40/argon-39 method for potassium–argon determination. *Geochim. Cosmochim. Acta*, **32**, 781–90.

Müller, H.W., Plieninger, T., James, O.B. and Schaeffer, O.A. (1977) Laser probe ^{39}Ar–^{40}Ar dating of materials from consortium breccia 73215. *Proc. 8th Lunar Planet. Sci. Conf.*, pp. 2551–65.

Onstott, T.C., Phillips, D. and Pringle-Goodell, L. (1991) Laser microprobe measurement of chlorine and argon zonation in biotite. *Chem. Geol.*, **90**, 145–68.

Perry, E.C. Jr and Bonnichson, W. (1966) Quartz and magnetite: oxygen 18–oxygen 16 fractionation in metamorphosed Biwabik Iron Formation. *Science*, **153**, 528–9.

Phillips, D. and Onstott, T.C. (1988) Argon zoning in mantle phlogopite. *Geology*, **16**, 542–6.

Phillips, D., Onstott, T.C. and Harris, J.W. (1989) ^{40}Ar/^{39}Ar laser-probe dating of diamond inclusions from the Premier kimberlite. *Nature*, **340**, 460–2.

Plieninger, T. and Schaeffer, O.A. (1976) Laser probe ^{39}Ar–^{40}Ar ages of individual mineral grains in Lunar basalt 15607 and Lunar breccia 15465. *Proc. 7th Lunar Sci. Conf.*, pp. 2055–66.

Ready, J.F. (1971) *Effects of High Power Laser Radiation*, Academic Press, London, 433 pp.

Ruffet, G., Feraud, G. and Amouric, M. (1991) Comparison of ^{40}Ar/^{39}Ar conventional and laser dating of biotites from the North Tregor Batholith. *Geochim. Cosmochim. Acta*, **55**, 1675–88.

Samson, S.D. and Alexander, E.C. Jr (1987), Calibration of the interlaboratory ^{40}Ar–^{39}Ar standard MMhb-1. *Chem. Geol. (Isotope Geosci.)*, **66**, 27–34.

Scaillet, S., Feraud, G., Lagabrielle, Y. *et al.* (1990) ^{40}Ar/^{39}Ar laser-probe dating by step heating and spot fusion of phengites from the Dora Maira nappe of the western Alps, Italy. *Geology*, **18**, 741–4.

Schaeffer, O.A., Müller, H.W. and Grove, T.L. (1977) Laser ^{39}Ar–^{40}Ar study of Apollo 17 basalts. *Proc. 8th Lunar Planet. Sci. Conf.*, 1489–99.

Steiger, R.H. and Jäger, E. (1977) Subcommission on geochronology: convention on the use of decay constants in geo- and cosmochronology. *Earth Planet. Sci. Lett.*, **36**, 359–62.

Sutter, J.F. and Hartung, J.B. (1984) Laser microprobe ^{40}Ar/^{39}Ar dating of mineral grains *in situ*. *Scann. Electr. Microsc.*, **4**, 1525–9.

Svelto, O. (1989) *Principles of Lasers*, 3rd edn, Plenum, New York.

Turner, G. (1971) Argon 40–argon 39 dating: the optimisation of irradiation parameters. *Earth Planet. Sci. Lett.*, **10**, 227–34.

Turner, G., Miller, J.A. and Grasty, R.L. (1966) The thermal history of the Bruderheim meteorite. *Earth Planet. Sci. Lett.*, **1**, 155–7.

Turner, G., Huneke, J.C., Podosek, F.A. and Wasserburg, G.J. (1971). ^{40}Ar–^{39}Ar ages and cosmic ray exposure age of Apollo 14 samples. *Earth Planet. Sci. Lett.*, **12**, 19–35.

Walter, R.C., Manega, P.C., Hay, R.L. *et al.* (1991) Laser-fusion ^{40}Ar/^{39}Ar dating of Bed 1, Olduvai Gorge, Tanzania. *Nature*, **354**, 145–9.

Wright, N., Layer, P.W. and York, D. (1991) New insights into thermal history from single grain ^{40}Ar/^{39}Ar analysis of biotite. *Earth Planet. Sci. Lett.*, **104**, 70–9.

York, D., Hall, C.M., Yanase, Y. *et al.* (1981) ^{40}Ar/^{39}Ar dating of terrestrial minerals with a continuous laser. *J. Geophys. Res.*, **8**, 1136–8.

CHAPTER NINE
Stable isotope ratio measurement using a laser microprobe

Ian P. Wright

9.1 Introduction

Many elements have two or more stable isotopes and can, therefore, be usefully studied for variations in their isotope ratios. These isotopes, unlike those of the radioactive variety, are completely stable and do not disintegrate with time. As such, when a non-volatile material such as a mineral is formed, the constituent elements should theoretically retain their isotopic integrity forever; subsequent measurement of these isotope ratios may assist with an effective reconstruction of the formation conditions. In reality, the isotopic compositions of a particular mineral may become modified as a result of an external influence, such as a heating process during hydrothermal activity. In this case, stable isotope measurements may be able to document something of the secondary activity which has befallen the sample of interest.

In its normal usage, the term 'stable isotope ratio' is restricted to measurements of the elements H, C, N, O and S. This grouping arises because (i) each of these elements can be readily converted into a gaseous form and then analysed using a gas-source mass spectrometer, and (ii) in a geological context, they are often analysed together in order to give complementary information (a factor which gives rise to the term 'light element geochemistry'). In regard of this latter point consider, for instance, that a sedimentary rock may contain carbonates, sulphates, hydrated minerals and organic compounds. The most complete description of the formation of these components may entail making stable isotope ratio measurements of all the elements mentioned above. The geochemist seeks to comprehend the extent to which the isotopes of an individual element, within a particular species, may have become separated from each other ('isotopic fractionation') by processes involving chemical reactions, or as a result of phenomena such as diffusion. Further information is derived from an appraisal of the degree to which coexisting species have equilibrated with respect to their isotopic compositions ('isotopic exchange'). The measurement of these parameters allows an insight into the nature of the conditions that prevailed during the formation of the materials under investigation. The light elements are particularly useful for these studies because the differences in mass between the isotopes are relatively large. Indeed, the relative mass difference between hydrogen (mass 1) and deuterium (mass 2) is the largest of any stable isotopes of any element. Since the magnitude of isotopic fractionation depends on relative mass differences, the light elements display the largest, and most easily measured, effects.

Traditionally, gases have been prepared for analysis by treating samples by heating/combustion techniques or, in the case of carbonate minerals, by reacting with acid. The use of a laser microprobe to release the gases of interest is an emergent and potentially revolutionary technique in the field of stable isotope ratio measurement. Since it is still in its infancy, it is difficult to assess for which types of investigation the laser microprobe will eventually be used. At present there are four areas where the laser has been successfully applied: (i) carbon and oxygen isotopic analysis of carbonates, (ii) sulphur isotope measurements of sulphides, (iii) oxygen isotope measurements of silicates and oxides and (iv) nitrogen isotope measurements of meteorite samples. All of these areas of application have been successfully carried out using the laser to make *in situ* isotopic measurements in a microprobe mode of analysis. However, there are still limitations to the technique; as such, further research and development is required before laser probe instrumentation can become a routine procedure used by non-specialists.

9.2 Stable isotope geochemistry

Excellent introductions to the breadth of investigations carried out within the field of stable isotope geochemistry can be found in Faure (1986) and Hoefs (1987). More advanced treatises by Valley, Taylor and O'Neil (1986) and Kyser (1987) are also available. The isotopic ratios of interest here are $^2H/^1H$ (D/H), $^{13}C/^{12}C$, $^{15}N/^{14}N$, $^{18}O/^{16}O$ and $^{34}S/^{32}S$. Other more specialized areas of investigation may require measurement of $^{17}O/^{16}O$, $^{33}S/^{32}S$ and $^{36}S/^{32}S$. The magnitudes of isotopic variations are normally given as δ values, where:

$$\delta = \left(\left(\frac{R_{sam}}{R_{ref}}\right) - 1\right) \times 1000 \qquad (9.1)$$

where R_{sam} is the ratio of the heavy to light (and, incidently, rare to abundant) isotope in an unknown sample and R_{ref} is the equivalent ratio in a reference material of known isotopic composition. The units of the δ value are per mil (‰), where 1‰ is equivalent to 0.1%. So that results from different laboratories can be compared, the δ values are measured relative to various internationally recognized standard materials. Thus:

$$\delta^{13}C_{PDB} = \left(\left(\frac{^{13}C/^{12}C_{sam}}{^{13}C/^{12}C_{PDB}}\right) - 1\right) \times 1000 \qquad (9.2)$$

where $^{13}C/^{12}C_{PDB} = 0.0112372$ (PDB stands for 'Peedee Belemnite'; this is also an oxygen standard with $^{18}O/^{16}O = 1.9452 \times 10^{-3}$). Other standards include 'Standard Mean Ocean Water', or SMOW (D/H = 1.6×10^{-4}, $^{18}O/^{16}O = 2.0052 \times 10^{-3}$, $^{17}O/^{16}O = 3.8288 \times 10^{-4}$), the terrestrial atmosphere, referred to as AIR ($^{15}N/^{14}N = 3.67647 \times 10^{-3}$) and 'Canyon Diablo Troilite', or CDT ($^{34}S/^{32}S = 4.50045 \times 10^{-2}$, $^{33}S/^{32}S = 8.09976 \times 10^{-3}$, $^{36}S/^{32}S = 1.79058 \times 10^{-4}$).

Stable isotope geochemistry involves the determination of isotope ratios in

1. an element partitioned between two coexisting species or;
2. different occurrences of an individual species.

The species concerned may ordinarily be minerals, but could equally well be fluids, gases, biological materials, etc. An example of (1), above, is oxygen isotope geothermometry where the distribution of isotopes between coexisting mineral phases is used to estimate the temperature of formation of an igneous rock. Geothermometry can also be applied to carbonate minerals precipitated from seawater since, at equilibrium, the oxygen isotopic distribution between these two entities is temperature dependent. For samples of a historical nature, or from the fossil record, the temperature of deposition of preserved carbonate minerals (palaeotemperature) is derived solely from the oxygen isotopic composition of the carbonate (an assumption is made regarding the isotopic composition of the seawater at the time of deposition). In the case of (2), above (different occurrences of the same species), since the oxygen and hydrogen isotopic composition of natural water samples from various sources show distinctive variations, the measurement of an unknown sample may help clarify its origin. During hydrous alteration at low temperatures, newly forming minerals, such as phyllosilicates, preserve an isotopic record of the fluids from which they were derived. Thus, from a knowledge of the hydrogen and oxygen isotopic compositions of hydrated minerals in a particular rock sample it may be possible to assess the source of the fluids which were present during an episode of weathering or alteration.

9.3 Conventional preparation of gases for stable isotope studies

The stable isotopic compositions of H, C, N, O and S are normally made on the following gaseous forms of the element: H_2, CO_2, N_2 and SO_2. More sophisticated analyses may be undertaken on O_2, NO or SF_6. The gases of interest are prepared in a number of ways. For instance, species which exist as trapped volatiles within a particular sample (e.g. CO_2 and N_2), may be removed simply by heating the material under vacuum. More refractory forms of the elements may be converted into gases by combustion. In this way, carbon in the form of organic materials, diamonds or graphite, for instance, is converted to CO_2; sulphur from sulphide minerals is combusted to form SO_2 etc. Hydrogen-bearing minerals are normally heated to liberate H_2O, which is subsequently reduced, using heated zinc or uranium, to give H_2 for isotopic measurement. Nitrogen from various minerals can be released by heating or combustion techniques. Carbonates are usually reacted with acid to produce CO_2 – in this way it is possible to obtain both carbon and oxygen isotopic compositions of the minerals.

Note that in all cases of sample preparation it is desirable to convert, quantitatively, the element under investigation to a single gaseous species. Failure to do this may result in erroneous isotopic measurements since the partitioning of an element between two or more product species may result in isotopic fractionation. Thus, in the case of a carbon combustion the extraction is carried out under an excess of oxygen to ensure the complete conversion of the element to CO_2 (too little oxygen would result in the partial production of CO). During the combustion of sulphides, excess oxygen results in the partial production of SO_3 which has to be reduced to SO_2 using heated copper. The production of nitrogen oxides during combustion must be suppressed by the catalytic reduction of oxides to N_2 gas. An example, which obviously violates the rule of quantitative conversion, is the production of CO_2 from carbonates; in this case three oxygen atoms, in the form of CO_3^{2-}, are converted to just two in the gaseous form. However, if the acid dissolution is conducted under reproducible experimental conditions and at a carefully controlled temperature, it is possible to apply a correction to establish the true oxygen isotopic composition of the carbonates.

Two more preparative techniques deserve mention here; both involve the use of fluorinating agents, such as F_2, ClF_3 or BrF_5. To convert the oxygen of silicate or oxide minerals into a gaseous form, it is necessary to use an oxidizing agent which is more powerful than oxygen itself. Thus, oxygen is displaced from minerals by reaction at high temperature with the fluorinating agent. Oxygen is liberated as O_2, which can be either analysed directly (rare) or subsequently converted, quantitatively, to CO_2. Treating sulphides with fluorinating reagents results in the formation of SF_6, a gaseous species which can be analysed for its sulphur isotopic composition.

9.4 Stable isotope ratio mass spectrometry

Conventional stable isotope ratio mass spectrometers represent an evolution of the original design by Nier (1947) and McKinney et al. (1950). Instruments of this type are not necessarily very accurate, but produce results of very high precision or reproducibility (i.e. δ values that have standard errors of ±0.05‰). It is the operational protocol which enables these high levels of precision to be obtained. The mass spectrometers are equipped with dual inlets, which allow sample and reference gases to flow continuously to a device known as a changeover valve (Figure 9.1). From this valve, one gas is selected for admission to the mass spectrometer while the other is bled to waste. If the pressures of gas in the two inlets are equalized, by adjusting the settings of variable volume reservoirs, the flow rates of both gases remain constant over relatively long periods of time (i.e. over the course of a single analysis, which may take typically 10–20 min). This factor ensures that any isotopic fractionation, which occurs when gas from the source reservoir is bled through the sampling capillary, occurs to the same degree for both gases. In this way, differential isotopic fractionation is avoided.

Mass spectrometers used for stable isotope analyses do not make absolute isotope ratio measurements. Rather, they are configured to make comparative measurements, i.e. δ values relative to the isotopic composition of the reference gas, which is known. In practice, δ values are made by repetitively changing the direction of gas flow from the changeover valve, i.e. by alternately admitting sample and then reference gases to the mass

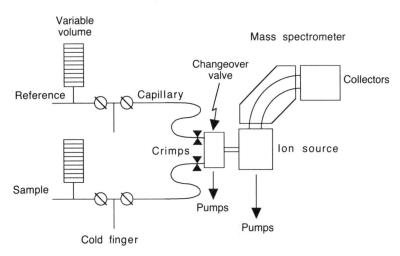

Figure 9.1 Schematic diagram of a conventional stable isotope ratio mass spectrometer. Sample and reference gases, stored in variable volume reservoirs, flow continuously through capillary tubes into a change over valve. This device selects one gas for analysis by the mass spectrometer and bleeds the other to waste. The δ value of interest is obtained by comparing the measured isotope ratio of the sample gas to that of the reference. In this way a differential isotope ratio is produced to a high degree of precision. Since the absolute isotopic composition of the working reference gas is always known, it is a straightforward matter to compare results from different laboratories.

spectrometer. A number of consecutive comparisons are used to attain a δ value. In this way any short-term changes in instrumental operating conditions, resulting from fluctuations in electrical supplies, changes in temperature, drift in electronic devices, etc., tend to affect the measurement of unknown and reference gases in the same way. The very nature of making a comparative, rather than an absolute, measurement leads to a result that has a high level of precision.

9.5 Historical development of the laser microprobe

Laser beams were first used in the early 1960s to vaporize atoms from solid targets, for the purpose of spectrographic determination (e.g. Honig and Woolston, 1963; Lincoln, 1965). By the 1970s the laser microprobe technique was beginning to gain acceptance as a quantitative analytical tool in the field of geological studies (e.g. Scott, Jackson and Strasheim, 1971; Erëmin, 1975). The further development of the laser microprobe is dealt with in reviews by, amongst others, Kovalev et al. (1978), Conzemius and Capellen (1980) and Conzemius et al. (1983). In addition, details of the nature of the laser interaction with solids can be found in Adrain and Watson (1984).

The development of the laser microprobe for gas analysis has effectively developed from the technique devised for noble gas determination (e.g. Megrue, 1967, 1971). As described in Chapter 8, one highly successful implementation of the laser microprobe heating technique is in the field of ^{40}Ar–^{39}Ar dating. The use of laser microprobes for the analysis of volatile species other than noble gases was first introduced in the 1970s, when the technique was applied to the release of organic materials. Preliminary efforts involved direct measurement of fragmented organic ions by mass spectrometry (e.g. Hillenkamp et al., 1975). A further technique involved using the laser to desorb organic compounds which were subsequently separated using gas chromatography (e.g. Vanderborgh, 1977). Subsequently, the laser microprobe was used to study fluid inclusions in various rock samples to determine the concentrations of elements pertinent to ore formation (e.g. Tsui and Holland, 1979; Bennett and Grant, 1980). A logical extension of this approach was to utilize the laser in conjunction with the analysis of the volatile constituents as well, thereby allowing *in situ* determination of the fluids in individual inclusions (e.g. Deloule and Éloy, 1982). This development represented a considerable advance on conventional fluid inclusion analysis in comparison with conventional heating–cooling techniques or bulk pyrolysis (e.g. review by Roedder, 1984). The use of laser decrepitation for the determination of H_2O/CO_2 ratios in individual fluid inclusions can now be considered as a viable, routine technique (e.g. Sommer et al., 1985).

The first reported attempts to use the laser microprobe to release gases (other than noble gases) for stable isotopic analyses were conducted by Norris, Brown and Pillinger (1981), Franchi et al. (1985, 1986) and Jones et *al.* (1986). These preliminary investigations concentrated on the design of the laser probe systems and associated vacuum chambers, and pointed out a number of potential problems which needed to be overcome before the technique could be sensibly employed for the analysis of unknown samples. In essence, the laser microprobe is used to heat selectively certain areas of a target, producing a gaseous species which is subsequently transferred to a mass spectrometer for stable isotope analysis. The technique relies on localized pyrolysis (heating in vacuum), combustion (heating in an atmosphere of oxygen gas), or fluorination (heating in the presence of a fluorinating agent).

9.6 Application of the laser microprobe to stable isotope studies

To obtain isotopic measurements of a suitable level of precision (i.e. small error compared to the anticipated range of values) requires a minimum quantity of material. This, in conventional techniques, corresponds to about $1-10\,\mu g$ in the case of carbon, and to $1-10\,mg$ for elements like oxygen and sulphur. The exact amount is dictated by factors such as the sensitivity of an individual instrument, the level of system blanks, whether cryogenic concentration of sample gas is used, etc. Depending upon the relative concentration, or form, of the element of interest, analysis may require the use of a whole-rock sample. On the other hand, to derive the maximum amount of information, it may also be possible to analyse separated components. Three examples are given below, where there is an obvious need to undertake stable isotope measurements on more than a single constituent of a particular sample.

1. In a hydrothermal alteration study, it may be desirable to measure the difference in isotopic compositions between coexisting sulphides and sulphates in a rock. In this case, direct analysis of a bulk sample is of no value. Various chemical/physical means can be used to extract the sulphide and sulphate minerals so that each can be analysed independently by conventional techniques (e.g. Alt, Anderson and Bonnell, 1989).
2. As described above, oxygen isotopic studies of carbonate minerals are used to derive the temperature of their formation. Note that for a marine organism, such as a coral, which has grown in a particular environment, the carbonate secreted as different growth bands will record local variations in seawater temperature or isotopic composition. Analysis of the bulk sample, or a fraction comprising many different growth bands, can only give information concerning average properties. To obtain the most highly resolved record of environmental fluctuations it is necessary, using conventional techniques, to use samples that have been mechanically removed from the coral at closely spaced intervals. This can be accomplished using a high-speed drill to extract material from individual growth

bands (e.g. Wefer and Killingley, 1980; Krantz, Williams and Jones, 1987). Figure 9.2 shows an example of the technique as applied to a recent mollusc shell collected off the coast of Virginia, USA.

3. Cathodoluminescence images of certain natural diamonds show that they consist of many concentric growth bands. To investigate the nature of the mantle fluids from which the diamonds formed, it would be desirable to measure carbon isotopic compositions of the individual growth bands.

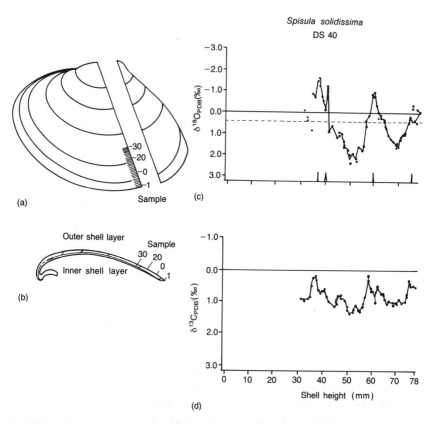

Figure 9.2 Carbon and oxygen isotopic zoning in a mollusc shell. (a) Example of a *Spisula* shell which has been sectioned along the axis of maximum growth and drilled incrementally to collect carbonate powders for isotopic analyses. (b) Cross-section of the same shell illustrating that powders are only drilled from the outer shell layer. (c) Oxygen isotopic profile, plotted as $\delta^{18}O_{PDB}$, from a *Spisula* shell. The data represent a growth profile (shell height) – small closed arrows on the horizontal axis represent annual internal growth increments. The solid line at 0‰ is for reference only; the dashed line represents the mid-point between highest and lowest $\delta^{18}O$ values. Note that isotopic temperatures become 'warmer' towards the top of the vertical scale. (d) Carbon isotopic profile, as $\delta^{13}C_{PDB}$, for the same shell. (Data from Krantz, Williams and Jones 1987).

Measurements of a bulk diamond can only yield information concerning the average composition of the mantle fluids. An innovative approach to the problem is to use a laser beam to cut individual diamonds into small pieces and then make carbon isotopic measurements on the individual fragments (Franchi et al., 1989). In this way it is possible to document the variation in carbon isotopic composition of the fluids that were in evidence over the time that a diamond grew (e.g. Boyd et al., 1987). Figure 9.3 shows carbon isotope data obtained by conventional mass spectrometry techniques, as well as additional information concerning nitrogen, obtained from a laser-dissected diamond.

The examples above show that, although innovations in sample preparation can be devised to permit analysis using conventional methods, there is a need for a microbeam technique capable of analysing stable isotopic compositions at high degrees of spatial resolution. The ion microprobe (Chapter 6) has already been used to very great effect for stable isotope analyses (e.g. Hervig, Thomas and Williams, 1989; Eldridge et al., 1989; Zinner, 1989; Harte and Otter, 1992). However, this instrument is a very expensive and complex piece of apparatus which, by virtue of its flexibility, is often used in a range of other applications. As such, the availability of analysis time for stable isotope work on an ion microprobe is generally very limited. The laser microprobe, on the other hand, is a relatively inexpensive and widely available option and there can be no doubt that in the future such devices will become commonplace in stable isotope laboratories.

9.7 Instrumentation

To interface a laser microprobe to a stable isotope ratio mass spectrometer, a sample chamber of some description is connected to the inlet of the instrument. Sample chambers can be constructed in many different ways, e.g. Figure 9.4 (also Chapter 8). Lasers of the appropriate specifications are available from a number of manufacturers, complete with viewing/focusing optics and a video camera to view the sample during laser extraction. Because of the inherent dangers associated with lasers it is necessary to make the systems conform to statutory safety guidelines. These standards can normally be achieved by enclosing the laser optics and sample chamber in a light-tight box, thereby avoiding the problem of stray radiation.

To create an effective extraction system, a number of valves (manual- or computer-controlled) are required in the system to enable opening, or removal, of the sample chamber without compromising the high vacuum of the mass spectrometer. Additional components may be required, such as copper oxide furnaces, to generate and readsorb oxygen gas, and cryogenic fingers for gas separations etc. Since the different areas of application have their own special requirements, each is dealt with separately below.

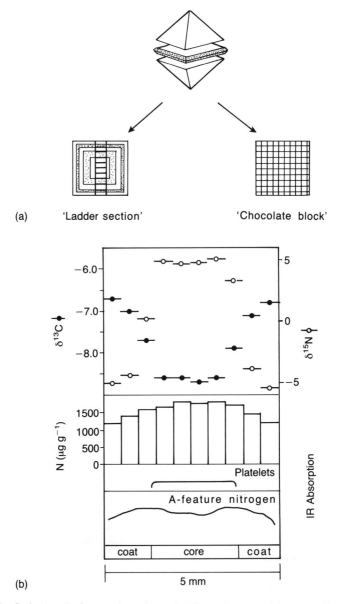

Figure 9.3 Carbon and nitrogen isotopic zoning in a diamond. (a) Sketch showing laser sectioning of diamond plates. A plate (about 5 × 5 × 0.3 mm) is cut from the centre of the stone by a diamond saw. This plate is then dissected by a Nd-YAG laser, according to patterns such as 'ladder sections' or 'chocolate blocks', to produce individual blocks of about 0.4 × 0.4 mm in size. Each one of these is then analysed for carbon and nitrogen isotopic composition. (b) An example of internal variations in carbon and nitrogen isotopic compositions from an individual diamond (known as a coated stone because of its rather obvious core and coat morphology). Also shown are the variations in nitrogen content and infrared absorption characteristics. (From Franchi et al., 1989).

CARBON AND OXYGEN ISOTOPIC ANALYSES OF CARBONATES

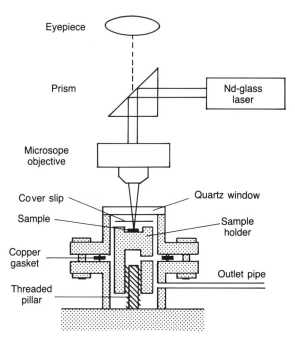

Figure 9.4 Schematic diagram of a typical sample chamber, laser and viewing optics. The laser chamber is constructed from two stainless steel flanges (70 mm diameter), one of which has a quartz window to allow the laser light to enter the vacuum system and to permit sample viewing.

Sample preparation generally involves placing either a powder or a thick polished section of the material under investigation into the sample chamber, which is then evacuated. Gas for subsequent analysis is extracted by irradiating the sample with laser light. Details of the interaction of laser light with sample materials are given in Chapter 8 (along with typical examples of laser pit morphologies). Unlike noble gas extraction, where the laser is simply used to heat the sample to a high enough temperature to liberate the gases of interest, the laser is generally used in stable isotope analysis to initiate a chemical reaction in a localized region. Clearly, there exists a variety of potential problems which can accompany this procedure, not least being uncontrollable isotopic fractionation in the gas phase species. A certain amount of effort has, therefore, been required to validate the technique using the laser to irradiate appropriate reference materials of various descriptions.

9.8 Carbon and oxygen isotopic analyses of carbonates

The first attempts to use laser microprobes to determine carbon and oxygen isotopic compositions of carbonates were undertaken by Franchi *et al.* (1986)

and Jones *et al.* (1986). In the former study, a Nd-glass pulsed laser (non-Q-switched) operating at 1064 nm wavelength was used to deliver 0.1–5 J of energy to the sample. The laser pits could be varied between 10 and 150 μm diameter (50–250 μm depth), resulting in power densities of 10^7 to 10^{10} W cm^{-2}. The laser light was focused onto the target through a microscope objective; the same optics were used to view the sample and position the laser beam (Figure 9.4). A number of different sample chambers were employed but in essence each one was made of stainless steel and comprised either one or two optical windows made of quartz (one was used for reflected light illumination only; two windows were necessary for reflected and transmitted light). In all cases a quartz coverslip was positioned between the sample and the optical window. This coverslip had the effect of preventing any ejected materials from coating the window, an important consideration since a dirty window reduces transmission, thereby making it harder to view the sample, and also risking the absorption of laser light, with attendant problems of heating the window.

Two different instrumental configurations were employed (Figure 9.5), one in which the sample chamber was not connected to a mass spectrometer and one in which it was connected to the extraction system of a static vacuum mass spectrometer. This type of mass spectrometer, which is highly sensitive, is found only in research laboratories. The principle of operation of a static mass spectrometer is similar to that used for noble gas analyses, but fundamentally different from the type normally used for conventional measurements. Resumés of the uses of static mass spectrometry in stable isotope analyses can be found in Wright (1984) and Pillinger (1992).

When the sample chamber was not connected directly to the mass spectrometer it was necessary to collect CO_2 gas in an evacuated vessel for subsequent transfer. Two carbonate minerals were analysed: calcite ($CaCO_3$) and siderite ($FeCO_3$). Because the calcite used was largely transparent with respect to the laser light employed, it was necessary to use of the order of 100 individual pulses to obtain enough gas for isotopic measurement, thereby negating some of the advantages of using a microbeam technique. In contrast, the laser light was readily absorbed into the siderite and so only 10 pulses were needed. Measured $\delta^{13}C$ and $\delta^{18}O$ values were higher that the true values (by 1–5‰ and 3–11‰ respectively) for both carbonate minerals, implying an isotopic fractionation of some description. For the experiments conducted with the sample chamber connected to the static mass spectrometer, it was possible to make measurements on CO_2 gas produced from individual pulses. In this case the $\delta^{13}C$ and $\delta^{18}O$ values were generally lower than expected. It also transpired that CO was produced in addition to CO_2 and that the carbon isotopic composition between the two gases was of the order of 25‰.

The technique of Jones *et al.* (1986) used a Nd-YAG laser operating in either continuous or pulsed modes. The sample chamber was not connected

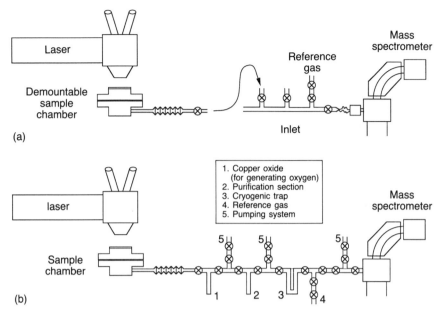

Figure 9.5 Two configurations of the laser microprobe used for stable isotope analysis of carbonates. (a) In this case the sample chamber was demountable. Prior to analysis the chamber, with the carbonate sample already loaded, was evacuated on a separate pumping system and then isolated by means of a valve. Upon subsequent laser interaction, gases such as CO_2 and CO were collected in the sample chamber itself. These gases were then transferred to the inlet of a stable isotope ratio mass spectrometer whereupon the carbon and oxygen isotopic compositions of CO_2 were determined. (b) A development of the demountable system whereby the sample chamber was connected directly to the inlet of a static mass spectrometer. The inlet, which is pumped to high vacuum, contains devices for generating oxygen (i.e. for laser combustions), converting and purifying the product gases, and a cryogenic trap to enable complete separation of CO_2. (From Franchi et al., 1986).

to a mass spectrometer and so analyses could only be conducted off-line. Apparently, using this system, it was possible to obtain the correct $\delta^{13}C$ and $\delta^{18}O$ values from calcite but, unfortunately, details of the procedure have not been forthcoming.

A number of other laser microprobe systems have been developed from these initial studies (e.g. Smalley et al., 1989; Powell and Kyser, 1991; Smalley et al., 1992). In each case the laser is a Nd-YAG system which can operate either in a continuous or pulsed (Q-switched) mode. From the results of Smalley et al. (1992), it would seem that there is no difference in the results obtained from either mode of operation. It should be noted that since the wavelength of radiation from the Nd-YAG laser is in the infrared (1064 nm), it is optically invisible. As such, it is impossible to assess the location of the laser beam, other than by trial and error as was used in the early techniques. Modern laser microprobe instrumentation includes a low-power He-Ne laser, which produces visible light coincident with the Nd-YAG and can therefore be used to position the analysis spot.

The sample chambers utilized with these second generation laser microprobe systems are variants of those used in the earlier studies. In each case, the gases from the sample chamber can pass directly to the mass spectrometer, without the use of demountable collection vessels. In the system of Powell and Kyser (1991), it is possible to determine the isotopic compositions of CO and CO_2, thereby allowing an appraisal of the isotopic fractionation between these gases. To analyse the isotopic composition of CO, it is necessary to convert it to CO_2, because this is the gas used by the mass spectrometer. The conversion is performed in one of two different ways: platinum at a high voltage ($\sim 3\,\text{kV}$) is used when the $\delta^{18}O$ of the original CO is required, while copper oxide at 450°C is used to make $\delta^{13}C$ measurements. Laser spot sizes are in the region of 25–35 μm, although it has been noted that the actual area of damage may extend to 50 μm.

Several different carbonate minerals have now been analysed using the laser microprobe. In each case it has been noticed that there is a degree of isotopic fractionation in the evolved CO_2 gas. Smalley et al. (1989) found that CO_2 liberated from calcite was 1.7‰ lower in $\delta^{18}O$ than calcite and 2.5‰ lower than aragonite. $\delta^{13}C$ was found not to be significantly altered during laser ablation. In contrast, Smalley et al. (1992) found that the $\delta^{18}O$ of CO_2 from calcite was 1.2‰ higher and $\delta^{13}C$ 0.8‰ lower. The two systems described by Smalley et al. (1989, 1992) are in different laboratories and can be considered to operate under different conditions. In neither case was the isotopic composition of the accompanying CO measured. However, it was noted by Smalley et al. (1992) that the CO/CO_2 ratio was fairly constant and independent of laser power. This observation means that the experimental conditions within a single system can be considered to be fairly reproducible and as such the analyses can be satisfactorily carried out by applying a correction to the measured results. Indeed, a laser microprobe system has been used by Dickson, Smalley and Kirkland (1991) to study aragonite of biogenic and abiogenic origin in sedimentary samples of Pennsylvanian age. The results of unknown samples, notwithstanding correction procedures, have 1σ errors of ± 0.4‰ for $\delta^{18}O$ and ± 0.2‰ for $\delta^{13}C$.

Powell and Kyser (1991) have conducted an in-depth study into the mechanism of isotopic fractionation during laser (i.e. Nd-YAG) microprobe sampling. They used a variety of different carbonate minerals: calcite, dolomite ($CaMg(CO_3)_2$), rhodochrosite ($MnCO_3$) and siderite. It was found that the CO_2 yield and CO/CO_2 ratio varied between the different mineral types; siderite and rhodochrosite gave the highest yields indicating a greater efficiency of laser light absorption. The CO/CO_2 ratios produced from siderite and rhodochrosite were lower than those from calcite and dolomite. The isotopic composition of the combined CO and CO_2 gases were found to be distinctly different from those obtained by acid dissolution (although reproducible in $\delta^{13}C$ and $\delta^{18}O$ to ±2‰ and ±3‰ respectively). The magnitude of the difference was related to the chemical composition of the carbonate. It was considered that the interaction of the laser with the sample proceeds in two stages. In the first instance, the laser beam excites the sample surface causing ablation of material; secondly, the laser beam interacts with the expanding ionized gas plume produced from stage one. The isotopic fractionation between CO and CO_2, which amounts to about 2‰ in the case of $\delta^{13}C$ and 5–10‰ in $\delta^{18}O$ (δ values are lower in CO than in CO_2, in both cases), occurs during the second stage. Clearly, there are a number of factors which need to be controlled adequately before reliable results can be obtained from a range of different carbonate minerals.

9.9 Sulphur isotopic measurements of sulphides

Laser microprobe measurements of $\delta^{34}S$ on sulphides can presently be made to precisions of about 0.25‰, at spatial resolutions of about 70 μm. Consequently, this technique fills the gap between high-precision orthodox combustion measurements of drilled samples with resultant limited spatial resolution ($\delta^{34}S$ ±0.2‰, 0.5 mm), and low-precision measurements at high spatial resolution by the ion microprobe ($\delta^{34}S$ ±1‰, <50 μm). The nature of the geological problem should thus dictate which method to adopt (see discussion in Kelley and Fallick, 1990).

Conventional gas source mass spectrometers use SO_2 or SF_6 for sulphur isotope ratio measurements, and both species have been produced by laser interaction with sulphides (Kelley and Fallick, 1990; Crowe, Valley and Baker, 1990; Rumble, Palin and Hoering, 1991). Amongst the advantages of SO_2 are the ready availability in many laboratories of suitable mass spectrometers and the convenience of simple oxidative combustion in an atmosphere of pure O_2. Drawbacks include the possibility of complex chemistry (e.g. production of SO_3) and the necessity for proper correction for isobaric interferences (e.g. the m/z 66 ion beam contains both $^{34}S^{16}O^{16}O^+$ and $^{32}S^{18}O^{16}O^+$). This latter problem can be addressed through empirical correction using minerals of known $\delta^{34}S$ (Kelley and Fallick, 1990) and by calibration with minerals synthesized with pure ^{32}S (as has been carried out at the Scottish Universities Research and Reactor Centre, for instance). Utilization of SF_6 obviates the requirement for correction of oxygen isotope interferences and should permit assessment of $^{33}S/^{32}S$ and $^{36}S/^{32}S$ in addition to $^{34}S/^{32}S$ (likely to be mainly of interest in analyses of extraterrestrial samples), but the chemistry involved is less convenient, requiring the use of a fluorinating reagent, and a specialized mass spectrometer must be able to cope with the higher m/z ions (e.g. $^{32}SF_5^+$ at 127). The inherent advantages of SF_6 mass spectrometry (Rees, 1978) make this the method of choice if very high precisions (e.g. ≤0.1‰) are required. However, since most work to date has employed SO_2, and because SF_6 is likely to be adopted only by a minority of sulphur isotope laboratories, attention is focused here on the SO_2 methodology.

Details of experimental techniques and gas handling vacuum lines have been given by Crowe, Valley and Baker (1990) and Kelley and Fallick (1990). It appears that the type of laser used is not critical; for example, argon ion, Nd-YAG and CO_2 lasers have all been successfully employed. Fallick et al. (1992) point out that since the laser beam is used merely as a heat source to promote a chemical reaction (i.e. combustion), the choice of irradiance (i.e. power density) is important to control the chemistry. In other words, for a given wavelength, it is imperative to select the right combination of laser output power and beam diameter (degree of focusing). Fallick et al. (1992) advocate an irradiance of the order of $10^9 \, W \, m^{-2}$. At such power densities, reaction of the hot sulphide with oxygen competes with thermal decomposition of the minerals giving gas-phase sulphur. It seems likely that sulphur produced during dissociation is less likely to react with oxygen since this would, presumably, require a gas-phase reaction. The existence of two product reservoirs of sulphur (S^0 and SO_2) may lead to sulphur isotope fractionation between starting mineral and product SO_2, and this might be expected to depend on the relative efficacy of thermal dissociation. In accord with this, Kelley and Fallick (1990) and Kelley et al. (1992) observed a smooth variation of sulphur isotope fractionation with the free energy of formation (a measure of relative bond strength) for a range of metal sulphides. Isotopic fractionation was small ($\leqslant 1‰$) for strong metal–sulphur bonds, and most importantly was independent of starting $\delta^{34}S$ over a range of almost 100‰. The implication of these findings is that for each laser microprobe configuration, a calibration program should be carried out and the inherent isotope fractionation carefully monitored.

Laser–sulphide mineral interactions have been studied by detailed investigation of surfaces and pit morphology after combustion, using SEM, electron probe and backscattered electron imagery; the results have been used to optimize conditions of analysis (Crowe, Valley and Baker, 1990; Fallick et al., 1992; Kelley et al., 1992). Again, it is important to tailor the experimental technique to the geological problem.

Because many natural sulphides are not only fine-grained but are also mineralogically complex intergrowths, the potential of laser microprobe $\delta^{34}S$ methods is particularly high. Crowe, Valley and Baker (1990) demonstrated isotopic zoning within individual mineral grains at the sub-millimetre scale and also assessed the degree of sulphur isotopic fractionation that exists between coexisting minerals in a variety of environments. Of special note are their results for variably metamorphosed sulphide ore deposits. Whereas pyrite–pyrrhotite isotopic equilibrium was not attained under greenschist facies conditions, pyrite and sphalerite were in $^{34}S/^{32}S$ equilibrium for amphibolite facies conditions. It was suggested that this result was perhaps related to the opportunity for extensive recrystallization.

Crowe and Valley (1992) reported laser microprobe δ^{34}S traverses along fluid conduits in a hydrothermal system from the Axial Seamount (Juan de Fuca Ridge, eastern Pacific). Average sphalerite δ^{34}S values of +3 to +4‰ were observed, with maximum variation of 5‰ within a given conduit and 7.4‰ between conduits. High mineral δ^{34}S values were used to identify short-lived excursions in fluid δ^{34}S, thought to be related to deep-seated changes in hydrothermal activity. For fossil hydrothermal chimneys from the Lower Carboniferous Silvermines (Eire), Fallick (1990) reported remarkably uniform δ^{34}S traverses despite the highly ^{34}S-depleted nature of the pyrite (δ^{34}S = -38 ± 1‰). Kelley and Fallick (1990) displayed a δ^{34}S profile for galena of up to 20‰ per millimetre. These results indicate that the laser microprobe technique is capable of providing surprising results, which in consequence have a strong impact on geological interpretation.

9.10 Oxygen isotopic analyses of silicates and oxides

A major problem with the conventional technique for analysing oxygen isotope ratios in silicates and oxides is that relatively large samples are required (typically 10–20 mg) to overcome the effects of blank contributions. This limitation has restricted investigations which require the measurement of isotopic ratios on a fine scale. Although the ion microprobe has been used to make δ^{18}O measurements on localized areas of rock samples (e.g. Giletti and Shimizu, 1989; Hervig, Thomas and Williams, 1989), the precision obtained by this technique (± 1–2‰) is not adequate for routine geochemical work. The use of laser microprobes for oxygen isotope investigations of silica-bearing minerals was first evaluated by Franchi et al. (1986), who used a laser to demonstrate that CO could be liberated from a mixture of graphite and quartz. This technique is a variant of the graphite-reduction method developed by Clayton and Epstein (1958). A more detailed study of the feasibility of using laser heating for this purpose was conducted by Sharp and O'Neil (1989). A finely powdered sample (1–3 mg) was mixed with excess graphite and then irradiated by a Nd-YAG laser. The CO produced in this way was subsequently converted to CO_2 and admitted to a mass spectrometer for isotopic analysis. The δ^{18}O values obtained in this fashion were in some cases found to agree rather well with the expected results (e.g. quartz, magnetite, diopside, garnet). However, forsterite gave rather erratic results and the data for feldspar were in error by about 3‰. The problem with a laser technique of this sort is that it offers no real hope of *in situ* microprobe applications, since samples have to be extracted and subsequently crushed. However, it should be pointed out that CO is a gas which is amenable to isotopic determination using static mass spectrometry (e.g. Gardiner and Pillinger, 1979) and so, theoretically, it is possible that the oxygen isotopic composition of sub-microgram samples could be obtained by preparation of this gas.

A far more encouraging use of the laser microprobe for oxygen isotope analyses has been described by Sharp (1990). In this case, laser heating is used in conjunction with BrF_5 or ClF_3 to produce oxygen gas from silicates and oxides. The success of this technique can be judged by the relatively rapid implementation of similar devices in other laboratories (e.g. Elsenheimer and Valley, 1992; Franchi, Akagi and Pillinger, 1992; Sharp, 1992; Mattey and Macpherson, 1993). A schematic diagram of a typical system is shown in Figure 9.6.

Two different types of laser have been used for fluorination purposes. Sharp (1990, 1992) and Franchi, Akagi and Pillinger (1992) used a CO_2 laser, while Elsenheimer and Valley (1992) and Mattey and Macpherson (1993) used a Nd-YAG device. The advantages and disadvantages of each type of laser have been expounded by Sharp (1992). Briefly, the Nd-YAG

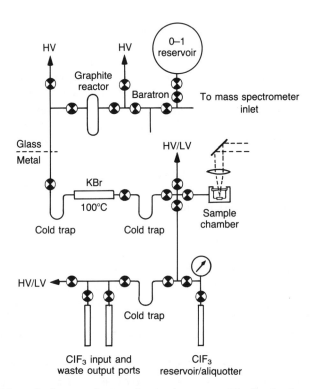

Figure 9.6 Schematic diagram of a laser-fluorination system. The fluorinating agent, ClF_3, is admitted to the sample chamber containing the minerals under investigation, whereupon the action of the laser causes oxygen gas to be formed. The sample gases are passed through a cold trap and heated KBr, to remove any unreacted ClF_3, or other fluorine-containing products. The oxygen gas is then reacted with heated graphite and converted to CO_2, which is subsequently admitted to the mass spectrometer for analysis. (From Mattey and Macpherson, 1993).

laser is capable of being focused to ~10 μm, but the emitted wavelength is not readily absorbed by all of the minerals of interest; furthermore, it is a relatively bulky piece of equipment. CO_2 lasers, on the other hand, emit radiation at 10.6 μm, which can be absorbed by all minerals. A typical device is small and inexpensive; however, the ultimate spot size of a CO_2 laser is about 100 μm. Another consideration is that whereas the radiation from a Nd-YAG laser can be transmitted through Pyrex, quartz or sapphire windows, it is necessary to use a material such as BaF_2 for optical elements used in conjunction with a CO_2 laser. Caution needs to be exercised when using BaF_2 windows since these have been known to fail during laser analysis. Because of the low weight and small size of a CO_2 laser, it is possible to move the laser itself, rather that the sample chamber, to position the analysis spot (Sharp, 1990). In the case of a large Nd-YAG laser, it is necessary to be able to move the sample chamber – this can be accomplished by connecting the chamber to the rest of the vacuum system by flexible stainless steel tubing.

The focal length of a typical Nd-YAG system is about 4 cm (Elsenheimer and Valley, 1992), whereas that for a CO_2 laser is 12.5 cm (Sharp, 1990). An advantage of a long focal length is that a relatively 'tall' sample chamber can be constructed, such that the optical window is far removed from the sample. This arrangement prevents materials ejected from the sample during laser irradiation from coating the window. The drawback of this approach is that the path length of the laser beam through the sample chamber is relatively long. It has been suggested that the laser beam may interact with a fluorinating agent such as BrF_5 to produce F_2. Obviously, a longer path length will result in the production of higher amounts of F_2. Because it is extremely difficult to separate totally F_2 and O_2 by cryogenic means, the presence of unwanted fluorine gas is considered undesirable. Various specialized techniques have had to be developed to cope with the separation of F_2 from O_2.

A typical sample chamber is constructed from stainless steel; the optical window is attached to the system using a fluorine-resistant material such as Kalrez (a perfluorinated polymer). Samples are generally introduced in two different forms – either as a single thick section, or as powders. In the latter case, the samples are placed into a number of separate blind holes drilled into a suitable nickel holder. Nickel is used because of its relative inertness towards the fluorinating agents.

Following the insertion of samples into the chamber, the vessel is evacuated and then F_2, BrF_5 or ClF_3, at a pressure of 0.2–0.6 bar, can be admitted in order to 'pre-fluorinate' the samples. This procedure, which is carried out at room temperature for several hours, serves to remove any adsorbed H_2O from the chamber walls and the samples. Inevitably there is some reaction of the fluorinating reagents with the samples themselves. In particular, phyllosilicates, carbonate and feldspars are attacked

(Elsenheimer and Valley, 1992), although ClF_3 at 60°C does not appear to react with ferromagnesian minerals (Mattey and Macpherson, 1993). An assessment of the blank level in the sample chamber is obtained by collecting those gases liberated when the fluorinating reagent is present in the system (ideally, with accompanying laser heating). Blanks range from $1\,\mu$mol O_2 (Elsenheimer and Valley, 1992) to $<0.02\,\mu$mol O_2 (Mattey and Macpherson, 1993).

After pre-fluorination, F_2, BrF_5 or ClF_3 at 0.1–0.5 bar, is admitted into the chamber. A He-Ne laser is then used to locate the target area on a thick section, or on the appropriate sample powder. The exact conditions required for a successful fluorination of a particular mineral or rock may have to be determined by trial and error. Sharp (1990) noted that powders need to be heated slowly to prevent violent reactions and consequent ejection of materials from the sample holder. As such, a defocused beam was used for the analysis of powdered samples – the laser power was then gradually increased during the analysis (which takes 30–240 s). In the system of Mattey and Macpherson (1993), which is used exclusively to analyse powdered samples and separated mineral grains, the laser beam size is of the order of $250\,\mu$m.

For *in situ* analyses, Elsenheimer and Valley (1992) found that even though a laser beam of $10\,\mu$m size was employed, the reaction pits in thick sections were typically 400–$1200\,\mu$m in diameter. Obviously, the reaction time in this case is an important factor in determining the size of the pits, and ultimately the spatial resolution of the technique. Sharp (1990) used a $100\,\mu$m laser beam for *in situ* analyses (continuous power, 1 s duration). This was found to give a pit of $300\,\mu$m diameter.

Following laser heating, unreacted BrF_5 or ClF_3 is removed from the system by cryogenic means. The oxygen gas is then purified, a process which can be accomplished using a mercury diffusion pump (Sharp, 1990), or KBr at 100°C for 5 min (Mattey and Macpherson, 1993). In the system of Franchi, Akagi and Pillinger (1992), this purified oxygen gas is then admitted to a mass spectrometer for isotopic analysis. Use of oxygen gas enables measurement of both $\delta^{17}O$ and $\delta^{18}O$. In other systems, the oxygen is converted to CO_2 by reaction of the gas with heated graphite (a process which takes several minutes). Unfortunately, it would seem that quantitative conversion of oxygen to CO_2 may not be achieved with small oxygen samples, resulting in erroneous $\delta^{18}O$ values (Mattey and Macpherson, 1993). However, careful appraisal of an individual laser microprobe system should enable appropriate correction factors to be derived.

The magnitude of the oxygen yield obtained from a particular specimen is dependent upon the minerals present. In the system of Sharp (1990), variable yields (80–120%) were obtained from a variety of different minerals. However, variations in yield did not appear to affect the measured $\delta^{18}O$ values. In contrast, Mattey and Macpherson (1993) found that $\delta^{18}O$ values from olivines were strongly dependent upon oxygen yield. In their system, oxygen yields from olivines were lower with a highly focused beam. Garnets, on the other hand, were found to react easily with ClF_3 giving 100% yields, regardless of laser beam diameter or reagent pressure. Clinopyroxenes melt very easily by laser heating, which can be a problem since the fluorination reaction proceeds vigorously, resulting in ejection of partially reacted materials (and thus, resulting in low yields). It seems that high reagent pressures are needed for the analysis of clinopyroxenes. However, orthopyroxenes give more satisfactory yields from low reagent pressures.

When the requirements of high yields are satisfied, the $\delta^{18}O$ values obtained from laser fluorination are good. Sharp (1990) measured $\delta^{18}O$ on <100 μg samples of quartz, feldspar, kyanite, olivine, diopside, garnet, muscovite and biotite to precisions of ±0.1‰. This level of precision is as good as that which can be obtained from a conventional fluorination system using 100 times more sample. Franchi, Akagi and Pillinger (1992), using samples of quartz in the range 500–1000 μg were able to measure both $\delta^{17}O$ and $\delta^{18}O$ to precisions of ±0.17‰ and ±0.15‰ respectively. Mattey and Macpherson (1993) have been able to obtain $\delta^{18}O$ values to precisions of <± 0.2‰ on samples as small as 20 μg, although it should be noted that a very strict operating protocol and correction procedure has to be implemented in order to accomplished this (see data in Figure 9.7). Replicate *in situ* analyses of quartz, magnetite and olivine (Sharp, 1990) gave $\delta^{18}O$ values within ±0.5‰ of the expected result. Similar levels of precision were obtained by Elsenheimer and Valley (1992) on *in situ* analyses of plagioclase.

The first results of oxygen isotope investigations using laser microprobe systems are now available. Schiffries and Rumble (1990) used a laser microprobe to analyse oxygen isotopic zoning in quartz grains from the Transvaal Supergroup. In recrystallized samples there was no apparent difference in $\delta^{18}O$ between cores and rims. In contrast, quartz grains from a low-grade sandstone had rims which were about 1‰ higher in $\delta^{18}O$ compared to cores; this is consistent with observations that quartz overgrowths on sand grains are generally enriched in ^{18}O compared to detrital quartz cores. Sharp (1991) analysed the oxygen isotopic compositions of magnetite in a calcite marble from the Grenville province. This study showed that $\delta^{18}O$ of magnetite varies systematically with grain size. By determining the difference in oxygen isotopic compositions between magnetite and calcite, the formation temperature was derived. Mattey, Macpherson and Harris (1992) and Macpherson, Mattey and Harris (1992) used the laser microprobe to analyse the oxygen isotopic compositions of syngenetic silicate inclusions in diamond samples. It transpires that the $\delta^{18}O$ values of peridotitic inclusions have a rather restricted range of +4.2 to +5.2‰, whereas the $\delta^{18}O$ of ecologitic inclusions is +5.8 to +7.4‰. These data are considered to show that the mantle was heterogeneous with respect to oxygen isotopic composition;

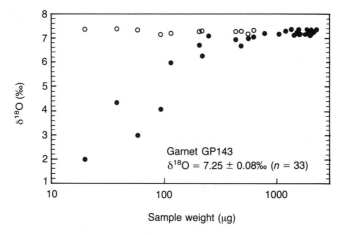

Figure 9.7 Results of oxygen isotope analyses of a garnet by a laser fluorination technique. The plot shows the variation of $\delta^{18}O_{SMOW}$ with sample weight. Note that there is a 'pressure effect', which gives rise to erroneous results below about 200 μg (closed symbols). However, the effect has been found to be reproducible and so a correction can be satisfactorily applied (open symbols). It can be seen that the minimum quantity of material that can be analysed in this case is 20 μg. (From Mattey and Macpherson, 1993).

furthermore, the heterogeneity was established before, or during, diamond formation in Archaean and Proterozoic times.

9.11 Nitrogen isotope analyses

The first attempt to use a laser microprobe to liberate nitrogen for stable isotopic measurement was undertaken by Norris, Brown and Pillinger (1981). In this system, a Nd-glass pulsed laser was used to pyrolyse nitrogen from ~100 μg quantities of lunar samples, contained in sealed quartz capillary tubes. The evolved nitrogen was subsequently released into the inlet of a static mass spectrometer by cracking open the tubes under vacuum. A refinement of the technique was introduced by Franchi *et al.* (1985, 1986) who interfaced the static mass spectrometer to a sample chamber. In this case, gases released by laser heating could be transferred directly to the purification section of the static mass spectrometer system. The obvious advantage of using a sample chamber, rather than capillary tubes, was that large samples (1 × 1 cm) could be analysed using the microprobe to make *in situ* determinations.

Samples are loaded into the sample chamber and then heated at 250°C for several hours to remove adsorbed atmospheric nitrogen. After this procedure, the outgassing rate is generally $<0.01\,\text{ng}\,N_2\,\text{min}^{-1}$. During the laser pyrolysis, various gases are released. Some of these are potentially troublesome for nitrogen isotope measurements; for instance, CO and C_2H_4 can interfere isobarically with N_2 in the mass spectrometer. Before any isotope ratio measurements can be made it is necessary to purify rigorously the nitrogen gas. Purification is accomplished using the techniques described by Boyd et al. (1988). Essentially, all of the carbon- and hydrogen-bearing species are converted to CO_2 and H_2O respectively, and then cryogenically removed, while NH_3 and NO_x are converted to N_2 gas.

To evaluate the laser microprobe technique for nitrogen isotope measurements, Franchi et al. (1986) analysed artificially produced titanium nitride. Individual laser pulses (300 μs duration, 50 μm diameter pits, 10^{10} W cm^{-2} power density) were found to liberate ~150 ng N_2, which is more than enough for analysis using static mass spectrometry. The $\delta^{15}N$ values recorded from five separate laser pulses gave a mean of -5.6 ± 2.1‰, which was within error of the value measured by conventional means (-3.8 ± 2.0‰). Franchi et al. (1986) also used the laser probe to investigate the nitrogen isotopic composition of chromium nitride in iron meteorites. It was necessary to use between 12 and 25 separate pulses to liberate enough nitrogen for measurement (each pulse produced a pit of the order of 100 μm in diameter). $\delta^{15}N$ values ranged from -66 to -76‰, roughly in agreement with the whole-sample value of -76.5‰. The microprobe data showed that most of the nitrogen in this sample is concentrated in nitride minerals.

Franchi et al. (1986) also conducted an investigation into the distribution of nitrogen in a carbonaceous chondrite. When laser pulses were directed into the dark matrix of the meteorite, the liberated gases had a $\delta^{15}N$ of $+42.4 \pm 2.2$‰, i.e. identical to that obtained from measurements of the bulk meteorite. Between five and ten pulses were needed to liberate enough nitrogen from the matrix. When high-temperature inclusions were analysed it was generally necessary to use more than ten pulses. $\delta^{15}N$ values of the inclusions showed a wide range ($+12.4$ to $+60.6$‰), indicating that isotopically distinctive nitrogen is associated with the inclusions.

Some of the potential problems for nitrogen isotope analyses using the laser microprobe were identified by Franchi et al. (1989), who conducted a study of a metal/silicate meteorite known to contain nitrogen of a highly anomalous isotopic composition ($\delta^{15}N$ c. $+1000$‰). Measured $\delta^{15}N$ values from laser probe extraction (6–20 pulses) of the metal fraction varied from $+420$ to $+960$‰, while silicate clasts gave $\delta^{15}N$ values of $+190$ to $+250$‰ (Figure 9.8). It was considered that the lowering of the $\delta^{15}N$ values, over those obtained from whole-sample measurements, was caused by the admixture of adsorbed atmospheric nitrogen. That adsorbed species are present in the sample had been demonstrated previously by Franchi, Wright and Pillinger (1986), using stepped heating. Quite clearly, adsorbed atmospheric nitrogen could be a problem for any investigation, but was only really obvious in this case because of the rather extreme values being investigated. It may be necessary, therefore, to use a variable power, continuous wave

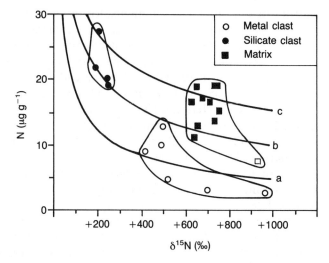

Figure 9.8 Results of laser microprobe analyses of nitrogen from an anomalous meteorite. The data are interpreted in terms of mixing of an indigenous component (δ^{15}N of + 1000‰) with a contaminant component (about 50 ppm N with δ^{15}N of 0‰). The metal clasts scatter along the curve having an assumed 5 ppm of isotopically heavy nitrogen (line a), whereas the silicate clasts and matrix require higher concentrations of the indigenous component of between 10 ppm N (line b) and 15 ppm N (line c). (From Franchi et al., 1989).

laser to provide variable heat to the sample. In this way the laser could be used to carry out a microbeam stepped heating extraction.

An alternative approach to liberating nitrogen (and also carbon and oxygen) from steel samples has been described by Shankai, Conzemius and Svec (1984). In their technique, a Nd-YAG laser microprobe was used to extract ions from the surface of steel samples. These ions were then admitted directly into a mass spectrometer for analysis. Although nitrogen isotope ratios were not measured, quantitative measurements were obtained on samples containing a few tens of parts per million nitrogen, using a laser beam of 15 μm diameter (rastered over a 300 × 300 μm area). It was concluded that the technique would ultimately be able to determine concentrations at the parts-per-billion level.

9.12 Conclusions

Without doubt, the laser microprobe has a future in the stable isotope laboratory. The equipment is not expensive to purchase and can be made to operate safely, quickly and routinely; furthermore, maintenance is fairly minimal. However, before the technique can be made available for use by geochemists at large, there are still a number of problems which need to be explored. Thus, it seems probable that, for the foreseeable future at least, laser microprobes are likely to remain in dedicated research laboratories where stable isotope specialists can continue to document analytical artefacts and refine procedures. Anybody wishing to use a laser microprobe as an adjunct to a stable isotope facility should not be deterred by this, but should be aware that a large amount of effort may be required in validating an individual system before it can be used for geological research. The benefits of microprobe research, however, cannot be overstated.

Acknowledgements

I gratefully acknowledge the help of Dr A.E. Fallick in the writing of this chapter, especially with regard to the application of the laser microprobe to sulphur isotope ratio analysis.

References

Adrain, R.S. and Watson, J. (1984) Laser microspectral analysis: a review of principles and applications. *J. Phys. D: Appl. Phys.*, **17**, 1915–40.

Alt, J.C., Anderson, T.F. and Bonnell, L. (1989) The geochemistry of sulfur in a 1.3 km section of hydrothermally altered oceanic crust, DSDP Hole 504B. *Geochim. Cosmochim. Acta*, **53**, 1011–23.

Bennett, J.N. and Grant, J.N. (1980) Analysis of fluid inclusions using a pulsed laser microprobe. *Mineral. Mag.*, **43**, 945–7.

Boyd, S.R., Mattey, D.P., Pillinger, C.T. et al. (1987) Multiple growth events during diamond genesis: an integrated study of carbon and nitrogen isotopes and nitrogen aggregation state in coated stones. *Earth Planet. Sci. Lett.*, **86**, 341–53.

Boyd, S.R., Wright, I.P., Franchi, I.A. and Pillinger, C.T. (1988) Preparation of sub-nanomole quantities of nitrogen gas for stable isotopic analysis. *J. Phys. E: Sci. Instr.*, **21**, 876–85.

Clayton, R.N. and Epstein, S. (1958) The relationship between O^{18}/O^{16} ratios in coexisting quartz, carbonate, and iron oxides from various geological deposits. *J. Geol.*, **66**, 352–73.

Conzemius, R.J. and Capellen, J.M. (1980) A review of the applications to solids of the laser ion source in mass spectrometry. *Int. J. Mass Spectrom. Ion Phys.*, **34**, 197–271.

Conzemius, R.J., Simons, D.S., Shankai, Z. and Byrd, G.D. (1983) Laser mass spectrometry of solids: a bibliography 1963–1982. *Microbeam Analysis–1983* (ed. R. Gooley), pp. 301–32.

Crowe, D.E. and Valley, J.W. (1992) Laser microprobe study of sulfur isotope variation in a sea-floor hydrothermal spire, Axial Seamount, Juan de Fuca Ridge, eastern Pacific. *Chem. Geol. (Isotope Geosci.)*, **101**, 63–70.

Crowe, D.E., Valley, J.W. and Baker, K.L. (1990) Microanalysis of sulfur-isotope ratios and zonation by laser microprobe. *Geochim. Cosmochim. Acta*, **54**, 2075–92.

Deloule, E. and Éloy, J.F. (1982) Improvements of laser probe mass spectrometry for the chemical analysis of fluid inclusions in ores. *Chem. Geol.*, **37**, 191–202.

Dickson, J.A.D., Smalley, P.C. and Kirkland, B.L. (1991) Carbon and oxygen isotopes in Pennsylvanian biogenic and abiogenic aragonite (Otero County, New Mexico): a laser microprobe study. *Geochim. Cosmochim. Acta*, **55**, 2607–13.

Eldridge, C.S., Compston, W., Williams, I.S. and Walshe, J.L. (1989) Sulfur isotopic analyses on the SHRIMP ion microprobe, in *New Frontiers in Stable Isotopic Research: Laser Probes, Ion Probes, and Small-sample Analysis* (eds W.C. Shanks and R.E. Criss), *US Geol. Surv. Bull.*, **1890**, 163–74.

Elsenheimer, D. and Valley, J.W. (1992) In situ oxygen isotope analysis of feldspar and quartz by Nd:YAG laser microprobe. *Chem. Geol. (Isotope Geosci.)*, **101**, 21–42.

Erëmin, N.I. (1975) Quantitative analysis by means of the laser microanalyser LMA-1. *Mineral. Mag.*, **40**, 312–14.

Fallick, A.E. (1990) High precision sulfur isotope ratio measurements by laser probe mass spectrometry. *Bull. Soc. fr. Mineral. Cristallog.*, **2/3**, 131.

Fallick, A.E., McConville, P., Boyce, A.J. et al. (1992) Laser microprobe stable isotope measurements on geological materials: some experimental considerations (with special reference to $\delta^{34}S$ in sulphides). *Chem. Geol. (Isotope Geosci.)*, **101**, 53–61.

Faure, G. (1986) *Principles of Isotope Geology*. John Wiley, New York, 589 pp.

Franchi, I.A., Akagi, T. and Pillinger, C.T. (1992) Laser fluorination of meteorites – small sample analysis of $\delta^{17}O$ and $\delta^{18}O$ (abstract). *Meteoritics*, **27**, 222.

Franchi, I.A., Boyd, S.R., Wright, I.P. and Pillinger, C.T. (1989) Application of lasers in small-sample stable isotopic analysis, in *New Frontiers in Stable Isotopic Research: Laser Probes, Ion Probes, and Small-sample Analysis* (eds W.C. Shanks and R.E. Criss), *US Geol. Surv. Bull.*, **1890**, 51–9.

Franchi, I.A., Gibson, E.K., Wright, I.P. and Pillinger, C.T. (1985) Nitrogen isotopes by laser probe extraction (abstract). *Lunar Planet. Sci.*, **XVI**, 248–9, Lunar and Planetary Institute, Houston.

Franchi, I.A., Wright, I.P., Gibson, E.K. and Pillinger, C.T. (1986) The laser microprobe: a technique for extracting carbon, nitrogen, and oxygen from solid samples for isotopic measurements. *J. Geophys. Res.*, **91**, D514–24.

Franchi, I.A., Wright, I.P. and Pillinger, C.T. (1986) Heavy nitrogen in Bencubbin – a light-element isotopic anomaly in a stony-iron meteorite. *Nature*, **323**, 138–40.

Gardiner, L.R. and Pillinger, C.T. (1979) Static mass spectrometry for the determination of active gases. *Anal. Chem.*, **51**, 1230–1236.

Giletti, B.J. and Shimizu, N. (1989) Use of the ion microprobe to measure natural abundances of oxygen isotopes in minerals, in *New frontiers in stable isotopic research: laser probes, ion probes, and small-sample analysis* (eds W.C. Shanks and R.E. Criss), *US Geol. Surv. Bull.*, **1890**, 129–36.

Harte, B. and Otter, M. (1992) Carbon isotope measurements on diamonds. *Chem. Geol. (Isotope Geosci.)*, **101**, 177–83.

Hervig, R.L., Thomas, R.M. and Williams, P. (1989) Charge neutralization and oxygen isotopic analysis of insulators with the ion microprobe, in *New frontiers in stable isotopic research: laser probes, ion probes, and small-sample analysis* (eds W.C. Shanks and R.E. Criss), *US Geol. Surv. Bull.*, **1890**, 137–43.

Hillenkamp, F., Unsöld, E., Kaufmann, R. and Nitsche, R. (1975) Laser microprobe mass analysis of organic materials. *Nature*, **256**, 119–20.

Hoefs, J. (1987) *Stable Isotope Geochemistry*. Springer-Verlag, Berlin, 241 pp.

Honig, R.E. and Woolston, J.R. (1963) Laser-induced emission of electrons, ions, and neutral atoms from solid surfaces. *Appl. Phys. Lett.*, **2**, 138–9.

Jones, L.M., Taylor, A.R., Winter, D.L. et al. (1986) The use of the laser microprobe for sample preparation in stable isotope mass spectrometry (abstract). *Terra Cognita*, **6**, 263.

REFERENCES

Kelley, S.P. and Fallick, A.E. (1990) High precision spatially resolved analysis of $\delta^{34}S$ in sulphides using a laser extraction technique. *Geochim. Cosmochim. Acta*, **54**, 883–8.

Kelley, S.P., Fallick, A.E., McConville, P. and Boyce, A.J. (1992) High precision, high spatial resolution analysis of sulfur isotopes by laser combustion of natural sulfide minerals. *Scann. Microsc.*, **6**, 129–38.

Kovalev, I.D., Maksimov, G.A., Suchkov, A.I. and Larin, N.V. (1978) Analytical capabilities of laser-probe mass spectrometry. *Int. J. Mass Spectrom. Ion Phys.*, **27**, 101–37.

Krantz, D.E., Williams, D.F. and Jones, D.S. (1987) Ecological and palaeoenvironmental information using stable isotope profiles from living and fossil molluscs. *Palaeogeogr. Palaeoclimat. Palaeoecol.*, **58**, 249–66.

Kyser, T.K. (ed.) (1987) *Short Course in Stable Isotope Geochemistry of Low Temperature Fluids*. Mineralogical Association of Canada, Short Course Handbook, **13**, 452 pp.

Lincoln, K.A. (1965) Flash-vaporisation of solid materials for mass spectrometry by intense thermal radiation. *Anal. Chem.*, **37**, 541–3.

McKinney, C.R., McCrea, J.M., Epstein, S. *et al.* (1950) Improvements in mass spectrometers for the measurement of small differences in isotope abundance ratios. *Rev. Sci. Instr.*, **21**, 724–30.

Macpherson, C., Mattey, D.P. and Harris, J. (1992) Oxygen isotope analysis of microgram quantities of silicate by a laser-fluorination technique: data for syngenetic inclusions in diamond (abstract). V.M. Goldschmidt Conference, May 8–10, 1992, Reston, Virginia, A-67.

Mattey, D.P. and Macpherson, C. (1993) High-precision oxygen isotope microanalysis of ferromagnesian minerals by laser-fluorination. *Chem. Geol. (Isotope Geosci.)*, **105**, 305–18.

Mattey, D.P., Macpherson, C.G. and Harris, J. (1992) Oxygen isotope analysis of syngenetic silicate inclusions in diamond by laser microprobe (abstract). *EOS, Trans. Am. Geophys. Union*, **73**, 336.

Megrue, G.H. (1967) Isotopic analysis of rare gases with a laser microprobe. *Science*, **157**, 1555–6.

Megrue, G.H. (1971) Distribution and origin of helium, neon, and argon isotopes in Apollo 12 samples by *in situ* analysis with a laser probe mass spectrometer. *J. Geophys. Res.*, **76**, 4956–68.

Nier, A.O. (1947) A mass spectrometer for isotope and gas analysis. *Rev. Sci. Instr.*, **18**, 398–411.

Norris, S.J., Brown, P.W. and Pillinger, C.T. (1981) Laser pyrolysis for light element and stable isotope studies (abstract). *Meteoritics*, **16**, 369.

Pillinger, C.T. (1992) New technologies for small sample stable isotope measurement: static vacuum gas source mass spectrometry, laser probes, ion probes and gas chromatography-isotope ratio mass spectrometry. *Int. J. Mass Spectrom. Ion Proc.*, **118/119**, 477–501.

Powell, M.D. and Kyser, T.K. (1991) Analysis of $\delta^{13}C$ and $\delta^{18}O$ in calcite, dolomite, rhodocrosite and siderite using a laser extraction system. *Chem. Geol. (Isotope Geosci.)*, **94**, 55–66.

Rees, C.E. (1978) Sulphur isotope measurements using SO_2 and SF_6. *Geochim. Cosmochim. Acta*, **42**, 383–9.

Roedder, E. (1984) *Fluid Inclusions*. Mineralogical Society of America, Reviews in Mineralogy, **12**, 644 pp.

Rumble, D., Palin, J.M. and Hoering, T.C. (1991) Laser fluorination of sulfide minerals with F_2 gas. *Annual Report to the Director of the Geophysical Laboratory, Carnegie Institution, Washington, 1990–1991*, pp. 30–4.

Schiffries, C.M. and Rumble, D. (1990) Oxygen isotopic zoning in quartz determined by laser microprobe–isotope ratio mass spectrometry. *Annual Report to the Director of the Geophysical Laboratory, Carnegie Institution, Washington, 1989–1990*, pp. 37–40.

Scott, R.H., Jackson, P.F.S. and Strasheim, A. (1971) Application of laser source mass spectroscopy to analysis of geological material. *Nature*, **232**, 623–4.

Shankai, Z., Conzemius, R.J. and Svec, H.J. (1984) Determination of carbon, nitrogen, and oxygen in solids by laser mass spectrometry. *Anal. Chem.*, **56**, 382–5.

Sharp, Z.D. (1990) A laser-based microanalytical method for the *in situ* determination of oxygen isotope ratios of silicates and oxides. *Geochim. Cosmochim. Acta*, **54**, 1353–7.

Sharp, Z.D. (1991) Determination of oxygen diffusion rates in magnetite from natural isotopic variations. *Geology*, **19**, 653–6.

Sharp, Z.D. (1992) *In situ* laser microprobe techniques for stable isotope analysis. *Chem. Geol. (Isotope Geosci.)*, **101**, 3–19.

Sharp, Z.D. and O'Neil, J.R. (1989) A laser-based carbon reduction technique for oxygen isotope analysis of silicates and oxides. *Annual Report to the Director of the Geophysical Laboratory, Carnegie Institution, Washington, 1988–1989*, pp. 72–8.

Smalley, P.C., Maile, C.N., Coleman, M.L. and Rouse, J.E. (1992) LASSIE (laser ablation sampler for stable isotope extraction) applied to carbonate minerals. *Chem. Geol. (Isotope Geosci.)*, **101**, 43–52.

Smalley, P.C., Stijfhoorn, D.E., Råheim, A. *et al.* (1989) The laser microprobe and its application to the study of C and O isotopes in calcite and aragonite. *Sediment. Geol.*, **65**, 211–21.

Sommer, M.A., Yonover, R.N., Bourcier, W.L. and Gibson, E.K. (1985) Determination of H_2O and CO_2 concentrations in fluid inclusions in minerals using laser decrepitation and capacitance manometer analysis. *Anal. Chem.*, **57**, 449–53.

Tsui, T.-F. and Holland, H.D. (1979) The analysis of fluid inclusions by laser microprobe. *Econ. Geol.*, **74**, 1647–53.

Valley, J.W., Taylor, H.P. and O'Neil, J.R. (eds) (1986) *Stable Isotopes in High Temperature Geological Processes Rev. Mineral.*, **16**, Mineralogical Society of America, 570 pp.

Vanderborgh, N.E. (1977) Laser induced pyrolysis techniques, in *Analytical Pyrolysis* (eds C.E.R. Jones and C.A. Cramers), Elsevier, Amsterdam, pp. 235–48.

Wefer, G. and Killingley, J.S. (1980) Growth histories of strombid snails from Bermuda recorded in their ^{18}O and ^{13}C profiles. *Marine Biol.*, **60**, 129–35.

Wright, I.P. (1984) $\delta^{13}C$ measurements of smaller samples. *Trends Anal. Chem.*, **3**, 210–15.

Zinner, E. (1989) Isotopic measurements with the ion microprobe, in *New frontiers in stable isotopic research: laser probes, ion probes, and small-sample analysis* (eds W.C. Shanks and R.E. Criss), *US Geol. Surv. Bull.*, **1890**, 145–62.

CHAPTER TEN
Micro-Raman spectroscopy in the Earth Sciences

Stephen Roberts and Ian Beattie

10.1 Introduction

When a monochromatic (i.e. single-frequency) beam of light traverses a medium (gas, liquid or solid) the majority of the scattered light will remain at the incident frequency. However, a small proportion of the scattered light will be at changed frequencies, above and below the incident frequency, and this is referred to as the Raman effect. The Raman effect was first observed by Raman and Krishnan (1928) using focused sunlight and filters and relied on the visual observation of colour changes in the scattered light. However, it was not until the advent of continuous wave visible lasers, during the 1960s, that the importance of Raman spectroscopy as a routine analytical technique was realized. Furthermore, the availability of this highly intense monochromatic light source, which could be focused to a narrow waist, allowed the analysis of small volumes of gas, liquid or solid.

Today several instrument manufacturers produce Raman instruments, with microscope attachments, which enable the routine application of Raman spectroscopy to small samples such as fibres or dust particles. Yet, within the Earth Sciences, micro-Raman spectroscopy represents a relatively new analytical technique, which to date is only established in a few geological laboratories. Nevertheless, the use of light to probe the vibrational behaviour of molecular systems is undoubtedly a 'growth area'. Particularly attractive aspects of the Raman technique include the small spot size, relative lack of sample preparation and the normally non-destructive nature of the analysis. These attributes have resulted in a variety of geological applications of which examples include the characterization and structural study of minerals, and the discrimination of the calcium polymorphs involved in the construction of foraminiferal tests.

10.2 Principles

If a gas, liquid or solid is illuminated with a monochromatic light source, most of the light will pass through the sample without undergoing any

Microprobe Techniques in the Earth Sciences. Edited by P.J. Potts, J.F.W. Bowles, S.J.B. Reed and M.R. Cave. Published in 1995 by Chapman & Hall, London. ISBN 0 412 55100 4

change (Figure 10.1). However, a small proportion will be scattered by the sample. Measurement of this scattered light by a spectrometer reveals that about 10^{-3} of the incident intensity has been scattered with the same frequency (v_0) as that of the incident light source. The process by which light is scattered at the same frequency as that of the incident beam is often called elastic or Rayleigh scattering.

In addition to the Rayleigh scattering, about 10^{-6} of the incident intensity is scattered at new frquencies above ($v_0 + \Delta v$) and below ($v_0 - \Delta v$) the incident frequency. This is referred to as Raman scattering or the Raman effect. The shifts in frequency ($\pm \Delta v$) from that of the incident radiation are independent of the exciting radiation v_0 and are characteristic of the species which gives rise to the scattering.

Another straightforward way to consider the Raman effect is through an energy level diagram (Figure 10.2). Here an incident photon (v_0), interacts with a vibrating molecule and is annihilated. As a result of this process a new photon at a lower frequency (v_r) than that of the incident photon is created, and the molecule undergoes a Raman active vibrational (rotational or, unusually, electronic) transition v_R to a higher energy level (Figure 10.2(a)). The difference in frequency ($v_0 - v_r = v_R$) is a vibrational frequency of the molecule under study. Alternatively, if the molecule is in an excited state (Figure 10.2(b)) and undergoes a transition to the ground state during the Raman process then ($v_r - v_0 = v_R$). This time the scattered photon (v_r) is shifted to high frequency, that is the photon is blue shifted relative to the incident beam.

Raman transitions to lower frequency, i.e. red shifted bands (Figure 10.2(a)), are referred to as Stokes lines, whereas transitions to higher frequency, blue shifted (Figure 10.2(b)), are referred to as anti-Stokes lines. Stokes lines are normally much more intense than anti-Stokes lines, as the population of the ground state is usually very much greater than that of the excited state of the molecule. However, this is not true if experiments are carried out at high temperatures or where Raman frequencies are very close

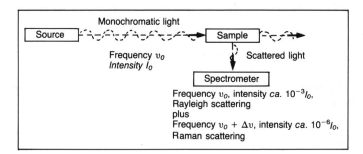

Figure 10.1 Schematic outline of a simple Raman experiment. The monochromatic light source is normally a laser operating in the visible region, typically at 488 or 514 nm.

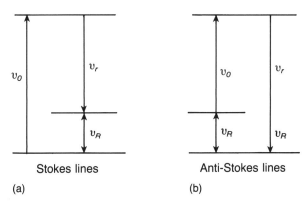

Figure 10.2 Energy-level diagram of the Raman effect. See text for explanation of the notation used.

to the incident frequency so that v_R is very small. The contrasting intensity of Stokes and anti-Stokes lines is clearly illustrated in Figure 10.3, which shows a Raman spectrum of CCl_4.

Not all molecular vibrations are Raman active, since the condition for obtaining a Raman spectrum for a given molecule depends on a change in the polarizability of the molecule during a vibration. When an oscillating electric field is applied to a molecule, the electrons migrate to follow the field. An incident beam of light has an oscillating electric field perpendicular to the direction in which the beam is propagating. Classically, the induced oscillating dipole of the molecule, or induced polarization (P) (i.e. the ease with which the electron cloud is deformed) is proportional to the applied field (E):

$$P = \alpha E \qquad (10.1)$$

where α is the polarizability. This polarizability must change during a molecular vibration if a band is to be Raman active. The polarizability of a molecule can be represented by a triaxial ellipsoid. Consider the linear molecule CO_2, with its three fundamental vibrations illustrated in Figure 10.4(a). For the symmetrical stretching vibration of CO_2 (Figure 10.4(a)(1)) the molecule clearly changes shape and so does its polarizability from (I) to (II) to (III) (Figure 10.4(b)). If we plot a graph of polarizability of the molecule (α) against some vibrational coordinate (Q) (Figure 10.4(c)) we can see that the change in shape results in a change in polarizability where the curve cuts the axis. Thus $(\partial \alpha / \partial Q) \neq 0$, and the vibration is therefore Raman active. If however, we consider the antisymmetric stretch (Figure 10.4(a)(2)), although the molecule changes shape during the vibration the two 'end-member' vibrational configurations give identical polarizability

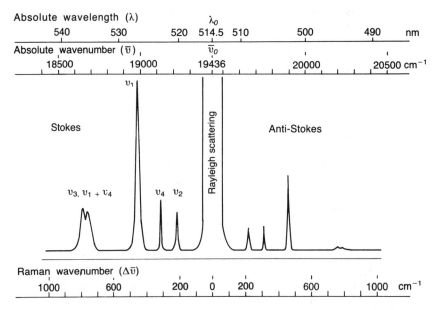

Figure 10.3 Raman spectrum of CCl_4 showing the more intense Stokes lines. Various units of light measurement are shown for comparison; relative wavenumbers (cm^{-1}) are most commonly employed as these are independent of the exciting radiation. The most intense band in the spectrum (v_1) is where the carbon remains stationary and the four chlorines move in and out in sinusoidal motion simultaneously in phase. This is sometimes called the 'breathing frequency' for obvious reasons.

ellipsoids (Figure 10.4(b) (IV and VI)). Thus, in this instance a plot of polarizability against the vibrational coordinate results in $(\partial a/\partial Q) = 0$ and the vibration is Raman inactive (Figure 10.4(c)). A similar argument would apply for the bending vibration (Figure 10.4(a)(3)). Although the bending vibration is Raman inactive, it is infrared active as the bending involves a change in the dipole moment of the molecule. Similarly, the antisymmetric stretching vibration is also infrared active. For infrared activity it is the change in the dipole that is important as the molecule need not have a permanent dipole. This contrasting infrared and Raman active nature of the vibrations serves to demonstrate the complementary nature of the two techniques. Thus, in the above example, only the symmetrical CO_2 vibration causes a change in the polarizability of the molecule and is Raman active (but is infrared inactive). It is generally found that symmetric vibrations tend to give rise to intense Raman lines whereas non-symmetric vibrations are usually weaker.

Nishimura, Hirawaka and Tsuboi (1978) have summarized some common observations about Raman spectral intensities into three main generalizations.

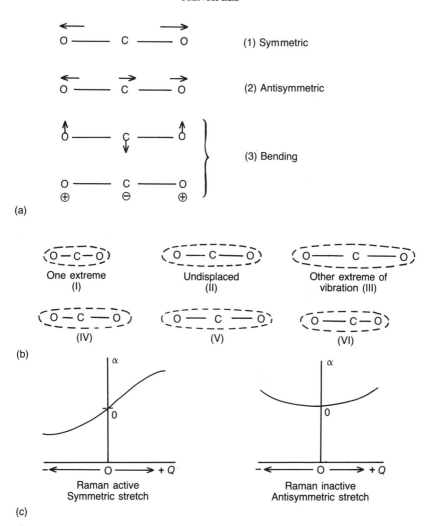

Figure 10.4 (a) Fundamental vibrations of the CO_2 molecule. (1) symmetric stretch, (2) antisymmetric stretch, (3) bending vibration. (b) Schematic outline of polarizability ellipsoids associated with the antisymmetric and symmetric stretch of the CO_2 molecule. (c) Polarizability (α) against a displacement vector (Q), for the Raman active symmetric stretch and the Raman inactive antisymmetric stretching vibration.

1. Stretching vibrations associated with chemical bonds should be more intense than deformational vibrations.
2. Multiple chemical bonds should give rise to more intense stretching modes, e.g. Raman lines due to a C=C vibration should be more intense than a C—C vibration.

3. Bonds involving atoms of large atomic mass are expected to give rise to stretching vibrations of high Raman intensity.

A vibrational spectrum is characteristic of a given sample and as individual peaks may be associated with the presence of particular structural groups within the sample, the spectra may be used to infer the presence of a particular phase or certain molecular groups. Further information can sometimes be obtained during Raman experiments if the degree of depolarization of the scattered light can be determined. The ability of minerals, e.g. calcite, to polarize normal light is familiar to geologists. The interaction of highly polarized laser radiation with molecules during Raman spectroscopy gives spectra with lines which are found to be polarized to differing extents; for example, the breathing frequency of CCl_4 (v_1) gives a line which is polarized exclusively in the direction of polarization of the incident laser beam. This information can be important in assigning bands to particular vibration modes. The degree of depolarization can be estimated by observing intensity variations when a piece of polarizer is rotated in the Raman beam. It also provides a method for 'removing' polarized bands from a spectrum which may contain overlapping bands.

10.3 Instrumentation

To complete a Raman experiment, the essential components required are a light source, a sample point, collection optic or optics, a dispersing optical element and a detection system system (Figure 10.5). The principal design criteria behind this instrumentation are the inherent weakness of the scattering phenomenon and the necessity to maximize the number of detected photon events.

To produce a detectable number of Raman scattered photons an intense light source is required. Continuous wave lasers offer sufficient power levels in the visible region of the spectrum. This property is important since the Raman scattering cross-section varies as the fourth power of the frequency. The laser source (having a typical output power of between 0.01 and 1 W) can be focused on the sample, either within a 'macro-chamber' or by a conventional microscope. With micro-Raman spectoscopy, the laser beam is directed into the microscope and onto a semi-reflecting mirror (beam splitter) (Figure 10.5) and a portion of the incident radiation is reflected down towards the sample. This reflected beam then passes through the microscope objective lens which focuses the beam onto the sample. It should be noted that when a laser has been focused through a microscope objective, 1 W of laser power focused to a $2\,\mu m$ spot size gives an intensity of the order of $10^8\,W\,cm^{-2}$ which in certain instances can result in damage to the sample. The scattered beam containing the Raman signal is then collected back

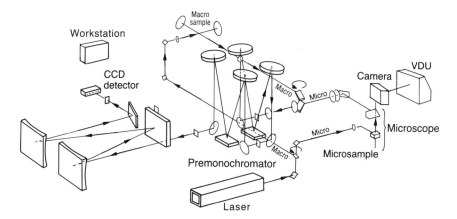

Figure 10.5 Schematic outline and optical diagram of the Jobin Yvon S3000 Raman microscope.

through the objective and is transmitted through the beam splitter and directed to the coupling optics at the entrance of the spectrometer. The coupling optics generally consist of an achromatic lens which refocuses the light onto the entrance slit of the monochromator.

The monochromator is required to disperse collected light across the exit slit for sequential presentation to a detector, or to disperse it across an array detector. In the instrument illustrated in Figure 10.5, a non-dispersive double monochromator rejects stray light prior to dispersion within the spectrometer. Since the scattered light also contains the elastically scattered (Rayleigh) component, the monochromator must be able to discriminate effectively against this large signal. The major problem associated with these systems is transmission loss owing to the many optical elements. In the system mllustrated, the spectrometer disperses the collected light across a multichannel detector (Figure 10.5).

Detector systems can be categorized as single and multichannel systems. Single-channel detection normally involves photon counting using photomultiplier tubes and is frequently applied where high resolution is required. Detector technology has developed rapidly during the past few years with multichannel detectors including diode array systems and, more recently still, multichannel charge coupled device (CCD) detectors. CCDs are integrated circuits within which discrete potential wells are capable of storing electrons created by incident photons. Multichannel detectors allow the simultaneous collection of data over a broad bandpass (typically $>400\,\text{cm}^{-1}$) and the extremely low noise of a cooled CCD allows for long integration times, which is a particularly attractive feature when measuring weak Raman signals.

10.4 Sample handling and routine operation

Sample preparation for micro-Raman spectroscopy can be extremely simple, the most convenient sample support being an ordinary glass microscope slide upon which the sample is placed, so that individual particles can be selected and brought into focus. By appropriate choice of high magnification and/or long working distance objectives, a range of sample sizes and shapes can be accommodated and analysed successfully.

Frequency calibration of the instrument is most readily achieved by checking the frequency of plasma emission lines from the laser source. During analysis, these lines are removed either by a filter or by the use of a laser pre-monochromator. Frequency response calibrations are less straightforward, with each of the various component parts of the instrument, optical elements (notably gratings), detector chip efficiency and linearity all contributing to the final spectrum. Clearly, standards need to be run which permit data from different laboratories to be compared. Typically, a standard sample such as sulphur or indene, both of which are good Raman scatterers, provides information on the relative intensity of the peak positions monitored. Alternatively, light from a quartz-tungsten halogen lamp may be introduced into the system for calibration.

The resolution of the instrument is dependent on the particular arrangement employed and is affected by the choice and combination of slit widths, gratings and use of mono- or multichannel detectors. In a multichannel system with a microscope, the instrument resolution (typically $1.0\,\text{cm}^{-1}$) is largely determined by the groove density of the grating employed and the separation of individual pixels within the detector chip.

In addition to 'routine' operation, Raman spectrometers are well suited to the attachment of variable temperature and/or pressure cells. For example, a heating/freezing stage can be readily attached which allows for the *in situ* measurement of substances through a range of temperatures. Similarly, diamond anvil cells may be fitted, allowing the effects of pressure to be monitored.

10.5 Advantages and disadvantages of the Raman technique

The principal advantages of the technique include the following.

1. Each scattering species gives its own characteristic vibrational spectrum, which can be used as a fingerprint for qualitative identification. In ideal circumstances, the intensity of the Raman scattering is proportional to the concentration of the scattering species. Both structural and compositional information can be obtained for the analysed material.
2. Relative lack of interference, with the vibrational bands of many molecules well separated and often narrower than the corresponding infrared absorption bands.

3. Spatial resolution. By using a microscope with high-power objectives (routinely ×100) spot sizes of down to $2\,\mu m$ are attainable.
4. No sophisticated sample preparation techniques are required and all phases – gas, liquid or solid, large or small – can be analysed, providing the host matrix transmits the incident and Raman radiation.

By contrast, the principal disadvantages of the technique include the following.

1. Inherent weakness of the Raman effect.
2. Fluorescence. As a phenomenon, fluorescence is 10^6–10^8 times stronger than the Raman radiation. Thus, in the presence of fluorescence, the much weaker Raman signal may prove impossible to detect.
3. Coloured samples may absorb the laser beam resulting in the heating of the sample which may consequently decompose.
4. In the region of an absorption band of the sample, the intensity of certain vibrations may be enhanced by orders of magnitude. Further, additional frequencies at $2v_R$, $3v_R$ and $4v_R$ may be observed. This is known as the resonance Raman effect which, although a disadvantage for quantitative work, has been used to detect S_2^- and S_3^- in ultramarine (Clark and Franks, 1975). Whenever a sample is coloured, resonance effects may be observed and to obtain quantitative data it is essential to obtain spectra with several excitation lines separated as widely as possible.

10.6 Some applications of micro-Raman spectroscopy in the Earth Sciences

10.6.1 Micro-Raman analysis of fluid inclusions

As crystals grow, they invariably contain imperfections in the form of occluded liquids, solids and/or vapours. Within a geological context, analysis of this trapped material can provide vital information to understand the physical and chemical conditions prevalent at the time of crystal growth. Unfortunately, the inclusions are typically rather small, of the order of a few tens of micrometres, and, as a crystal may endure a long residence time within the Earth, more than a single generation of inclusions may be present. Thus, an analytical method capable of analysing individual inclusions would be highly desirable.

As Raman spectroscopy is carried out within the visible region of the spectrum, the incident laser beam can be focused by normal light optics to give spatial resolution in the region of micrometres. It therefore provides a rapid, non-destructive means of analysing the molecular species of fluid inclusions (Delhaye and Dhamelincourt, 1975; Rosasco, Roedder and Simmons, 1975; Dhamelincourt et al., 1979). This technique has proven to be particularly successful for the qualitative analysis of species such as CO_2, CO, CH_4, C_2H_6, N_2, H_2O, H_2S, HS^-, O_2, SO_4^{2-}.

Characterization of palaeofluids, observed in fluid inclusions, has enabled significant advances in our understanding of ore genesis and the P–T evolution of metamorphic terranes. The melting or dissolution points of solids measured during microthermometric analysis provide information on the dissolved gas species in a given system. For example, the triple point of pure CO_2 is significantly depressed to lower temperatures in the presence of additional volatiles such as CH_4, N_2 and H_2S. However, microthermometric data alone do not permit either unambiguous identification or quantification of the additional species present. Micro-Raman is the technique that offers a non-destructive method capable of resolving these additional species and has evolved into a routine tool for fluid inclusion analysis. An example of a two-phase aqueo-carbonic fluid inclusion, and its Raman spectrum indicating the presence of CO_2 and CH_4, is shown in Figure 10.6.

Quantitative analysis of these species is possible, although great care must be taken over the analysis and interpretation of the results. Clearly, an inclusion within a substrate does not constitute anything like the ideal optical system for quantitative measurements. Not only will the refractive

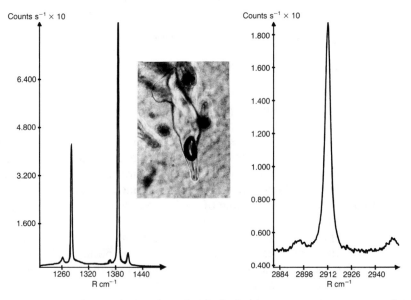

Figure 10.6 Raman spectra of carbon dioxide (with peaks at ~1280 and 1385 cm^{-1}) and methane (at about 2917 cm^{-1}) obtained from a two-phase aqueo-carbonic inclusion within quartz from an auriferous vein, a photomicrograph of which is shown in the insert.

index of the substrate be dependent on the wavelength of the exciting light but, in the region of an absorption band, attenuation of the incident beam or Raman radiation may also occur, together with significant changes in refractive index. Furthermore, the spectra of the constituents of the inclusions will be dependent on the physical conditions within the sample. Frequency shifts, changes in peak half-widths and a change of scattering cross-section are inevitable consequences of factors such as pressure, change of phase (including solution of gases in liquids), interaction with other species etc. Apart from the nature of the substrate, additional factors that will affect the accurate measurement of intensities include grating transmission and detector response, which are frequency (and hence excitation frequency) dependent. Factors which affect quantification, accuracy and precision of the analysis of gas mixtures of geological interest have been discussed by Wopenka and Pasteris (1986) and Pasteris, Wopenka and Seitz (1988).

Polyatomic ions such as CO_3^{2-}, NO_3^-, HCO_3^-, SO_4^{2-}, PO_4^{3-} and HS^- exhibit Raman spectra and can be detected within inclusions at concentrations greater than 1000 ppm (Dubessy et al., 1982). Monatomic ions such as Na^+, K^+, Ca^{2+} and Mg^{2+} cannot be detected in inclusions. However, salt hydrates of various cationic species do show characteristic spectra (Dubessy et al., 1982) and these can be used to identify the major cation species present, a technique which involves cooling the sample under the Raman microscope to cause crystallization. Another approach is to study the O—H stretching region (2800–3800 cm^{-1}) of aqueous solutions as this region is sensitive to changes in salinity (Georgiev et al., 1984). Systematic changes in spectral data permit the determination of the salinity of fluid inclusions at room temperature from Raman microprobe spectra (Mernagh and Wilde, 1989).

Another aspect of fluid inclusion research, facilitated by the ability to obtain spectra at differing temperatures and pressures, has involved the analysis of gas hydrates (clathrates) during microthermometric analysis. Raman analysis of clathrates enables estimates of the variable partitioning of gas species during the freezing of aqueo-carbonic inclusions (Seitz, Pasteris and Wopenka, 1987). The results of such studies are crucial to our interpretation of microthermometric data and suggest that if these data are not taken into account, significant errors can be made in the estimate of additional volatile species present.

A further application of micro-Raman in fluid inclusion research involves the identification of 'daughter' and trapped mineral phases within fluid inclusions, which cannot be measured by conventional microbeam techniques. In certain instances, these phases can be characterized by Raman spectroscopy. Carbonate and sulphate daughter minerals and a variety of silicate minerals can be identified by comparison with suitable standard spectra.

10.6.2 Microanalysis of carbon

The Raman analysis of carbonaceous material provides a good example of the molecular and structural information which vibrational spectra can offer. As Raman scattering of carbon material is sensitive to the structure within the sample, micro-Raman provides a useful non-destructive technique for characterization of carbon materials.

Tuinstra and Koenig (1970) demonstrated that high-quality single crystals of natural graphite show a single Raman line at $1575\,\text{cm}^{-1}$. However, carbonaceous material shows an additional line at $1355\,\text{cm}^{-1}$, whose intensity increases as a function of the amount of 'unorganized' carbon in the sample and a decrease in the graphite crystal size. Thus, an evaluation of the crystallinity in terms of the (001) in-plane crystallite size of carbonaceous material in natural samples is possible.

As organic matter is metamorphosed, chemical changes occur which involve the driving off of volatiles which include hydrogen, nitrogen and oxygen and cause the resulting material to become enriched in carbon. In conjunction with these changes, the basic structural units, which make up the material and which are randomly ordered at low temperatures, become increasingly ordered (Beny-Bassez and Rouzaud, 1985). Pasteris and Wopenka (1991) developed this work into a potential geothermometer by comparing carbonaceous material from a variety of metamorphic grades. Variations in the intensity ratio of the $1360/1572\,\text{cm}^{-1}$ bands (Figure 10.7(a)) could be correlated to a change of grade (Figure 10.7(b)). Recent analysis of Chitinozoa (organic-walled microfossils) suggest that the poor correlation evident amongst the low rank material may be the result of maturation following a power law curve, an observation which is in agreement with spectra determined for a variety of carbonaceous materials by Beny-Bassez and Rouzaud (1985).

Work on carbonaceous material also reveals two practical aspects which arise during analysis of single crystals and dark materials. First, the spectroscopist should be aware of any orientation effects which may result in dramatic changes in the relative intensities of certain bands (Wang *et al.*, 1989), as observed in the case of graphitic material (Figure 10.8). Second, dark and coloured samples can readily absorb the laser beam, resulting in rapid heating and possible decomposition of the sample. In the studies outlined above, *in situ* maturation of the sample under the laser beam can be achieved if the laser power is not kept to a minimum. Whenever a sample is coloured (and this includes black), resonance Raman effects may be observed.

10.6.3 Raman spectroscopy of minerals

The potential of Raman spectroscopy for the identification and structural analysis of minerals has long been recognized (Griffith, 1974; White, 1974;

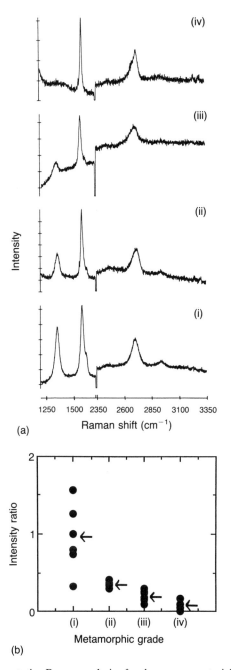

Figure 10.7 (a) Representative Raman analysis of carbonaceous material from four contrasting metamorphic grades. (i) chlorite zone, (ii) garnet zone, (iii) staurolite zone and (iv) sillimanite zone. (b) Plot of intensity ratio of Raman peaks $I(\sim 1360\,\mathrm{cm}^{-1}):I(1582\,\mathrm{cm}^{-1})$ of graphite samples grouped according to their spectrally recognized metamorphic categories. Numerals (i)–(iv) refer to metamorphic grade, as for Figure 10.7(a). (Both diagrams after Pasteris, and Wopenka, 1988).

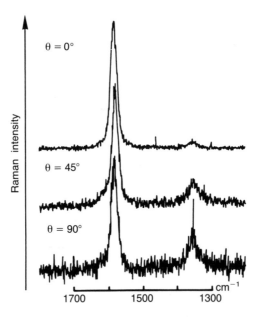

Figure 10.8 Evolution of the first-order Raman spectrum of a graphitic compound as a function of the angle θ between the c-axis of the crystal and the optical axis of the microscope. (From Wang et al., 1990).

McMillan, 1985; McMillan and Hofmeister, 1988). Raman spectra of minerals can provide important information on the type of chemical bonding between atoms and on the presence of specific compounds within a sample. However, the situation for a crystal is complicated in comparison to a gas and to a lesser extent a liquid, because we are dealing with an extended array in three dimensions. The carbonates are used as the main example in this brief outline as they represent one of the most abundant minerals in the Earth's crust and tend to be studied by the whole gamut of Earth Science specialists, from micropalaeontologists, interested in the construction of calcareous tests, through to metamorphic and experimental petrologists, interested in the phase transitions/high-pressure polymorphs which calcite demonstrates.

Any crystal can be characterized by the primitive cell, this being the smallest unit which, if repeated in three dimensions, would generate the crystal. For infrared and Raman activity to be observed, all atoms related by a primitive translation must move identically. Identical atoms which are related by a translation along or parallel to one of the axes of the primitive cell, and are separated by a distance equal to that cell edge, are 'primitively related'. In addition (as with discrete species), if the structure has a centre of symmetry, only vibrations which preserve that centre can be Raman active. Although this may seem complicated, the value of such an approach can be seen in Figure 10.9, where the primitive cell of caesium chloride is shown together with the only possible vibration. This vibration can occur in the x, y or z direction, and obviously all have the same frequency; such a vibration is said to be triply degenerate. If the symmetry is lowered from cubic to orthorhombic then the x, y and z directions are no longer equivalent and three frequencies result instead of the one observed for cubic symmetry. These vibrations are referred to as lattice modes as they result from translational movements of the atoms/ions within the crystal lattice. Note that the vibration shown in Figure 10.9 is infrared active as translational movements generate an oscillating dipole. Thus, caesium chloride does not have a first order Raman spectrum. In general, lattice vibrations occur to low frequency except in the case of very light atoms such as Be, O, Li and F.

If we now consider a material such as calcite ($CaCO_3$), then vibrational modes can be categorized as internal modes, here internal to the CO_3^{2-} ion, and external modes based on Ca^{2+} and the CO_3^{2-} (considered as a unit). As we have seen, the lattice or external modes tend to be at much lower frequencies than the vibrational frequencies of the carbonate ion, especially those based on stretching vibrations. Carbonates of divalent cations of intermediate size usually crystallize in the calcite structure, with two CO_3^{2-} ions in the primitive cell. Placing carbonate ions in the calcite crystal will clearly perturb the vibrations to some extent. The first feature that may occur in the crystal is that the crystal symmetry of the carbonate ion sites may be lower than that of the free ion. This effect can cause resolution of the degeneracy of a vibration so that v_3 and v_4 (the anti-symmetric stretch and in-plane bend) could appear as doublets, not single lines. In the case of CO_3^{2-}, the free ion has a threefold axis of symmetry which is retained in calcite. In this case, the degeneracy is not removed and from this point of view the CO_3^{2-} should display three Raman active fundamentals, of which v_1

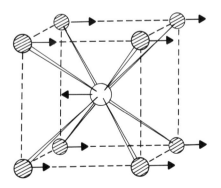

Figure 10.9 Primitive cell and vibrations of caesium chloride.

(symmetric stretch) will be by far the most intense. A complication is that because there are two CO_3^{2-} in the primitive cell, each vibration can couple in-phase and out of phase. This is illustrated for the v_1 vibration in Figure 10.10. Note that this gives one Raman active band.

Taking into account these vibrational modes, the Raman spectrum of calcite (Figure 10.11) has bands at $1088\,cm^{-1}$ (the very intense breathing frequency in which the carbon atom does not move); $1443\,cm^{-1}$ (clearly the antisymmetric stretching mode, the high frequency being due to the movement of the light carbon atom); and $714\,cm^{-1}$ (a low-frequency mode based on angle deformation). These vibrations are thus assigned to v_1, v_3 and v_4 respectively, v_2 being Raman inactive. Note that CO_3^{2-} unlike CO_2 does not have a centre of symmetry. The low-frequency vibrations at 159 and $286\,cm^{-1}$ are lattice modes.

Turning now to carbonates of divalent elements of large ionic radius which crystallize in the aragonite structure (having orthorhombic symmetry). There are now four molecules in the unit cell and the site symmetry is very low – one mirror plane only passes through the carbon atoms. The loss of the threefold axis means that any degeneracy will be resolved, and it can be predicted that doublets for the degenerate vibrations (v_3 and v_4) will be present in the Raman spectra. Depending on the magnitude of this splitting, the doubling may or may not be observed. Further, because the crystal structure is so different, the lattice vibrations will be radically changed. The distinction between aragonite and calcite using micro-Raman techniques will be explained in the following section.

(a) Phase transformations

Although aragonite should be the stable high-pressure form of calcite (Jamieson, 1953), calcite invariably transforms to calcite-II and calcite-III at

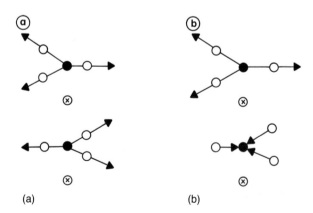

Figure 10.10 (a) In-phase and (b) out-of-phase coupling of carbonate modes within the calcite unit cell. (After White, 1974).

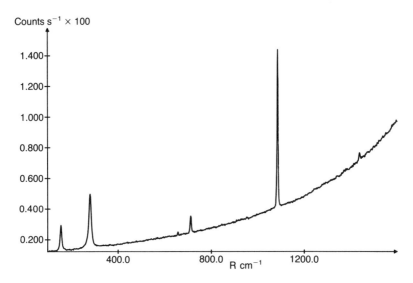

Figure 10.11 Raman spectrum of a calcite crystal. See text for explanation of the bands observed.

high pressures and at temperatures below 600°C. Calcite inverts to calcite-I at around 1.5 Gpa and from calcite-II to III at around 2.0 Gpa (Fong and Nicol, 1971). Good Raman spectra of calcite-II and III have been reported by Fong and Nicol (1971) and Liu and Mernagh (1990); the results of Fong and Nicol (1971) are illustrated in Table 10.1. It can be seen that all frequencies increase at high pressures, as expected from compression of the crystal. Superimposed on the blue-shifted vibrations are other notable spectral effects, including the splitting of the v_3 and v_4 bands in calcite-II and v_1 band in calcite-III. These changes are reversible and disappear when the sample is returned to atmospheric pressure. The presence of a v_2 vibration within the Raman spectra is probably due to the pressure-induced distortion of the parent calcite structure (White, 1974).

These experimental data confirm the possibility of stable carbonate minerals within the upper mantle and therefore provide a means by which carbon may be retained within this region of the Earth. These data have implications for the genesis of carbon-containing magmas and thus an understanding of the high pressure stability and compressional behaviour of these phases is of considerable geological importance.

(b) Recognition of carbonates with aragonite structure
As noted earlier, carbonates of divalent elements of large ionic radius crystallize in the aragonite structure. Micro-Raman has been used to distinguish calcite from aragonite unambiguously and rapidly in samples prepared as standard petrographic thin sections. The two polymorphs show

Table 10.1 Raman spectra of calcite II and calcite III (cm^{-1})

Modes	Calcite-II 14 kbar	Calcite-III 18 kbar	Calcite-III 38 kbar
v_1	1096	1104	1108
		1087	1093
v_2	866	870	870
	1471	1540	1540
v_3	1445	1515	1520
	721	741	745
v_4	715	733	739
Lattice modes			
	–	–	695
	–	333	351
	319 (p)	314 (p)	318 (p)
	292	299	312
	240	269	270
	–	–	234 (p)
	–	221	224 (p)
	–	208 (p)	224 (p)
	–	204 (n)	206 (n)
	204	202 (p)	205 (p)
	155	161	173
	133	137	139
	–	131	–
	–	105	109
	99	99	99

All data from Fong and Nicol, 1971. The notation: (p) refers to crystals oriented with the optical axis parallel to the pressure axis. (n) refers to natural orientation in which the thrust of the pistion was against a cleavage face.

contrasting low-frequency (<300 cm^{-1}) lattice vibrations: aragonite (~150, 180, 190, 206 cm^{-1}); calcite (~153, 283 cm^{-1}). In metamorphic rocks aragonite is representative of blueschist metamorphic conditions, usually occurring with jadeite, quartz and lawsonite, which confirms its stable, high-pressure, low-temperature origins. Raman analysis has been employed to confirm the presence of micrometre-sized crystals of aragonite and therefore contribute to an evaluation of the $P-T$ evolution of a metamorphic terrane (Gillet and Goffe, 1988).

At the other end of the metamorphic spectrum, the use of micro-Raman to distinguish aragonite–calcite on a very small spot size has resulted in a detailed description and classification of foraminiferal tests. For example, Raman spectra have been used to confirm the calcite nature of *Ammonia beccarii* compared to the aragonite nature of *Hoeglundina elegans* (Venec-Peyre and Jaeschke-Boyer, 1979).

10.7 Other applications of micro-Raman spectroscopy in the Earth Sciences

As remarked in the introduction, Raman and micro-Raman spectroscopy is a growth area in the Earth Sciences with a burgeoning literature and a wide variety of applications. This section now very briefly describes other applications which have produced, or are in the process of producing, significant contributions to our understanding of key processes in the Earth Sciences.

Raman spectra have been used extensively in the structural studies of silicate glasses and melts. Spectra from glasses of variable composition have been assigned to various structural groups based on observed band position and symmetry character, examination of systematic changes in spectra with composition and corresponding glass and crystal spectra; studies which have been excellently reviewed by McMillan (1984(a)).

One goal of the research into silicate glasses has been to attempt to rationalize properties such as viscosity, density and heat content of magmas as a function of temperature, pressure and composition. For example, it has been noted that the isothermal viscosity of silicate liquids and glass decreases as alkali and alkali-earth oxides are added to pure silica, an observation interpreted in terms of depolymerization of the silicate melt or glass network (Matson, Sharma and Philpotts, 1983; McMillan, 1984(b)). Other workers (Mysen, Virgo and Seifert, 1982; Seifert, Mysen and Virgo, 1982) have shown that Raman spectra can be related to changes in the bulk physical properties of silicate glasses and melts, which may not be quenchable in certain instances (Hemley *et al.*, 1986). For example, Kubicki, Hemley and Hofmeister (1992) showed that the dominant change in silicate glasses with increased pressure is a major shift in the mid-frequencies ($625-650\,\text{cm}^{-1}$) which is reversible. They concluded from the Raman data that the main compression mechanism is a decrease in the average inter-tetrahedral Si–O–Si angles.

Raman spectra of H_2O, CO_2 and H_2 have all been reported for silicate glasses and have enabled insights into the volatile dissolution mechnisms in melts (Mysen and Virgo, 1980, 1986; McMillan, 1984(a), 1989, for review).

Owing to the weak Raman spectrum of H_2O, Raman offers an ideal analytical technique with which to study solutes in hydrothermal solutions, especially given the increasing availability of $P-T$ cells suitable for the collection of Raman spectra at pressures and temperatures well above ambient. For example, the behaviour of gold in chloride-bearing hydrothermal systems is of considerable interest to economic geologists and hydrothermal geochemists. Recently, Pan and Wood (1991) were able to provide direct spectroscopic evidence for the existence of $AuCl^-$ in hydrothermal solutions. Moreover, Peck *et al.* (1991), using Raman and resonance Raman techniques, showed that with increasing pH, chloride ligands are successively replaced by hydroxide ligands, and thus reported for the first time the potential importance of mixed chloro-hydroxo species in a $P-T$ range over which most epithermal Au deposition is believed to take place.

10.8 Other Raman techniques and closing remarks

In addition to Raman spectroscopy as described above, an array of Raman-based technologies now exists which have yet to find their way into the geological laboratory to a significant extent. Amongst these are included Fourier transform (FT)-Raman, hyper-Raman and coherent anti-Stokes Raman (CARS) (see McMillan and Hofmeister, 1988, for more details). Of these, FT-Raman appears the most attractive to the geologist as it offers a potential major advantage over conventional dispersive Raman spectroscopy in its ability to give spectra which are effectively free of fluorescence interference. Commercial micro-FT-Raman systems are now available.

In closing, although Raman spectroscopy remains a relatively underemployed technique in the geological laboratory, this situation will undoubtedly change in the near future as the potential of the technique to contribute to a wide variety of problems within the Earth Sciences becomes increasingly recognized.

Acknowledgements

We extend our thanks for thoughtful and thorough reviews of the manuscript to Phil Potts, Stephen Reed, Dave Alderton, Ian Croudace, Trevor Gilson and John Murray. Barry Marsh and Anthea Dunkley are thanked for plates and cartographic work respectively. One of us (IRB) thanks the Leverhulme Trust for a fellowship.

References

Beny-Bassez, C. and Rouzaud, J.N. (1985) Characterisation of carbonaceous materials by correlated electron and optical microscopy and Raman microspectroscopy. *Scann. Electron. Microsc.*, 1985, 119–32.

Clark, R.J.H. and Franks, M.L. (1975) The resonance Raman spectrum of ultramarine blue. *Chem. Phys. Lett.*, **34**, 69–72.

Delhaye, M. and Dhamelincourt, P. (1975) Raman microprobe and microscope with laser excitation. *J. Raman Spectrosc.*, **3**, 33–43.

Dhamelincourt, P., Beny, J.M., Dubessy, J. and Poty, B. (1979) Analyse d'inclusions fluides à la microsonde MOLE à effet Raman. *Bull. Mineral.*, **102**, 600–10.

Dubessy, J., Audeoud, D., Wilkins, R. and Kosztolanyi, C. (1982) The use of the Raman microprobe in the determination of electrolytes dissolved in the aqueous phase of fluid inclusions. *Chem. Geol.*, **37**, 137–50.

Fong, M.Y. and Nicol, M. (1971) Raman spectrum of calcium carbonate at high pressures. *J. Chem. Phys.*, **54**, 579–85.

Georgiev, G.M., Kalkanjiev, T.K., Petrov, V.P. and Nickolov, Z. (1984) Determination of salts in water solutions by a skewing parameter of the water Raman band. *Appl. Spectrosc.*, **38**, 593–5.

REFERENCES

Gillet, P. and Goffe, B. (1988) On the significance of aragonite occurrences in the Western Alps. *Contrib. Mineral. Petrol.*, **99**, 70–81.

Griffith, W.P. (1974) Raman spectroscopy of minerals, in *The infrared spectra of minerals* (ed. V.C. Farmer), *Min. Soc. Monograph*, **4**, 119–35.

Hemley, R.J., Mao, H.K., Bell, P.M., Mysen, B.O. (1986) Raman spectroscopy of SiO_2 glass at high pressure. *Phys. Rev. Lett.*, **57**, 747–50.

Jamieson, J.J. (1953) Phase equilibria in the system calcite–aragonite. *J. Chem. Phys.*, **21**, 1385–90.

Kubicki, J.D., Hemley, R.J. and Hofmeister, A.M. (1992) Raman and infrared study of pressure induced structural changes in $MgSiO_3$, $CaMgSiO_6$, and $CaSiO_2$ glasses. *Am. Mineral.*, 258–69.

Liu, L.G. and Mernagh, T.P. (1990) Phase transitions and Raman spectra of calcite at high pressures and room temperature. *Am. Mineral.*, **75**, 801–6.

Matson, D.W., Sharma, S.K. and Philpotts, J.A. (1983) The structure of high sillica alkali-silicate glasses: a Raman spectroscopic investigation. *J. Non-Cryst. Solids*, **58**, 323–52.

Matson, D.W., Sharma, S.K. and Philpotts, J.A. (1986) The structure of high-silica alkali-silicate glasses along the orthoclase–anorthite and nepheline–anorthite joins. *Am. Mineral.*, **71**, 694–704.

McMillan, P.F. (1984a) Structural studies of silicate glasses and melts: applications and limitations of Raman spectroscopy. *Am. Mineral.*, **69**, 622–44.

McMillan, P.F. (1984b) A Raman spectroscopic study of glasses in the system $CaO-MgO-SiO_2$. *Am. Mineral.*, **69**, 645–59.

McMillan, P.F. (1985) Vibrational spectroscopy in the mineral sciences, in *Microscopic to Macroscopic* (eds S.W. Kieffer and A. Navrotsky, *Rev. Mineral*), **14**, Mineralogical Society of America, 9–63.

McMillan, P.F. (1989) Raman spectroscopy in mineralogy and geochemistry. *Ann. Rev. Earth Planet. Sci.*, 255–83.

McMillan, P.F. and Hofmeister, A.M. (1988) Infrared and Raman spectroscopy, in *Spectroscopic methods in mineralogy and geology* (ed. F.C Hawthorne), *Rev. Mineral.*, **18**, Mineralogical Society of America, 99–159.

Mernagh, T.P. and Wilde, A.R. (1989) The use of the laser Raman microprobe for the determination of salinity in fluid inclusions. *Geochim. Cosmochim. Acta*, **53**, 765–71.

Mysen, B.O. and Virgo, D. (1980) Solubility mechanisms of carbon dioxide in silicate melts: a Raman spectroscopic study. *Am. Mineral.*, **65**, 885–99.

Mysen, B.O. and Virgo, D. (1986) Volatiles in silicate melts at high pressure and temperature 2. Water in melts along the join $NaAl_2O_2-SiO_2$ and a comparison of solubility mechanisms of water and fluorine. *Chem. Geol.*, **57**, 333–58.

Mysen, B.O., Virgo, D. and Seifert, F.A. (1982) The structure of silicate melts: implications for chemical and physical properties of natural magma. *Rev. Geophys. Space Phys.*, **20**, 353–83.

Nishimura, Y., Hirawaka, A. and Tsuboi, M. (1978) Resonance Raman spectroscopy of nucleic acids. *Adv. Infrared Raman Spectrosc.*, **5**, 217–75.

Pan, P. and Wood, S.A. (1991) Gold-chloride complexes in very acidic aqueous solutions at temperatures 25–300 °C: a laser Raman spectroscopic study. *Geochim. Cosmochim. Acta*, **55**, 2365–71.

Pasteris, J.D. and Wopenka, B. (1991) Raman spectra of graphite as indicators of degree of metamorphism. *Can. Mineral.*, **29**, 1–9.

Pasteris, J.D., Wopenka, B. and Seitz, J.C. (1988) Practical aspects of quantitative laser Raman microprobe spectroscopy for the study of fluid inclusions. *Geochim. Cosmochim. Acta*, **52**, 979–88.

Peck, J.A., Tait, C.D., Swanson, B.I. and Brown, G.E. (1991) Speciation of aqueous gold (III) chlorides from ultraviolet/visible absorption and Raman/resonance spectroscopies. *Geochim. Cosmochim. Acta*, **55**, 671–76.

Raman, C.V. and Krishnan, K.S. (1928) A new type of secondary radiation. *Nature*, **121**, 501.

Rosasco, G.J., Roedder, E. and Simmons, J.H. (1975) Laser-excited Raman spectroscopy for non-destructive partial analysis of individual phases in fluid inclusions in minerals. *Science*, **190**, 557–60.

Seifert, F., Mysen, B.O. and Virgo, D. (1982) Three-dimensional network structure of quenched melts (glass) in the systems SiO_2 – $NaAlO_2$, SiO_2–$CaAl_2O_4$ and SiO_2–$MgAl_2O_4$. *Am. Mineral.*, **67**, 697–717.

Seitz, J.C., Pasteris, J.D. and Wopenka, B. (1987) Characterization of CO_2–CH_4–H_2O fluid inclusions by microthermometry and laser Raman microprobe spectroscopy: inferences for clathrate and fluid equilibria. *Geochim. Cosmochim. Acta*, **51**, 1651–64.

Tuinstra, F. and Koenig, J.L. (1970) Raman spectrum of graphite. *J. Chem. Phys.*, **53**, 1126–30.

Venec-Peyre, M.-T. (1980) Microanalyseur ionique et microsonde moléculaire à laser mole: application à l'étude chimique et minéralogique du test d'ammonia becarii (Linne) Foraminifère. *Bull. Centre Rech. Explor.-Prod. Elf-Aquitaine*, **4**, 55–79.

Venec-Peyre, M.-T. and Jaeschke-Boyer, H. (1979) Interet de la microsonde moléculaire à laser Mole en systematique: Etude du foraminifère. *C. R. Acad. Sci. Paris*, **288**, 819–21.

Wang, A., Dhamelincourt, P., Dubessy, J. *et al.* (1989) Characterization of graphite alteration in an uranium deposit by micro-Raman spectroscopy, X-ray diffraction, transmission electron microscopy and scanning electron microscopy. *Carbon*, **27**, 209–18.

White, W.B. (1974) The carbonate minerals, in *The infrared spectra of minerals*, *Mineral. Soc. Monograph*, **4**, 227–79.

Wopenka, B. and Pasteris, J.D. (1986) Limitations to quantitative analysis of fluid inclusions in geological samples by laser Raman microprobe spectroscopy. *Appl. Spectrosc.*, **40**, 144–51.

Index

Absorption coefficients 84–5
Absorption corrections 82
  phi-rho-z models 83–4
Absorption spectrometry
  point-projection X-ray microscope 32
  and X-ray microscopy 30–3
ACPS (area counts per second) 303, 308, 310, 313
ALCHEMI 107
Alkali feldspars
  LA-ICP-MS analysis 309
  quantitative analysis 318
Alpha coefficient 82–3
Analytical electron microscopy 26–7, 91–139
  choice of kV 106
  historical note 93
  instrument 91–3
  mineralogical applications 96
  nanoprobe mode 94
  probe-forming system 93–4
  spurious X-rays 103–5
  thin specimens 96–124
    specimen preparation 94–6
  X-ray detectors 99–101
  see also X-ray analysis of thin specimens
Anhydrite, EELS spectrum 129
Apatite, LA-ICP-MS analysis 309, 311
APFU (atom per formula unit) 307–8
Ar, interfering reactions 333
Ar–Ar dating 327–58
  applications 349–56
    laser stepped heating 350–1
    unique samples 349–50
  'chrontours' 352–3
  estimation of errors 334–5
  historical note 327

  instrumentation 336–46
    gas handling and mass spectrometry 336–8
    lasers and laser/solid interaction 338–44
    microscopes and laser ports 344–6
  and K–Ar dating 328–9
  laser technique
    instrumentation 336–46
    interfering reactions 333
    laser function 340
    methodology 347–9
    reactor-produced isotopes (^{37}Ar and ^{39}Ar) 329
    technique 329–34
    interference factors 334
Arctica islandica, quantitative analysis 317–18
Area counts per second (ACPS) 303, 308
Area selection 2–3
Argon plasma torch 291–3
Atmospheric pollutants, particle-induced X-ray emission 29–30
Atom per formula unit (APFU) 307–8
Atom-beam milling 95
Atomic number 55
Atomic structure
  inner electron shells 53
  X-ray spectroscopy 52–7
Auger electrons 91

Backscattering
  coefficient, defined 71
  electron (BSE) signal 70
  Rutherford backscattering 157–9
Backscattering correction 82
Beam calibration, proton microprobe 149–50

INDEX

Beam milling 95–6
Beam scanning, historical note 8–9
Bence–Albee coefficients 82–3
Biotite
  AEM analyses 110
  laser analysis 350–1
  laser stepped heating compared with bulk stepped heating 350–1
  standards 334
Bragg reflection, wavelength-dispersive spectrometers 57–8
Bragg's law 58
Bremsstrahlung 144–6, 173
  synchrotron storage rings 173

Calcite, LA-ICP-MS analysis 307
Calcium
  interfering reactions 333
  ^{42}Ca (n, α) ^{39}Ar and ^{40}Ca (n, α) ^{39}Ar ratios 331–2
Cambridge
  Cavendish Laboratory
    point-projection X-ray microscope 30–3
    scanning electron microscope 8–9
    X-ray scanning microscope 8–10
  Geoscan electron probe 11
Carbon
  analysis, micro laser–Raman spectroscopy 398
  particle-induced X-ray emission 29–30
Carbonates
  with aragonite structure, micro laser–Raman spectroscopy 403–4
  carbon and oxygen analyses, stable isotope ratio measurement 369–73
CARS (coherent anti-Stokes Raman) 406
Castaing, R.
  approximation, quantitative analysis 81–2
  'sonde electronique' 7–8
Cathodoluminescence 28–9
CCD camera 270
Cerium hexaboride emitter 66
Chloride, interfering reactions 333
Chromite, polymineralic standards 309

'Chrontours', Ar–Ar dating 352–3
Cliff–Lorimer factor 108
Coal, XRF measurements 213–14
Coherent anti-Stokes Raman (CARS) 406
Compton scattering 205
Contact microradiography 31
Convergent-beam electron diffraction (CBED) 117
Cosmic dust, XRF measurements 214–15
Cosmochemistry, electron probe microanalysis 51
Counting statistics
  and error prediction 15–17
  X-ray intensities 75–6
Critical excitation energy, defined 54
Crookes, W., cathodoluminescence 28
Cyanoacrylate mounts 306

DAFS (diffraction absorption fine structure) 172
Depth profiling 270–4, 321
Diffraction absorption fine structure 172
Diffusion studies, depth profiling 270–4, 321
Diopside 119, 121
Dolomite
  proton and electron microprobes 147
  zoning 151–4
Drift correction software 121
Dust, airborne, particle-induced X-ray emission 29–30

EDS, *see* Energy-dispersive spectrometers
Elastic recoil detection analysis (ERDA) 159
Electron beam instruments, integration with EDs 70
Electron bombardment
  history 5–11
  resulting processes 3–5
Electron column 65–70
  astigmatism 67
  beam diameter and current 67–8
  computer control 69–70
  optical microscope 68–9

probe-forming system 66–7
specimen chamber 69
vacuum system 69
see also Analytical electron microscopy; Scanning electron microscopy
Electron energy-loss spectroscopy (EELS) 124–35
  vs EDS 135
  energy loss near-edge structure (ELNES) 126
  extended energy loss fine structure (EXELFS) 126
  log-ratio measurement efficiency of specimen thickness 118
  operational conditions 131–3
  PEELS 126–7, 129
  principles 124–8
  quantification of the EELS spectrum 129–31
  spatial resolution 133
Electron gun 65–6
Electron images 70
Electron microprobe analysis 49–90, 164, 305–6
  accuracy and sensitivity 49–50
  applications 51–2
  with beam scanning 8–9
  electron column 65–70
  energy-dispersive spectrometers 60–5
  ion analogue 35–9
  measured signal 6
  principles 6, 49
  probe size and spatial resolution 98–9
  qualitative analysis 73–5
  quantitative analysis
    data reduction 81–7
    experimental 75–81
  for reference standards 303, 305
  sample preparation 50–1
  scanning and mapping 70–3
  sensitivity 164
  spatial resolution 50
  specimen damage 123–4
  standardization and matrix factors 12–25
  standards, list 80, 81
  summary of performance 26

valence state determination 80
see also Analytical electron microscopy; Ion probe; Quantitative analysis; Scanning electron microscopy
Electron probe
  historical note 5–11
  probe size and spatial resolution 98–9
Electron probe microanalysis (EPMA), see Electron microprobe analysis
Element partitioning and diffusion couples, XRF measurements 215
EMPA, see Electron microprobe analysis
Energy filtering, ion microprobe 253, 257
Energy loss near-edge structure (ELNES) 126
Energy-dispersive spectrometers 60–5
  collection efficiency 65
  dead time and throughput 63–4
  vs EELS 135
  energy resolution 61–3
  entrance windows 64
  filtering, 'top-hat' function 78–9
  integration with electron beam instruments 70
  'light' elements 79–80
  lithium-drifted Si detectors 61
  qualitative analysis 73–5
  quantitative analysis 77–8
  synchroton techniques 196–200
Environmental applications, synchrotron X-ray microanalysis 216–17
EPMA, see Electron microprobe analysis
Error prediction, combination of errors 18
Extended energy loss fine structure (EXELFS) 126
Extended X-ray absorption fine structure (EXAFS) 171–2, 201
  vs XANES 209–13

Fayalite, LA-ICP-MS analysis 307–8
Field-emission guns (FEGs) 93, 120, 133
Figures of merit, synchrotron storage rings 185–6

INDEX

Fluid inclusions
  micro laser–Raman spectroscopy 395–8
  PIXE 154–6
  synchrotron X-ray microanalysis 217–18
  XRS 206
Fluorescence correction 82
  see also X-ray fluorescence analysis (XRFA)
Forsterite, LA-ICP-MS analysis 307–8
Fourier-transform–Raman 406

Gaussian function, FWHM and FWTM 98–9
Gaussian (normal) distribution, and standard deviations 16–17
Geoscan electron probe 11
Germanium detectors, AEM 101
Glass reference standards
  k_{AB} factors 109
  LA-ICP-MS analysis 305, 310
  NIST glasses 109, 310–11
Glass shards, LA-ICP-MS analysis 319–20
Greenstone, laser stepped heating compared with bulk stepped heating 350–1

Hafnium, historical note 7
Hole count, spurious X-rays 103–4
Hornblende
  diffusion constant and temperature 354–6
  laser stepped heating compared with laser spot analysis 352–6
  standards 334
Hydrides 253

ICP-MS, *see* Inductively coupled plasma mass spectrometry
Image-plan selection 1–3
Incident probe
  diameter 22–4
  and measured signal, microanalysis techniques 6
Inductively coupled plasma atomic emission spectrometry (ICP-MS) 41–2

Inductively coupled plasma mass spectrometry 291–300
  ion detection 297
  ion extraction interface 293–5
  ion lenses 295–6
  ion source 291–3
  laser operation 298–300
  mass analyzer 296
  sample introduction 297
    laser ablation 297
Infrared spectrometry 42
Interference factors 334
Interfering reactions 333
Ion counting, ion microprobe 259
Ion detection, single channel electron multiplier 297
Ion lenses 292, 295–6
Ion mass spectrometry, secondary 34–9
Ion microprobe analysis 34–9, 235–89
  analytical procedures
    contamination 259–61
    energy filtering 257
    high mass resolution 257
    ion counting 259
    molecular ion corrections 258–9
    quantitative analysis, limits 261–2
    specimen charging 256
    specimen preparation 255–6
  applications, elemental analysis 262–73
    depth profiling 270–3
    imaging 270
    light elements 262–3
    rare-earths 264–70
    trace elements 263–4
  applications, isotopic analysis 273–81
    Pb–Pb and U–Pb dating of zircon 279–81
    radiometric dating 279
    stable isotopes 275–9
  characteristics 35–6
  comparison with electron probe 35–9
  formation of secondary ions 235–9
  instrumentation 235–46
  ion microscopy 246
  mass spectrometry 34–9, 242–5
  measured signal 6
  primary ion source 239–40

probe-forming system 240
resolution 35
secondary ion detection 245–6
secondary ion production
  energy distribution 251–4
  extraction system 240–2
  influence of primary ion beam 254–5
  molecular ions 250–1
  multiply charged species 251
  sputter rates 246–7
  yields 247–50
selected-area ion mass spectrometry
  direct excitation 34–9
  indirect excitation 39–40
  SHRIMP and ANU instruments 38–9, 280
summary and future developments 281–2
Ion microscopy 246, 270
Ion sputtering 34, 246–7
Isotopes
  ion microprobe analysis 273–81
    radiometric dating 279
  stable isotope ratio
    ion microprobe measurement 275–9
    laser microprobe measurement 359–86
      applications 365–7
      carbon and oxygen analyses of carbonates 369–73
      geochemistry 360–1
      historical note 364–5
      instrumentation 367–9
      mass spectrometry 362–4
      nitrogen isotope analyses 380–2
      oxygen isotopic analyses of silicates and oxides 375–80
      preparation of gases 361–2

Jadeite, ED spectrum 62

K, interfering reactions 333
K spectra, typical 53, 55–6
K–Ar dating
  and Ar–Ar dating 328–9
  constants for decay of ^{40}K 330
K–Ca dating 328

Kapton film 3–4
Kirkpatrick–Baez systems 193–5
k_{AB} factors 109–12, 115, 117, 121–2

L spectra, typical 53, 55–6
LA-ICP-MS, see Laser ablation inductively coupled plasma mass spectrometry
Lanthanum emitter 66
Laser(s)
  heating systems 40–2
    laser ablation inductively coupled mass spectrometry 41–2
    localized release of gases 41
  laser spot analysis 347–8
  and laser/solid interaction, Ar–Ar dating 338–44
  measured signal 6
  microprobe technique
    argon diffusion studies 351–6
    stepped heating 348–51
    see also Ar–Ar dating
  Nd-YAG 297–8
    CW lasers 341–4
    ICP-MS 300–2
    increase of frequency 306
    transition wavelength 299
  Q-switching 339
  ruby 297, 338–9
  spot analysis 347–8
Laser ablation inductively coupled mass spectrometry, heating systems 41–2
Laser ablation inductively coupled plasma mass spectrometry 300–23
  ablation volume 314
  analytical methodology 304–13
    calibration 307–8
    examples 308–9
    qualitative calibration 311–13
    sample preparation 305–6
    synthetic glass standards 310–11
  analytical rationale 302–3
  applications
    depth profiling 321
    quantitative analysis 316–19
    semiquantitative analysis 319–21

stable isotope ratio measurement
359–86
bulk sampling 304
detection limits 313–16
LLDs 313–16
historical note 303–4
summary and future prospects 321–2
typical system 301
Laser probe microanalysis (LPMA), see
Laser ablation inductively coupled
plasma mass spectrometry
Light elements
EELS 96, 125
electron probe analysis 79–80
elemental analysis, ion microprobe
262–3
energy-dispersive spectrometers
79–80
k_{AB} factors 109
Line scans 72
Lithium fluoride, wavelength-dispersive
spectrometers 58
Lithium-drifted Si detectors 61–2,
101–3, 148
collection efficiency 65
Luminescence micrographs, first 28

M spectra, typical 56
Magnesium, direct analysis in terms of
molecular formula 308
Mantle materials 154
synchrotron X-ray microanalysis 218
Mass analyzer, ICP-MS 296
Mass defect, defined 37
Mass resolution, defined 38
Mass spectrometry
Ar–Ar laser technique 336–8
electron energy-loss spectroscopy
(EELS) 126–35
inductively coupled plasma atomic
emission spectrometry (ICP-MS)
41–2
ion microprobe 34–9, 242–5
laser heating 40–2
mass resolution 38
metals and non-metals, contamination
problems 34
non-metallic elements 34

secondary ion mass spectrometry
(SIMS) 242–5
selected-area
direct excitation 34–9
indirect excitation 39–40
stable isotope ratio measurement
362–4
thermal ionization 34
Mass spectroscopy, see Selected-area
mass spectroscopy
Matrix corrections 82
Matrix factors 12–25
calculation 13–14
Meteorites, synchrotron X-ray
microanalysis 219
Micro laser–Raman spectroscopy 42–3,
387–408
advantages/disadvantages 394–5
applications
carbon analysis 398
fluid inclusions 395–8
minerals 398–404
instrumentation 392–3
other Raman techniques 406
principles 387–92
sample handling 394
Microanalysis
quantitative analysis 11–12
standardization and matrix factors
12–25
techniques 1–48
comparisons 164–5
history 5–11
see also Quantitative analysis
Microprobe methods (general)
area selection 3
electron probe, with beam scanning
8–9
incident probe and measured signal 6
see also Ion microprobe
Microradiography, see X-ray microscopy
Minerals
micro laser–Raman spectroscopy
398–404
carbonates with aragonite structure
403–4
phase transformations 402–3
silicate glasses 405

solutes 405–6
synchrotron X-ray microanalysis
  moon minerals 219–20
  ores 220
  soils 221
  see also specific substances
Minimum detectable mass (MDM) 123
MMHb1 standard 334
Molecular ions 236
  energy distribution 252–3
  ion microprobe analysis 250–1
  ion microprobe, corrections 258–9
Molluscs, quantitative analysis 316–18
Moon, minerals, synchrotron X-ray microanalysis 219–20
Moseley, H.G.J.
  law 55
  X-ray emission spectra 5–7
Muscovite, standards 334

Nanoprobe mode (AEM) 94
Nitrogen isotope analyses 380–2
Nixon, W.C., point-projection X-ray microscope 8
Nomenclature, characteristic X-rays in X-ray spectroscopy 54
Normal distribution, and standard deviations 16–17
Nuclear microprobe analysis 141–61
  elastic recoil detection analysis (ERDA) 159
  principles 141–2
  proton microprobe 141–4
  Rutherford backscattering (RBS) 157–9
  see also Particle-induced X-ray emission (PIXE)
Nuclear reaction analysis (NRA) 156–7

Olivines
  analysis 118–19
  LA-ICP-MS analysis 307–8
Optical absorption microspectrometry 42
Optical microscope
  electron column 68–9
  image-plane selection 1–3
  limitations 2
  in microprobe instrument 4
Orthopyroxene, AEM analyses 110
Oxygen, isotopic analyses of silicates and oxides 375–80

Particle-induced gamma-ray emission (PIGE) 165
Particle-induced X-ray emission (PIXE)
  applications
    geology 151–6
    thick specimens (TTPIXE) 148
  historical note 142
  principles 29–30, 141–51
    beam calibration 149–50
    detection limits 150–1
    detectors and filters 148–9
    sample thickness 146–8
    X-ray background 144–6
  see also Scanning proton microprobe
Pauli exclusion principle 52
Pb–Pb dating of zircon 279–81
Pentaerythritol, wavelength-dispersive spectrometers 58
Phengite, 'chrontours' 353, 354
Phi-rho-z models 83–4
Phlogopite, argon diffusion 352
PIGE (particle-induced gamma-ray emission) 165
PIXE, *see* Particle-induced X-ray emission
Plasmon scattering 124
Pockels cell 298, 299
Point-projection X-ray microscope 8, 30–3
Poisson statistics, applications 16
Precision, accuracy and sensitivity 14–19
  spatial resolution 20–2
  standards 19–20
Primary beam
  interactions 3–5
  penetration and scattering 20–1
Proton beam 141–4
Proton bremsstrahlung 144–6, 173
Proton microprobe 29–30, 141–4
  principles 29–30, 144–51
  see also Scanning proton microprobe
Protons, measured signal 6

INDEX

Qualitative analysis
  calibration, LA-ICP-MS 311–13
  energy-dispersive spectrometers 73–5
  standardization and matrix factors 25
  wavelength-dispersive spectrometers 75
Quantitative analysis
  data reduction 81–7
    absorption coefficients 84–5
    accuracy 85
    Bence–Albee coefficient 82–3
    Castaing's approximation 81–2
    detection limits 86
    iteration 85
    matrix corrections 82
    phi-rho-z models 83–4
    results 86–7
    water content 87
    ZAF corrections 83
  experimental 75–81
  ion microprobe 261–2
  principles 11–12
  thick samples, synchrotron 207–13
  wavelength-dispersive spectrometry 75–7

Radiometric dating, ion microprobe 279
Raman spectroscopy, *see* Micro laser–Raman spectroscopy
Rare-earth elements (REE), analysis
  ion microprobe 264–70
  L spectra 200
Rayleigh scattering 205
RBS (Rutherford backscattering) 157–9
Resin mounts 306
Resolution
  defined 20–1
  EELS vs EDS 133
  electron probe 50, 98–9
  energy-dispersive spectrometers 61–3
  ion microprobe analysis 35, 257
  mass resolution, defined 38
  scanning images 24–5
  standardization and matrix factors 20–2
  X-ray analysis of thin specimens 98–9
Resonance ion mass spectrometry 40
Ring and bending magnets, synchrotron
  storage rings 173–8
Rutherford backscattering (RBS) 157–9

Saha factor 313
Sample preparation
  electron microprobe analysis 50–1
  LA-ICP-MS 305–6
Sampling
  bulk, laser heating systems 40–2
  choice of technique 43–6
  initial measurements 43
  measurement efficiency 44–6
  measurement efficiency of sample thickness 115–18
  optical absorption microspectrometry 42
  selected-area ion mass spectrometry, direct/indirect excitation 34–40
Sanidine, standards 334
Scanning electron microscope
  accelerating voltage, choice 76
  electron images 70–1
  historical note, Cambridge 8–9
  scanning and mapping 70–3
    digital image storage, display and analysis 72–3
    line scans 72
    selection of points 76
    X-ray images 71–2
  with X-ray spectrometry 70
  see also Electron microprobe analysis
Scanning proton microprobe 29–30
  principles 144–51
  see also Particle-induced X-ray emission
Scanning transmission electron microscope (STEM) 93–4, 101
Scanning X-ray microscopy, zone-plate 31
Secondary beam, interactions 3–5
Secondary ion mass spectrometry (SIMS) 34–5, 235
  analytical procedures 255–62
  applications 262–81
  mass spectrometry 242–5
  primary ion source 239–40
  probe-forming system 240
  problems 237–9

secondary ion detection 245–6
secondary ion energy distribution 251–4
secondary ion extraction system 240–2
secondary ion production 246–55
see also Ion microprobe analysis; Ion probe
Secondary ion spectra 34–7
Selected-area mass spectroscopy
  with direct excitation 34–9
  with indirect excitation 39–40
  measurement efficiency 45–6
Selected-area X-ray fluorescence analysis 27–8
Selenium in soils 221
Sensitivity
  and limit of detection 18–19
  precision, accuracy and sensitivity 14–19
  summary 19
Shelled animals, quantitative analysis 316–18
Silicates
  measurement efficiency of specimen thickness 117
  and oxides, oxygen isotopic analyses of silicates and oxides 375–80
  X-ray analysis 108–9
Silicon
  direct analysis in terms of molecular formula 308
  see also Lithium-drifted Si detectors
SIMS, see Secondary ion mass spectrometry
Slodzian, G., selected-area ion mass spectrometry 34–5
Small particles, analysis 29–30, 118–19
Sphalerite, analysis 118
Spiking 12
SPM, see Scanning proton microprobe
SRIXE (synchrotron radiation-induced X-ray emission) 167–8
Standard deviations, counting statistics 16
Standardization and matrix factors 12–25
  calculation of matrix factors 13–14
  calibrated microstandards 14

diameter of incident probe 22–4
matching standards 12
precision, accuracy and sensitivity 14–19
qualitative applicability 25
resolution in scanning images 24–5
spatial resolution 20–2
spiking 12
STEM (scanning transmission electron microscope) 93–4, 101
Stopping power correction 82
Sulphur, stable isotope ratio measurement 373–5
SXAS (synchrotron X-ray absorption spectroscopy) 171–2
SXRD (synchrotron X-ray diffraction) 171
SXRFA (synchrotron X-ray fluorescence microanalysis) 164–72
Synchrotron radiation-induced X-ray emission (SRIXE), available techniques 167–8
Synchrotron X-ray microanalysis 163–233
  abbreviations and notation 222–4
  advantages and limitations of synchrotron sources 166–71
  applications and future advances 213–21
  beamline and experimental stations analysis 204–13
  general aspects 186–91
  XAS 172–3, 186–8, 201–2, 204, 209–10
  XRD 171, 188, 200–2, 209
  XRF 187–8, 191–200, 202–9
  comparison of microanalytical techniques 164–5
  distance operation 168–9
  finance and success 169
  insertion devices (IDs)
    undulator type 181–5
    wiggler type 179–81
  specimen preparation 202–4
  storage rings
    bremsstrahlung 173
    figures of merit 185–6

first and second generation rings
176–7
multi-magnet insertion devices
179–85
ring and bending magnet 173–8
X-ray diffraction 171, 188, 200–2, 209
X-ray fluorescence 187–8, 191–200, 202–9

Take-all disease of wheat 221
TEM, *see* Analytical electron microscopy
Tephra, glass shards, LA-ICP-MS analysis 319–20
Tephroite, LA-ICP-MS analysis 307–8
Thallium acid phthallate, wavelength-dispersive spectrometers 58
Thermal ionization, mass spectrometry 34
Thick samples, quantitative analysis 207–13
Thin films
  standards 207
  thin film criterion 112
Thin specimens, *see* X-ray analysis of thin specimens
Trace element analysis
  ion microprobe 263–4
  PIXE 151–6
Transmission electron microscopy, *see* Analytical electron microscopy
Tremolite 119, 121
TTPIXE, *see* Particle-induced X-ray emission (PIXE)
Tungsten filament 65–6

U-Pb dating of zircon 279–81

Valence state determination 80
van de Graaff accelerator 142
Visible light, measured signal 6

Water analysis, LA-ICP-MS analysis 319
Water content
  quantitative analysis 87
  vacuum system 260–1
Wavelength-dispersive spectrometers 57–60
  Bragg reflection 57–8
  crystals used 58
  focusing 58–9
  multiple spectrometers 59
  proportional counters 59–60
  qualitative analysis 75
  quantitative analysis 75–7
  synchrotron techniques 196–9
Wheat, take-all disease 221
White X-rays 179–81

X-ray(s)
  absorption coefficients 84–5
  continuum 105
  measured signal 6
  spurious, hole count 103–4
  synchrotron sources 166–71
X-ray absorption near-edge structure 171
X-ray analysis of thin specimens 96–124
  choice of kV 106
  contamination of specimen 106
  limits of X-ray microanalysis 121–4
    accuracy of quantification 121–3
    specimen damage 123–4
  operational conditions 101–3
  principles 96–8
  probe size and spatial resolution 98–9
  quantitative analysis 107–8
    analysis of small particles 118–19
    background subtraction 107–8
    breakdown of thin-film criterion 112–15
    fluorescence in specimen 115
    k_{AB} factor 109–12
    measurement of sample thickness 115–18
    ratio technique 108–9
    X-ray mapping 119–21
  spurious X-rays 103–5
  X-ray detectors 99–101
  see also Synchrotron X-ray microanalysis
X-ray diffraction 188, 200–2, 209
  synchrotron X-ray microanalysis 171, 202
X-ray emission
  particle-induced (PIXE), *see* Particle-induced X-ray emission

particle-induced (proton microprobe) 29–30
spectra, first apparatus 7
X-ray fluorescence analysis (XRFA) 187–8, 191–200, 204–9
  advantages and limitations 163–4
  selected-area analysis 27–8
  synchrotron X-ray microanalysis 202–3
  see also Synchrotron X-ray microanalysis
X-ray images 71–2
X-ray microscopy
  and absorption spectrometry 30–3
  contact microradiography 31–3
  point-projection X-ray microscope 30–3
  scanning microscope, Cavendish Laboratory, Cambridge 8–10
  zone-plate scanning microscopy 31
X-ray spectroscopy 52–7
  atomic structure 52
  continuous spectrum 57
  energy-dispersive spectrometers 60–5
  origin of characteristic X-rays 52–4
    critical excitation energy 54
    nomenclature 54
  with SEM, *see* Electron microprobe analysis
  wavelength-dispersive spectrometers 57–60
  wavelengths, energies and intensities 55–7
  see also Wavelength-dispersive spectrometers
XANES (X-ray absorption near-edge structure) 171, 201–2
  vs EXAFS 209–13
XAS (X-ray absorption spectrometry) 172–3, 186–8, 201–2, 204, 209–10
Xenoliths 154–5
XRD (X-ray diffraction) 171, 188, 200–2, 209
XRF (X-ray fluorescence analysis) 187–8, 191–200, 204–9
  NRLXRF program 208–9
XRS, *see* X-ray spectroscopy

ZAF corrections 83
Zero-loss peaks 124–5
Zircon
  ion microprobe analysis 269–70, 281
  LA-ICP-MS analysis 309
    quantitative analysis 318–19
  Pb-Pb and U-Pb dating 279–81